Signals and Systems

(As per JNTU Syllabus)

Dr. K. PADMANABHAN

B.E., M.Sc. (Engg.), Ph.D., FIEF, FIETE, SMIEEE

Dr. S. ANANTHI

B.E., M.Tech. (I.I.Sc.), Ph.D., MIEEE, SMIEEE

PUBLISHING FOR ONE WORLD

NEW AGE INTERNATIONAL (P) LIMITED, PUBLISHERS

(formerly Wiley Eastern Limited)

New Delhi • Bangalore • Chennai • Cochin • Guwahati • Hyderabad
Jalandhar • Kolkata • Lucknow • Mumbai • Ranchi

Visit us at www.newagepublishers.com

Published by New Age International (P) Ltd., Publishers
First Edition : 2007

ISBN : 81-224-1962-3

Rs. 150.00

C-06-08-970

Printed in India at Mohan Lal Printers, Delhi.
Typeset at Kalyani Computers, Delhi.

PUBLISHING FOR ONE WORLD

NEW AGE INTERNATIONAL (P) LIMITED, PUBLISHERS
(formerly Wiley Eastern Limited)
4835/24, Ansari Road, Daryaganj, New Delhi - 110002
Visit us at **www.newagepublishers.com**

PREFACE

Today's advances in communication techniques are all based on the developments in the theoretical methods propounded by Fourier, Shannon, Nyquist and others on discrete signal processing during the past twenty years. Digital techniques have increased communication throughput very much over the early day analog methods. All this has been due to the understanding of the mathematics of Fourier Series, transforms and functions by the Electronic Engineers well enough so as to deploy the techniques using the presently available digital processing chips. Simultaneous with this study, the growth of the hardware industry in the area of microprocessors and digital signal processing chips has also resulted in increasing the communication capabilities of any channel.

The subject of Signals and Systems is basic to the development of such Modern Electronic Communication. In this volume, the essentials of the theory of analysis and synthesis of signals in the continuous time and discrete time modes is presented.

In the first chapter, the concept of signal space is introduced, with practical examples. How this signal space is transformed into the frequency domain is brought out in the second and third chapters. The passage of signals through system components in the process of communication is dealt with in Chapter 4. The important topics of convolution and correlation are explained in Chapter 5. Chapter 6 deals with the theoretical background of taking samples of a signal and also introduces the important concept of bandpass sampling and re-construction from such samples.

The most important transforms used for every design in digital signal processing are the Laplace and z-transformations for continuous and discrete signals. A thorough understanding of these is very much required for a further study and use in practical digital signal processing methods for communication or even instrumentation applications. These are given in chapters 7 and 8.

In all these topics, quite many illustrative examples are worked out with ample figures portraying the concepts and calculation techniques. A good number of exercises are provided in each topic, with answers and hints whenever necessary.

In writing a volume of this kind, preciseness, conciseness and essentiality are all kept in mind. Within the span of one semester classroom study, it is felt that more than this size for the book will be too much of a load for the time available. Anyway, since further extensions of this subject in the form of DSP, Filter Design, Digital Communication etc., are to be faced by the students, the material provided here will be a solid foundation to all such studies.

Suggestions for improvement are welcome.

K. PADMANABHAN
S. ANANTHI

PROLOG

Sources of signals in the world are many and varied. A knowledge of the signals we get from various sources helps us to realise the events or basic phenomena behind them. We have signals from Earthquake tremors, sunspot observations, tidal waves, blood blow in heart chambers and so on.

Today, the subject on the systematic analysis of signals right from Fourier to Laplace transform is a built-in component in many an Engineering discipline. We now have a set of powerful methods to observe, record and analyse the signals from any source. The use of powerful computers has even paved the way for very fast on-line complex signal processing. Many sophisticated instruments in various fields have exploited the techniques to provide information about them.

In this book, the fundamentals of signals and systems is given for a fresher to understand the mathematical background required for analysing signals after digitising the same through samples.

The several chapters lead from the basic understanding of the linear system through Fourier, Laplace and the Z-transform methods.

More details on these such as the fast Fourier transform, the chirp-z-transform etc. are covered in another book "A Practical Approach to DSP" by the same author.

A number of worked examples with illustrative write-ups on practical signals are included in the book wherever necessary.

Hope this volume will cover the rudiments of signals and be a forerunner to the study of DSP.

K. PADMANABHAN
S. ANANTHI

SYLLABUS

JAWAHARLAL NEHRU TECHNOLOGY UNIVERSITY HYDERABAD

II Year B.Tech. ECE-I Semester **T P C**

4+1 0 4

(EC 05517) SIGNALS AND SYSTEMS

UNIT I

Signal Analysis : Analogy between vectors and signals, Orthogonal signal space, Signal approximation using orthogonal functions, Mean square error, Closed or complete set of orthogonal functions, Orthogonality in complex functions, Exponential and sinusoidal signals, Concepts of impulse function, Unit step function, Signum function.

UNIT II

Fourier Series Representation of Periodic Signals : Representation of Fourier series, Continuous time periodic signals, properties of Fourier series, Dirichlet's conditions, Trigonometric Fourier series and Exponential Fourier series, Complex Fourier spectrum.

UNIT III

Fourier Transforms : Deriving Fourier transform from Fourier series, Fourier transform of arbitrary signal, Fourier transform of standard signals. Fourier transform of periodic signals, properties of Fourier transforms, Fourier transforms involving impulse function and Signum function, Introduction to Hilbert Transform.

UNIT IV

Signal Transmission Through Linear Systems : Linear system, impulse response, Response of a linear system, Linear time invariant (LTI) system, Linear time variant (LTV) system, Transfer function of a ITI system. Filler characteristics of linear systems. Distortion loss transmission through a system, Signal bandwidth, system bandwidth, Ideal LPF, HPF and BPF characteristics. Causality and Poly-Wiener criterion for physical realization, relationship between bandwidth and rise time.

UNIT V

Convolution and Correlation of Signals : Concept of convolution in time domain and frequency domain, Graphical representation of convolution, Convolution property of Fourier transforms. Cross corrolation and auto corrolation of functions, properties of corrolation function, Energy density spectrum, Parseval's theorem, Power density spectrum, Relation between auto correlation function and energy/power spectral density function. Relation between convolution and correlation, Detection of periodic signals in the presence of noise by correlation, Extraction of signal from noise by filtering.

UNIT VI

Sampling : Sampling theorem—Graphical and analytical proof for Band Limited Signals, impulse sampling, Natural and Flat top Sampling, Reconstruction of signal from its samples, effect of under sampling, Aliasing Introduction to Band Pass sampling.

UNIT VII

Laplace Transforms : Review of Laplace transforms, Partial fraction expansion, Inverse Laplace transform, Concept of region of convergence (ROC) for Laplace transforms, constraints on ROC for various classes of signals. Properties of L.T's relation between L.T's and FT of a signal Laplace.

UNIT VIII

Z-Transforms : Fundamental difference between continuous and discrete time signals, discrete time signal representation using complex exponential and sinusoidal components, Periodicity of discrete time using complex exponential signal, Concept of Z-Transform of a discrete sequence. Distinction between Laplace, Fourier and Z-transforms. Region of convergence in Z-Transform, constraints on ROC for various classes of signals, Inverse Z-transform, properties of Z-transforms.

CONTENTS

1

SIGNAL ANALYSIS

1.1 ANALOGY OF SIGNALS WITH VECTORS

Alternating current theory describes the continuous sinusoidal alternating current by the equation

$$i = I_{max} \sin \omega t \qquad ...(1.1)$$

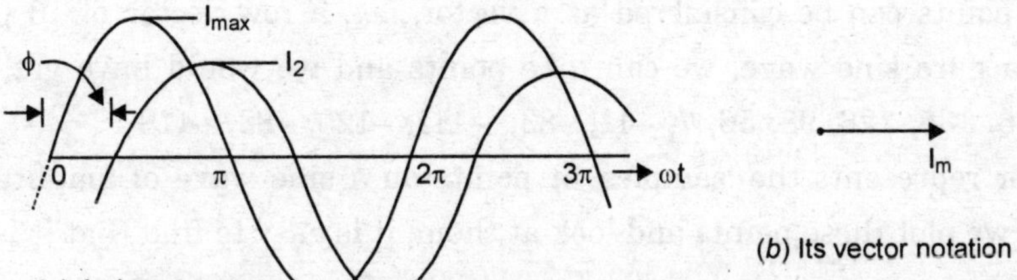

(a) A sine wave a.c.

(b) Its vector notation

Fig. 1

The same is usually represented by a vector indicating the maximum value by the length of the vector.

Another current, say i_2 which lags the first current by a phase angle ϕ is similarly denoted as :

$$i_2 = I_{2m} \sin (\omega t - \phi) \qquad ...(1.2)$$

Its vector notation is given to below the first by the lagging angle ϕ.

This notation of sinusoidal signals by vectors is useful for easy addition and subtraction.

The addition is done by the parallelogram law for vector addition.

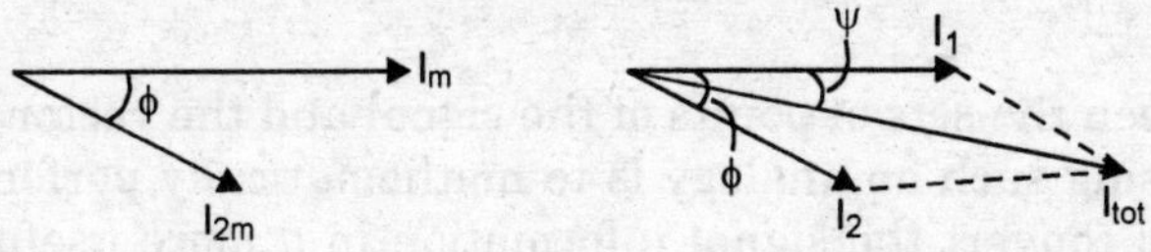

(a) (b)

Fig. 2

The total current $(i_1 + i_2)$ lags behind the first current is by the angle ψ.

This method of representing waveforms of a.c. voltage or current by vectors is also known as phasor representation.

The above vector represents a time varying signal. There are signals which are from image data. These are two dimensional signals when a plane image is considered. A three dimensional object can be taken as a three vector signal.

In short, a vector of the kind used in geometrical applications can be used to indicate a time-varying sinusoidal signal or a fixed two dimensional image or a three dimensional point.

But when a signal, such as arising from a vibration or a seismic signal or even a voice signal, is to be represented, how can we use a compact vector notation? Definitely not easy.

Such signals change from instant to instant in a manner which is not defined as in the pure sine save. If we take the values from instant to instant and write them down, we will get a large number even for a short time internal. If there are a thousand points in one second of the signal, then there will be a thousand numbers which would represent the signal for that one second interval.

For example, a variation with time of a certain signal could be as in :

–5 0 1 5, 7 4 3 2, 0 –2.4 –7, –5 –3 –1 0 ...(1.3)

These 16 points can be considered as a vector, *i.e.*, a row vector of 16 points.

Even for a pure sine wave, we can take points and we would have got, for *e.g.*,

56, 98, 126, 135, 126, 98, 56, 7, –41, –83, –111, –120, –83, –418 ...(1.4)

This vector represents the 'samples' or points on a sine wave of amplitude 128.

But when we plot these points and look at them, it is easy to find that it is a sine signal.

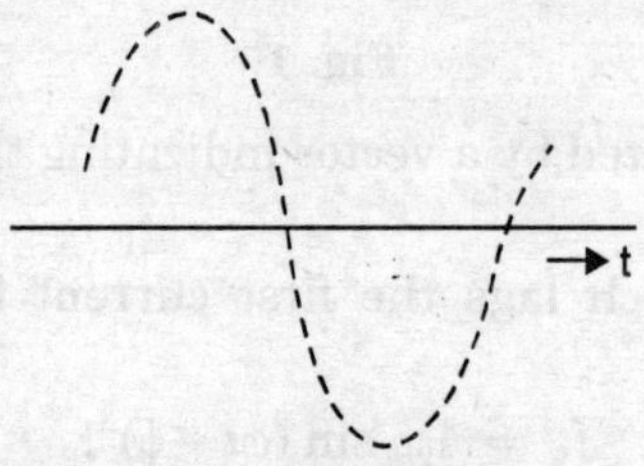

Fig. 3. Points of a sine wave make a vector.

In this case, the representation of the entire set of points is done by simply stating that the signal is a sine wave and denoted as A sin ωt, where A is the peak value and ω is the frequency.

The analogy between the sets of points of the signal and the vector of points is now clear. The purpose of visualising such an analogy is to mathematically perform such operations on the signal which would convert the signal information in a more useful form. A filter which operates on the signal does it. A Fourier transform which operates on the vector of data tells the frequencies present in the signal.

As an example, we present below the signal obtained by nuclear magnetic resonance of a chemical, dichloro-ethane from such a spectrometer instrument. The sample points of the signal are shown in Fig. 4(*b*).

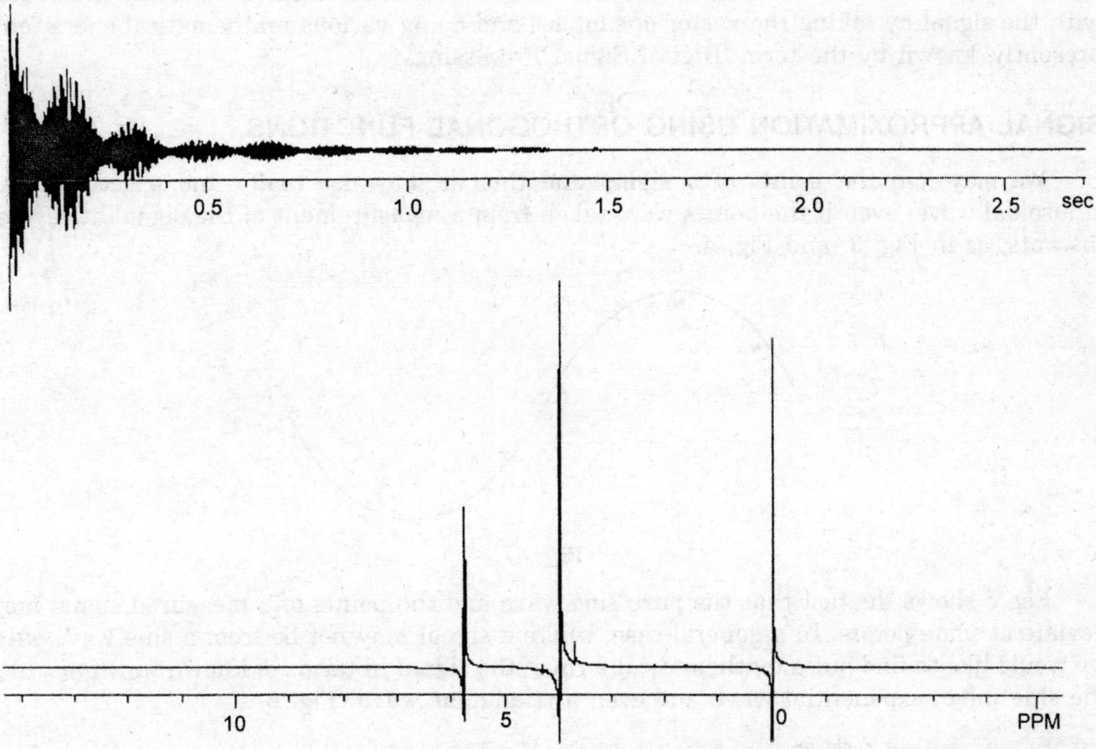

Fig. 4 (*a*) Shows the free induction decay signal of 1-2-3 dichloroethane and its Fourier transform spectrum shown in terms of parts per million of the Carrier frequency of the NMR instrument.

Sweep Width = 6024 Spectrometer freq. 300 ref. shift 1352 ref pt. 0

Acquistions = 16, 1-1-2-Dichloroethane

Fig. 4 (*b*) Shows the actual signal points. These points form the vector. There are two signals, one is the in-phase detected signal and the other is the 90°-phase detected (qudarature) signal. These two are considered as a complex signal *a+jb*.

Like this, there are hundreds of examples in scientific and engineering fields which deal with the signal by taking the vector of samples and doing various mathematical operations, presently, known by the term 'Digital Signal Processing'.

SIGNAL APPROXIMATION USING ORTHOGONAL FUNCTIONS

We may join the points of a signal and then it may not really and exactly fit the sinusoidal wave, even if the points were taken from a measurement of the signal at several instants, as in Fig. 3. and Fig. 4.

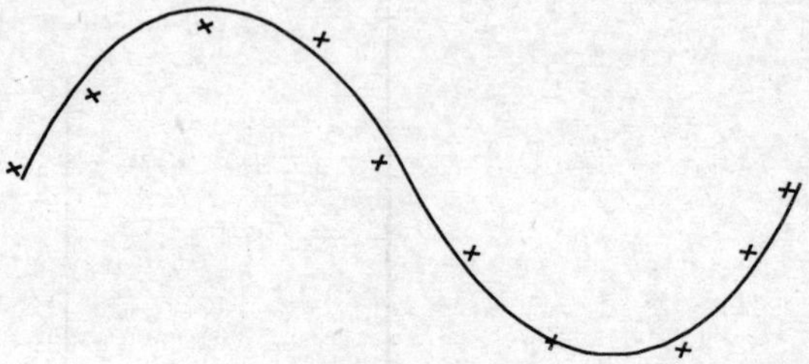

Fig. 5

Fig. 5 shows the fact that the pure sine wave and the points of a measured signal may deviate at some points. In a general case, where a signal may not be from a sine wave only, we would like to find out a mathematical form of the signal in terms of known functions, like the sine wave, exponential wave and even a triangular wave (Fig. 6).

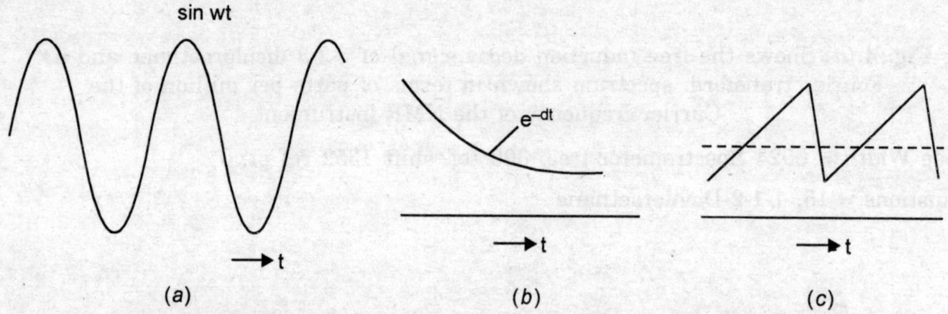

Fig. 6 *(a)* A sine signal *(b)* A decaying signal *(c)* A Ramp signal.

These signals and many others are amenable for mathematical formulation to represent and operate on the signal vector.

Thus, we have signals of the form

$$\sin \omega t, \sin (\omega t \pm \phi) \quad \text{...(1.5)}$$

$$\cos (\omega t), \cos (\omega t \pm \phi) \quad \text{...(1.6)}$$

$a \sin (m\omega t \pm \phi) \pm b \cos (n \cos \omega t \pm \phi)$ (for various values of m and n),

$$e^{-\alpha t} \sin(\omega t + \phi) \quad \text{...(1.7)}$$

$$e^{-\alpha_1 t} \pm e^{-\alpha_2 t} \quad \text{...(1.8)}$$

$$kt \text{ for } t = 0 \text{ to } 1 \quad \text{...(1.9)}$$

$$-kt \text{ for } t = 1 \text{ to } 2 \text{ and so on.} \quad \text{...(1.10)}$$

But the question arises about the error or differences between the vector points and the function curve (Fig. 7).

In Fig. 7, the differences at the points are A_1 A_2, B_1 B_2,...............

To find the function that best fits the points is a mathematical problem. This is done by the principle which states that if the squares of the errors at all points of the vector and the function are summed up, it will be a minimum.

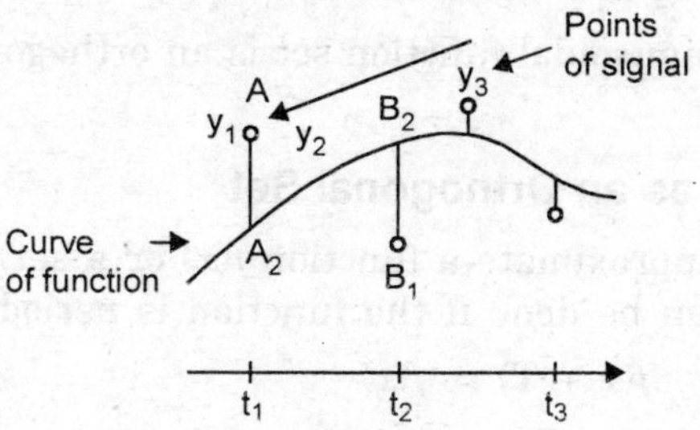

Fig. 7

DERIVATION OF LEAST SQUARE ERROR PRINCIPLE

Given a vector of r points of time samples of a signal as (t_1, y_1) (t_2, y_2), (t_3, y_3) ... (t_r, y_r).

We find such a function $f(t)$ which best fits this vector, where

$$f(t) = a_0 + a_1 t + a_2 t^2 + \dots + a_n t^n \dots \qquad \text{...(1.11)}$$

Here $f(t)$ is expressed as a series. If $r = n$, then the given r points can be substituted in the above equation and then solve them (r equations). That will give the values of a_0, a_1, ..., a_r.

But if $r > n$, then we have more points than a values to determine. In that case $f(t)$ will not satisfy all the r points. The deviation is

$$v_i = f(t_i) - y_i \qquad \text{...(1.12)}$$

which is called the 'residual' or error.

The probability of obtaining the n observed values is, according to Gauss' law of errors, for n points :

$$P = \left(\frac{h}{\sqrt{\pi}}\right)^n e^{-h^2 \Sigma v_i^2} \qquad \text{...(1.13)}$$

If P is to be a maximum, then Σv^2_i must be a minimum, which is the principle of the least square error.

The average of the sum of errors will also be a minimum, which is called the 'least mean square error.'

ORTHOGONAL FUNCTIONS

A set of so-called 'orthogonal' functions are useful to represent any signal vector by them. These functions possess the property that if you represent a set of points (or signal vector) by the combinations of an orthogonal function set, then the error will be the least. In other words, using a set of orthogonal functions, we can represent a signal vector as approximately as possible.

What are these 'Orthogonal' Functions ?

There is nothing special about them. Even the sine and cosine functions belong to this category and we are familiar with them.

We can therefore represent a signal as a sum (or combination, in other words), of a number of sine and cosine waves sin θ, sin 2θ, sin 3θ... cos θ, cos 2θ, cos 3θ... etc. Here θ denotes the time variable, by the usual $\theta = \omega t$ relationship.

Let us show how the sinusoidal function set is an orthogonal one and how it makes the MSE as least.

Sine and cos Functions as an Orthogonal Set

Suppose we want to approximate a function $f(x)$ or a set of discrete values y_i by trigonometric functions. This can be done if the function is periodic of period T, so that

$$f(x + T) = f(x).$$

By a change of variable $x = \frac{T}{2\pi} x^1$, the above becomes

$$f(x^1 + 2\pi) = f(x^1)$$

and the period is 2π. (Rewrite x^1 as x again.)

Even if a function is not periodic, it can be represented by a trigonometric approximation in a certain interval.

We take a trigonometric sum

$$\Sigma a_m \cos mx + b_m \sin mx.$$

Then by the principle of least squares,

$$\int_{-\pi}^{\pi} \left[f(x) - \overset{n}{\Sigma} (a_m \cos mx + b_m \sin mx) \right]^2 dx \qquad \text{...(1.14}a\text{)}$$

would be a minimum.

Differentiating partially w.r.t. a_k for $0 < k < n$, and equating to zero, we get, after rearranging terms,

$$\Sigma a_m \int_{-\pi}^{\pi} \cos mx \cos kx\, dx + b_m \int \sin mx \cos kx\, dx = \int_{-\pi}^{\pi} f(x) \cos kx\, dx$$

But we know that the definite integrals

$$\left.\begin{aligned} &\int_{-\pi}^{\pi} \cos mx \sin kx\, dx = 0 \\ &\int_{-\pi}^{\pi} \cos mx \sin kx\, dx = 0 \end{aligned}\right\} \text{for } m \neq k \qquad \text{...(1.14}b\text{)}$$

$$= \pi \text{ if } m = k \neq 0$$

$$= 2\pi \text{ if } m = k = 0$$

The above relation is what is meant by 'an orthogonal' function.

So, $$\pi a_k + 0 = \int f(x)\cos kx\,dx, \text{ giving}$$

$$a_k = \frac{1}{\pi}\int f(x)\cos kx\,dx, \text{ (between limits} -\pi \text{ and } +\pi)$$

$$a_0 = \frac{1}{2\pi}\int f(x)\,dx, \text{ (between limits} -\pi \text{ and } +\pi) \quad ...(1.15)$$

similarly, by differentiating (1.14) w.r.t. b_k,

$$b_k = \frac{1}{2\pi}\int f(x)\sin kx dx \text{ between same limits} \quad ...(1.16)$$

(1.15) and (1.16) are the co-efficients of the well-known Fourier Series (see later in chapter 2).

If the function is an odd function, $f(x) = -f(-x)$ and so, from (1.15),

$$\pi\, a_k = \int_0^{\pi} f(x)\cos kx\ dx + \int_{-\pi}^{0} -f(x)\cos kx\ dx$$

Put $x = -x^1$ in the second integral, $\cos k\,(-x^1) = \cos kx^1$,

$$\pi\, a_k = \int_0^{\pi} f(x)\cos kx\ dx + \int_{\pi}^{0} -f(x^1)\cos kx^1(-dx^1)$$

$$= 0$$

So if $f(x)$ is odd, terms a_k are absent. Similarly even functions such that $f(x) = f(-x)$, have no cosine terms. Also for odd functions

$$b_k = \frac{1}{\pi}\int_{-\pi}^{\pi} f(x)\sin kx\,dx \quad ...(1.17$$

$$= \frac{2}{\pi}\int_0^{\pi} f(x)\sin kx \quad \text{if } f(x) \text{ is odd}$$

[For *even* functions, a_k is got by a similar formula involving cos kx inside the interga in place of sin kx.]

If a certain function is neither odd nor even, we can write it as

$$f(x) = \tfrac{1}{2}[f(x) - f(-x)] + \tfrac{1}{2}[f(x) + f(-x)]$$

$$= f_1(x) + f_2(x) \quad ...(1.18$$

Note that $f_1(x)$ is odd and $f_2(x)$ is even. So, a given function can thus be split and thei a_k terms of $f_2(x)$ and b_k terms of $f_1(x)$ be found separately. For large problems, this artific will save computer time.

Example: The magnetising current waveform of a transformer is given by the functio $i = f(t)$ and the values for 12 points are :

t	0	2	4	6	8	10	12	14	16	18	20	22	24
i	1	2	2.5	4	5	3	0	–3	–5	–4	–2.5	–2	–1

Find the least squares approximation to a trigonometric series of 3 terms (or, find the Fourier series upto the third harmonic).

We first choose the origin at $t = 12$ as we notice odd symmetry to the left and right of $t = 12$.We have to change the variable to x so that $-\pi < x < \pi$, and so

$$x = \frac{2\pi}{24}t - 12$$

x	–180°	–150°	–120°	–90°	–60°	–30°	0°	30°	60°	90°	120°	150°	180°
i	1	2	2.5	4	5	3	0	–3	–5	–4	–2.5	–2	–1

Notice that the function is odd, so that cosine terms are absent. The b_k terms are got by :

x	j	i	*sin x*	*sin 2x*	*sin 3x*
0	0	0	0	0	0
30°	1	–3	0.5	0.866	1
60°	2	–5	0.866	0.866	0
90°	3	–4	1	0	–1
120°	4	$-2\frac{1}{2}$	0.866	–0.866	0
150°	5	–2	0.5	–0.866	1
180°	6	–1	0	0	0
			12.995	–3.031	–0.167

The integrals (1.17), for discrete number of points, are to be interpreted as representing

$$b_k = \text{Twice Average value of } f(x) \sin kx$$

since division by π after integration from 0 to π means only averaging.

$$\therefore \quad b_1 = \frac{2}{6}[0.0 + (-3)0.5 + (-5)0.866 + (-4)1]$$

$$+ (-2.5)(0.866) + (-2)0.5 + (-1)0]$$

$$= -12.995 \times 2/6 = -4.33.$$

The value (12.995) is entered usually at the end of the column 'sin x', (and it is not sum of that column, of course).

Similarly, $b_2 = \frac{2}{6}(-3.031) = -1.01$ $\quad b_3 = \frac{2}{6}(-1) = -0.333$

So the function $i(x) = b_1 \sin x + b_2 \sin 2x + b_3 \sin 3x$

$$\therefore \quad i(x) = b_1 \sin(\pi t/12 - 12) + \ldots$$

COMPLETE ORTHOGONAL SYSTEM OF FUNCTIONS

A set of orthogonal functions $\{\phi_n(x)\}$ is termed as complete in the closed interval $x \in [a, b]$ if, for every particular function which is continuous, $f(x)$, say, we can find the error term between the actual function f and its equivalent representation in terms of these orthogonal functions as a square ingegral.

$$E_n = |(f - (c_1\phi_1 + c_2\phi_2 + ... + c_n\phi_2))|^2 \qquad ...(1.17)$$

For more terms included in the $c_n\phi_n$ functions, E_n will become or tend to zero. This means that when we use a large number of sine terms in a sine series, the equivalence to the actual function is tending to be exact.

'Completeness' of a set of functions is given by the integral below, which should tend to zero also for large n.

$$\underset{n\to\infty}{\text{Lt}} \int_a^b [f(x) - \sum_{m=0}^{n} a_m\phi_m(x)]^2\, w(x)\,dx \to 0 \qquad ...(1.18)$$

In this, $w(x)$ is a 'weighting function' which is requires to satisfy the convergence of the limit.

The above limit integral just finds the area under the squared error curve (the error meaning the difference between the original signal or function and its approximation in terms of sum of sine, cos or any orthogonal set of function). This area is found only between $x = a$ and $x = b$. The $w(x)$ is a function of x, which is used to limit the area value and enable the integral to tend to zero.

The above integral is called 'Lebesgue Integral'.

1. $\{\sin(nx), \cos(nx)\}$ is an example of complete biorthogonal set of functions. The limits are $-\pi$ to $+\pi$.
2. The Legendre polynomials $\{P_n(x)\}$.
3. The Bessed polynomials $\{\sqrt{x}J_o(\alpha_n x)\}$, limits being 0 to 1, $J_0(x)$ = Bessel function of the first kind

$$\alpha_n = n\text{th root of the function.}$$

These systems lead to

(1) Fourier series

(2) Legendre series

(3) Fourier Legendre series

(4) Fourier Bessel series.

Thus all the above functions can be used to approximate a signal. We are usually familiar with sine-cosine series of a signal.

Fig. 8 shows how the signal (a square wave) is approximated better and better by adding more and more terms of the sine series; when n tends to be large, the approximation is perfect and matches the square wave itself.

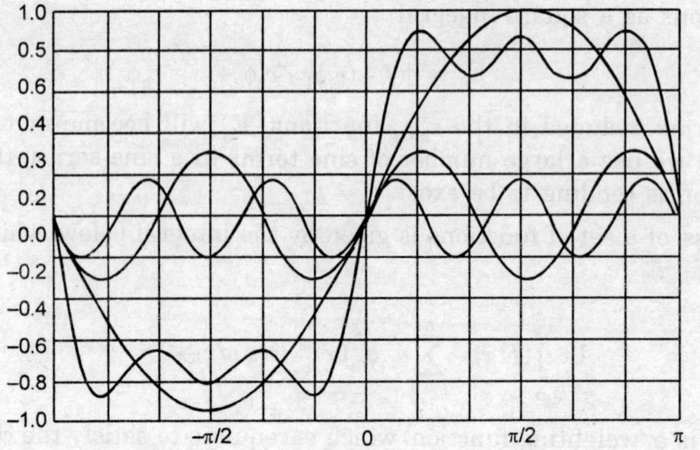

Fig. 8 (*a*) A square wave and its sinusoidal function approximation.

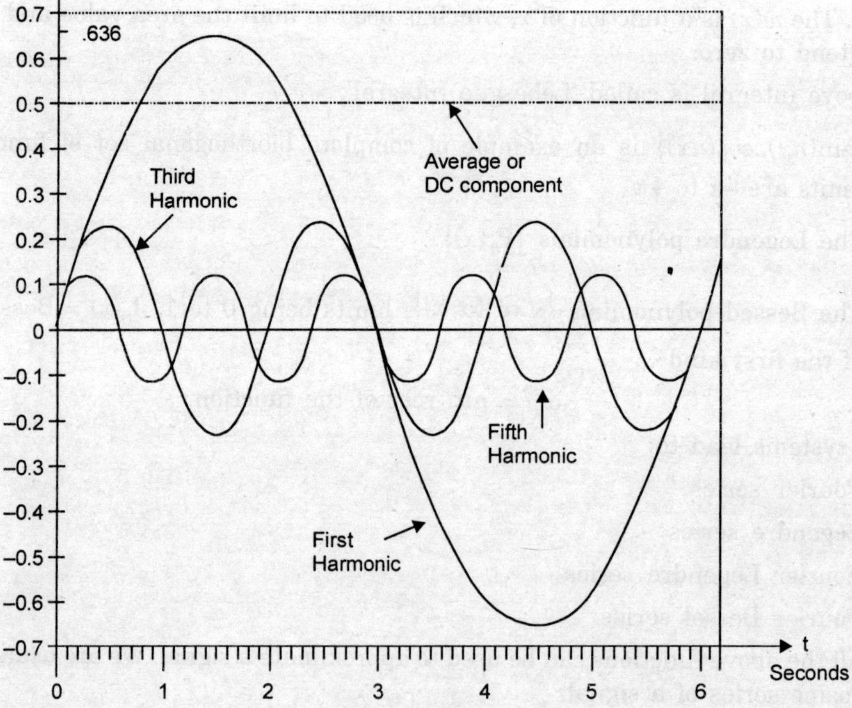

Fig. 8 (*b*)

Adding up these four curves we obtain the approximation of **f(t)** as:

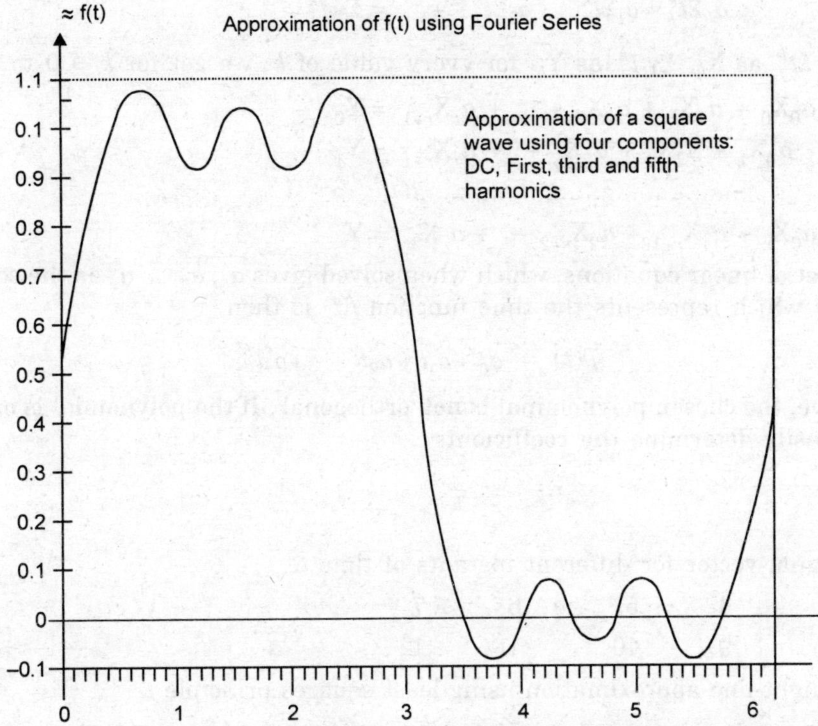

Fig. 8 (*c*) Four compnent approximation for square wave signal.

FITTING ANY POLYNOMIAL TO A SIGNAL VECTOR

$$E = |(f - k_1\phi_1 - K_2\phi_2 - \cdots)|^2 \qquad ...(1.19)$$

for a polynomial function, we can represent *f*(*t*) as any polynomial,

$$f(t) = a_o + a_1 t + a_2 t^2 + \cdots a_n t^n \text{ upto } n \text{ terms.}$$

If the *f*(*t*) signal points *r* in number are just equal to *n*, then we get a unique value for each of the coefficients a_0 to a_n which is obtainable by a solution of the *n* equations, one for each point.

But if $r > n$, then we can fit the *r* points to pass through the function (polynomial function).

Since E is a minimum,

$$\frac{\partial E}{\partial a_0} = \frac{\partial E}{\partial a_1} = \frac{\partial E}{\partial a_2} = \ldots = 0 \qquad ...(1.20)$$

Here,

$$E = (f(t) - \Sigma k\phi)^2 = (f(t) - \Sigma a_k x^k)^2 \qquad ...(1.21)$$

Taking the partial derivative, we get

$$2\sum_{j=0}^{j=n} [f(t_j) - a_0 - a_1 t_1 - a_2 t_2^2 - \ldots]\, t_j^k = 0 \qquad ...(1.22)$$

This gives

$$a_0\Sigma t_j^k + a_1\Sigma t_j^{k+1} + a_2\Sigma t_j^{k+2} + \ldots = \Sigma y_i t_j^k \qquad ...(1.23)$$

Rewriting Σt_j^k as X_k, $\Sigma y_i t_j^k$ as Y_k, for every value of k, we get for $k = 0$ to r,

$$a_0X_0 + a_1X_1 + a_2X_2 + \ldots + a_rX_{r+1} = Y_0$$
$$a_0X_1 + a_1X_2 + a_2X_3 + \quad + a_rX_{r+1} = Y_1$$
$$\ldots\ldots\ldots\ldots\ldots\ldots\ldots\ldots \qquad ...(1.24)$$
$$a_0X_r + a_1X_{r+1} + a_2X_{r+2} + \quad + a_rX_{2r} = Y_r$$

Here is a set of linear equations, which when solved gives a_0, a_1, ... a_r as the coefficients. The polynomial which represents the time function $f(t)$ is then

$$f(t) = a_0 + a_1 t + a_2 t^2 + \cdots + a_r t^r. \qquad ...(1.25)$$

In the above, the chosen polynominal is not 'orthogonal'. If the polynomial is orthogonal, we can more easily determine the coefficients.

EXERCISE

Given a signal vector for different instants of time t,

t	3	4	5	6	7
$f(t)$	6	9	10	11	12

Find a straight-line approximation using least squares principle.

Let $y = f(t)$. Put $x = t - 5$.

Let us fit $y = a_0 + a_1x$

x	y	xy	x_2
–2	6	–12	4
–1	9	–9	1
0	10	0	0
1	11	11	1
2	12	24	4
0	48	14	10

Since $E = \sum_0^5 (y_j - a_0 - a_1x_j)^2$ is the squared error, which has to be minimised, its partial derivatives w.r.t. a_0 and a_1 will vanish.

$$2\sum_0^5 (y_j - a_0 - a_1x_j) = 0 \qquad \text{from } \frac{\partial E}{\partial a_0} = 0$$

$$2\sum_0^5 (y_j - a_0 - a_1x_j)x_j = 0 \qquad \text{from } \frac{\partial E}{\partial a_1} = 0$$

The above equations can be rewritten as

$$5a_0 + a_1\Sigma x_j = \Sigma y_j$$

$$a_0\Sigma x_j + a_1\Sigma x_j^2 = \Sigma x_j y_j$$

Substituting values from table above,

$$5a_0 + 0 = 48$$

$$0 + 10a_1 = 14$$

Thus $$y = \frac{48}{5} + \frac{7}{5}x$$

$$= \frac{48}{5} + \frac{7}{5}(t-5) = \frac{13}{5} + \frac{7}{5}t$$

This is the straight line approximation for the signal vector given.

Note that the approximation holds good with minimum total error between the points given (for the 5 time values) and the points on the straight line (at these 5 time values).

From the straight line approximation, we connot find the actual points at any time.

INNER PRODUCT AND NORM OF A SIGNAL VECTOR

Let us consider the inner product of two signals $x_1(t), x_2(t)$. We form the product at any t.

$$x_1(t)\, x_2(t)$$

Then we find the area under this curve with Δt.

$$x_1(t)\, x_2(t)\, dt$$

Then, let this area span the entire range of real values of t, from $-\infty$ to ∞. Then divide by this range. This can only be done using limits.

$$\underset{T\to\infty}{\text{Lt}} \frac{1}{2T}\int_{-T}^{T} x_1(t) x_2(t)\, dt$$

This is called the inner product; represented as (x_1, x_2). Also as $< x_1, x_2 >$ in some cases, by another notation. Let us take an example.

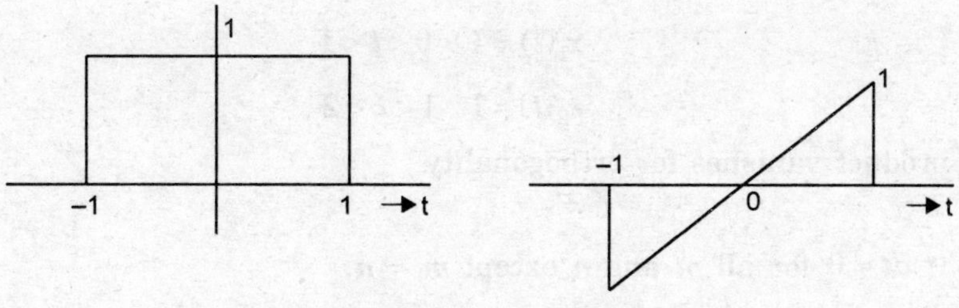

Fig. 9. Inner product of function–example.

Let us find the inner product of the above two funcitons. This gives

$$(x_2, x_1) = \int_{-1}^{1} x_2\, x_1{}^*\, dt$$

In general, the conjugate of the first function is used for the inner product. In case the function is real, the conjugate is the same as the function itself.

$$\int_{-1}^{0}(-t)(1)dt + \int_{0}^{1}(t)(1)dt = -\left[\frac{t^2}{2}\right]_{-1}^{0} + \left[\frac{t^2}{2}\right]_{0}^{1} = -\left[0-\frac{1}{2}\right]+\left[\frac{1}{2}-0\right] = 1.$$

Norm

If we do the operation of multiplication of a function $x(t)$ by itself in forming the product integral average, then we get the **Norm.** Denoted as $\|x\|$.

Example: Find the norms of the above two functions. The norm is got as the square root of the product integral.

$$\|x_1\|^2 = \int_{-1}^{1}[1]^2 dt = 2$$

(Since the function is 1 between $-1 < t < 1$).

$$\|x_1\| = \sqrt{2}$$

Likewise,

$$\|x_2\|^2 = \int_{-1}^{0}[-t][-t]dt + \int_{0}^{1}[t][t]dt$$

$$= \frac{1}{3}+\frac{1}{3} = \frac{2}{3}$$

$$\|x_2\| = \sqrt{2/3}\cdot$$

Example: There are two waveforms given below. Show that these time functions are orthogonal in the range 0 to 2.

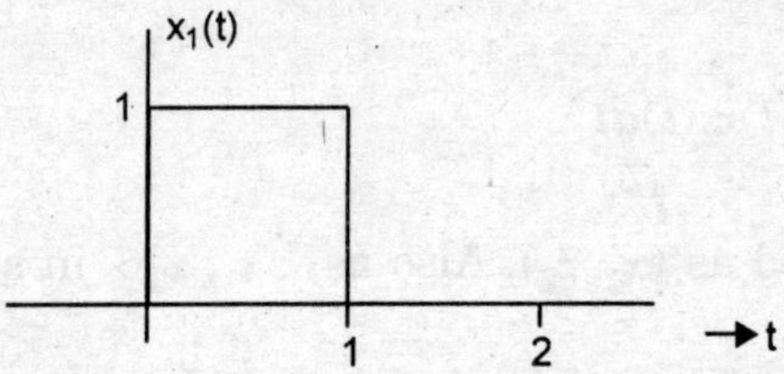

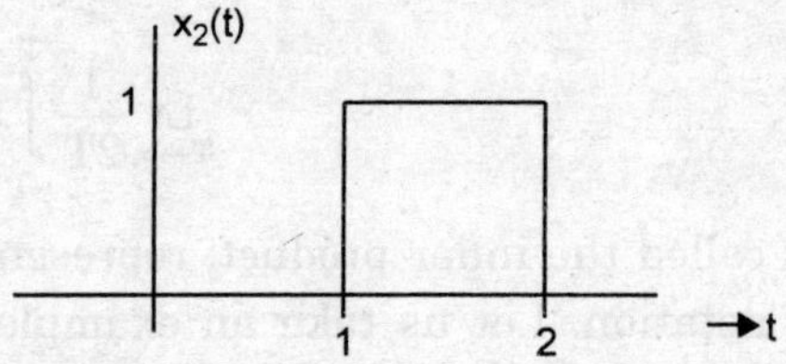

$$x_1(t) = 1 \quad 0 < t < 1$$

$$x_2(t) = 1 \quad 1 < t < 2$$

The inner product vanishes for orthogonality.

$$\int_{a}^{a+T} x_m(t)\, x^*{}_n(t)dt = 0 \text{ for all } m \text{ and } n \text{ except } m = n.$$

So,

$$\int_{a}^{a+T} x_1(t)x_2(t)dt = \int_{0}^{1}(1)(0)dt + \int_{1}^{2}(0)(1)dt = 0$$

So, the two functions are orthogonal (only in the range (0, 2)).

Example: There are four time functions, shown as wave forms in Fig (a). Then write down the function shown in Fig. (b) below as a sum of these orthogonal functions.

That the functions in Fig. (*a*) are orthogonal to each other in the range is clear because they do not at all overlap.

So, we can represent any other time function (in that range 0 to 4) using the combinations of these four functions.

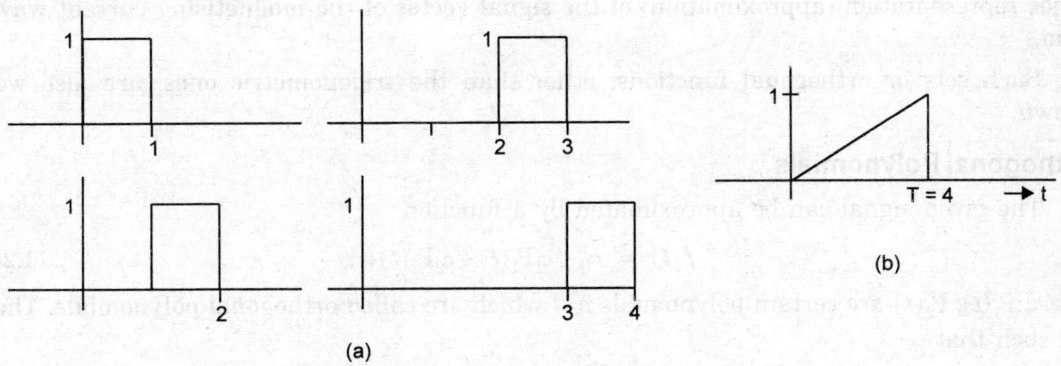

Fig. 10. Example–Orthogonal functions.

The given function is a ramp (uniformly increasing, triangular) function, truncated at t = T. Here T = 4.

Then it can be approximated by the combination of the four functions ϕ_1, ϕ_2, ϕ_3 and ϕ_4 in Fig. 10 (*a*), as

$$x(t) = c_1\phi_1(t) + c_2\phi_2(t) + c_3\phi_3(t) + c_4\phi_4(t)$$

$$C_1 = \int_0^1 x(t)\phi_1(t) = \int_0^1 \frac{t}{T}\,dt = \frac{1}{T}\left(\frac{t^2}{2}\right) = \frac{1}{4}$$

$$C_2 = \int_1^2 x(t)\phi_2(t) = \frac{3}{8}$$

$$C_3 = \int_2^3 x(t)\phi_3(t) = \frac{5}{8}$$

$$C_4 = \int_3^4 x(t)\phi_4(t) = \frac{7}{8}$$

Hence $x(t)$ is approximated as

$$x(t) = \frac{1}{4}\phi_1(t) + \frac{3}{8}\phi_2(t) + \frac{5}{8}\phi_3(t) + \frac{7}{8}\phi_4(t)$$

ORTHOGONAL FUNCTIONS FORM A CLOSED SET

From the relations 1.14(*b*). We can infer how the sine and cosine functions form a closed set exhibiting the property of 'orthogonality'.

A sine function of n-degree will integrate with a cosine or sine function with any other degree m, ($n \neq m$) only to yield a zero result. The integration can be either for one wave cycle (0 to 2π) or for entire range of (0 to ∞) in R.

If m and n are identical, then only the integral will yield a value.

This means that each sine or cosine component (as in sin t, sin $2t$... or cos t, cos $2t$...) will be **uniquely** determinable for a given signal vector.

That was how, in the previous example, we obtained the three cosine terms of the 3-term series representation (approximation) of the signal vector of the magnetising current waveform.

Such sets or orthogonal functions, other than the trigonometric ones, are also well known.

Orthogonal Polynomials

The given signal can be approximated by a function

$$f(t) = c_0 + c_1 P_1(t) + c_2 P_2(t) + \ldots \qquad \text{...(1.26)}$$

where $P_1(t)$; $P_2(t)$ are certain polynomials in t which are called orthogonal polynomials. They are such that

$$\sum_{t=0}^{n} P_1(t)P_2(t)$$

or

$$\Sigma P_2(t)P_3(t),\ \Sigma P_1(t)P_3(t) = 0 \qquad \text{...(1.27)}$$

In general, $\Sigma P_r(t)P_s(t) = 0$ unless $r = s$. ...(1.28)

This is the orthogonality relation.

In the light of the least squares error,

$\Sigma\{y_j - c_0 - c_1P_1 - c_2P_2 \ldots\}^2$ is a minimum.

The necessary condition that partial derivatives with represent to c_0, c_1, c_2 be zero, yields

$$\Sigma - 2P_j(t)[y_j - c_0 - c_1P_1(t) - \ldots c_jP_j(t) - \ldots] = 0 \qquad \text{...(1.29)}$$

Using the orthogonality relation, only the j-th term product sum does not vanish, so, this becomes

$$\Sigma P_j(t)y_j + \Sigma c_j P_j^2(t) = 0 \text{ so that}$$

$$c_j = \frac{\Sigma y_j P_j(t)}{\Sigma P_j^2(t)} \qquad \text{...(1.30)}$$

The c_j^0 values are independently determined provided we know $p_j(t)$ for all values of t. (We don't have to solve a set of equations as in 1.24).

It can be shown using algebraic methods that $p_j(t)$ satisfying the orthogonality relation is given by

$$P_1(t) = 1 - 2\frac{t}{n}$$

$$P_2(t) = 1 - 6\frac{t}{n} + \frac{6t(t-1)}{n(n-1)}$$

$$P_3(t) = 1-12\frac{t}{n}+30\frac{t\ (t-1)}{n\ (n-1)}-20\frac{t(t-1)(t-2)}{n(n-1)(n-2)}$$

$$P_4(t) = 1-20\frac{t}{n}+90\frac{t\ (t-1)}{n\ (n-1)}-140\frac{t(t-1)(t-2)}{n(n-1)(n-2)}$$

$$+70\frac{t(t-1)...(t-3)}{n(n-1)...(n-3)} \quad ...(1.31)$$

These values are generally available for different t, n values as tables. For 6 points *i.e.*, t = 0, 1, 2, 3, 4, 5 the values of $P_1(t)$, $P_2(t)$ appear as in table.

t	$P_1(t)$	P_2	P_3	P_4
0	5	5	5	1
1	3	–1	–7	–3
2	1	–4	-4	2
3	–1	–4	4	2
4	–3	–1	7	–3
5	–5	5	–5	1
S	70	84	180	28

These numbers have to be divided by the respective top entries –5 for P_1, 1 for P_4 etc. For example, $P_2(3)$ = –4/5, $P_4(4)$ = –3/1.

The values $P_1^2(t)$ are given by $70/5^2$, $P_2^2 = 84/5^2$, $P_3^2 = 180/5^2$,

$P_4^2(t) = 28/1^2$. In general = S/(term on top of column)^2.

Example: Using orthogonal polynomials, approximate the following simple signal vector to a third degree.

t	5	8	11	14	17	–	20
y	1	2	3	4	5	–	7

First we have to convert the variable t so that, instead of samples at 5, 8, 11, 14, we have 0, 1, 2, 3. Put t' = (t – 5)/3 Then t' varies as 0, 1, 2, 3.

t'	y	P_0	P_1	P_2	P_3	yP_1	yP_2	yP_3
0	1	1	5	5	5	5	5	5
1	2	1	3	–1	–7	6	–2	–14
2	3	1	1	–4	–4	3	–12	–12
3	4	1	–1	–4	4	–4	–16	16
4	5	1	–3	–1	7	–15	–5	35
5	7	1	–5	5	–5	–35	35	–35
S	22	10	70	84	180	–40	5	–5

$$C_1 = \frac{\Sigma y P_1}{\Sigma P_1^2} = \frac{-40/5}{70/5^2} = \frac{-40\times 5}{70} = -\frac{20}{7}$$

$$C_0 = \frac{\Sigma y P_0}{\Sigma P_0^2} = \frac{22}{10} = 2.2$$

$$C_2 = \frac{\Sigma y P_2}{\Sigma P_2{}^2} = \frac{5/5}{85/5^2} = 25/84$$

$$C_3 = \frac{\Sigma y P_3}{\Sigma P_3^2} = \frac{-5/5}{180/5^2} = -25/180$$

So the polynomial approximation to the signal vector is

$$f(t) = 2.2 - \frac{20}{7} P_1(t') + \frac{25}{84} P_2(t') - \frac{25}{180} P_3(t') \quad ...(1.32)$$

We can leave at just three terms as above. For further computations, the further P's can be got from tables or from computer storage.

One main advantage of orthogonal functions is the co-efficients (c's) are independent and so, if a fit has been made with an mth-degree polynomial in P, and it is decided later to use a higher degree, giving more terms, only the additional coefficients are required to be calculated and those already calculated remain unchanged.

LEGENDRE POLYNOMIALS

The general set of Legendre polynomials $P_n(x)$ (for $n = 0, 1, 2...$) form an orthogonal set.

$$P_n(t) = \frac{1}{2^n n!} \frac{d^n}{dt^n} (t^2 - 1)^n \quad ...(1.33)$$

Thus $P_0(t) = 1$

$$P_1(t) = t$$

$$P_2(t) = \frac{3}{2} t^2 - \frac{1}{2} \quad ...(1.34)$$

$$P_3(t) = \frac{5}{2} t^3 - \frac{3}{2} t \text{ etc.}$$

These are actually orthogonal, which can be verified by taking the products of any two of the above and integrating between –1 and +1.

$$\int_{-1}^{+1} P_m(t) P_n(t) dt = 0 \quad ...(1.35)$$

But if $m = n$, then the value of the integral becomes non-zero.

$$\int_{-1}^{+1} P_m(t)P_m(t)dt = \frac{2}{2m+1} \qquad \text{...(1.36)}$$

So, it is possible to express a signal function $f(t)$ as an approximation by the Legendre series (or Legendre-Fourier series).

$$f(t) = C_0P_0(t) + C_1P_1(t) + C_2P_2(t) + \ldots \text{ upto as many terms as we want.} \qquad \text{...(1.37)}$$

Here
$$C_k = \frac{\int_{-1}^{+1} f(t)P_k(t)dt}{\int_{-1}^{+1} P_k^2(t)dt} = \frac{2k+1}{2}\int_{-1}^{1} f(t)P_k(t)dt$$

By a change of variable, the limits –1 to 1 can be made useful for any other range for which time the signal exists.

Exericse: Find the Legendre-Fourier series of the square-wave given below.

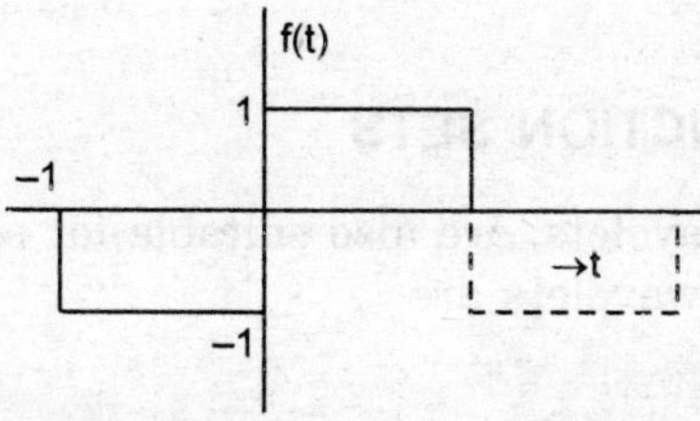

Given: $f(t)$ is +1 for $t > 0$ and $t < 1$

$f(t)$ is –1 for $t < 0$ and $t < -1$

$$C_0 = \frac{2.0+1}{2}\int_{-1}^{1} f(t)\,dt = 0$$

This first coefficient is the average function; it is zero.

$$C_1 = \frac{3}{2}\int t\,f(t)dt = \frac{3}{2}\left(\int_{-1}^{0} -tdt + \int_{0}^{1} tdt\right)$$

$$= \frac{3}{2}\left[\left(\frac{-t^2}{2}\right)_{-1}^{0} + \left(\frac{t^2}{2}\right)_{0}^{1}\right] = \frac{3}{2}\left(\frac{1}{2}+\frac{1}{2}\right) = \frac{3}{2}$$

$$C_2 = \frac{5}{2}\int_{-1}^{1} f(t)\left(\frac{3}{2}t^2 - \frac{1}{2}\right)dt$$

$$= \frac{5}{2}\left[\int_{-1}^{0} -\frac{3}{2}t^2 - \frac{1}{2}dt - \int_{0}^{1}\left(\frac{3}{2}t^2 - \frac{1}{2}\right)\right]dt$$

$$= 0$$

Here C_2 is an even value. The function is having even symmetry. So all even values are zero only.

$$C_3 = \frac{7}{2}\int_{-1}^{1} f(t)\left[\left(\frac{5}{2}t^3 - \frac{3}{2}\right)dt\right]$$

$$= \frac{7}{2}\int_{-1}^{0} -\frac{5}{2}t^3 + \frac{3}{2}t\ dt + \int_{0}^{1}\left(\frac{5}{2}t^3 - \frac{3}{2}t\right)dt$$

$$= -\frac{7}{8}$$

Hence the function, upto four terms, is $f(t) = \frac{3}{2}t - \frac{7}{8}\left(\frac{5}{2}t^3 - \frac{3}{2}t\right) + \ldots$

OTHER ORTHOGONAL FUNCTION SETS

A class of signals, called wavelets, are also suitable for signal vector approximation by functions. The commonly used wavelets are

1. Daubechies wavelet
2. Morlet wavelet
3. Gaussian wavelet
4. Maxican Hat wavelets
5. Symlet

The Daubechies wavelet is an orthogonal function. The **Morlet wavelet** is the function:

$$f(x) = \exp\left(-x^2/2\right).\ \cos(5x) \qquad \ldots(1.39)$$

Gaussian

It is not completely orthogonal. It is given by

$$f(x) = C_n \text{ diff}\left(\exp(-x^2), n\right) \qquad \ldots(1.40)$$

Where 'diff' denotes the symbolic derivative; C_n is such that the 2-norm of the Gaussian wavelet $(x, n) = 1$. This is not orthogonal.

Maxican Hat

This is a function

$$f(x) = c\, e^{-x^2/2}\left(1 - x^2\right) \qquad \ldots(1.41)$$

where $$c = \frac{2}{\pi^{1/4}\sqrt{3}}.$$

This is also not fully orthogonal.

Coiflet and Symlet wavelets are orthogonal.

Application

Now-a-days, a large number of applications make use of signed processing of data vectors using the several wavelets.

Wavelets based approximation helps in de-noising the signal or for compressing the signal with minimal data for storage. Feature extraction from unknown signals is possible using wavelet approximations.

Complex Functions

Fourier transforms involve complex numbers, so we need to do a quick review. A complex number $z = a + jb$ has two parts, a real part x and an imaginary part jb, where j is the square-root of –1. A complex number can also be expressed using complex exponential notation and Euler's equation:

$$z = a + jb = \mathrm{A}e^{j\phi} = \mathrm{A}[\cos(\phi) + j\sin(\phi)] \quad ...(1.42)$$

where A is called the amplitude and ϕ is called the phase. We can express the complex number either in terms of its real and imaginary parts or in terms of its amplitude and phase, and we can go back and forth between the two :

$$a = \mathrm{A}\cos(\phi), \quad b = \mathrm{A}\sin(\phi) \quad ...(1.43)$$

$$\mathrm{A} = \sqrt{a^2 + b^2}, \quad \phi = \tan^{-1}(b/a)$$

Euler's equation, $e^{j\phi} = \cos(\phi) + j\cos(\phi)$, is one of the wonders of mathematics. It relates the magical number e and the exponential function e^{ϕ} with the trigonometric functions, $\sin\phi$ and $\cos(\phi)$. It is most easily derived by comparing the Taylor series expansions of the three functions, and it has to do fundamentally with the fact that the exponential function is its own derivative:

$$\frac{d}{d\phi}e^{\phi} = e^{\phi}.$$

Although it may seem a bit abstract, complex exponential notation, $e^{j\phi}$, is very convenient. For example, let's say that you wanted to multiply two complex numbers. using complex exponential notation,

$$\left(\mathrm{A}_1e^{j\phi_1}\right)\left(\mathrm{A}_2e^{j\phi_2}\right) = \mathrm{A}_1\mathrm{A}_2e^{j(\phi_1+\phi_2)} \quad ...(1.44)$$

so that the amplitudes multiply and the phases add. If you were instead to do the multiplication using real and imaginary $a + jb$, notation, you would get four terms that you could write using sin and cos notation, but in order to simplify it you would have to use all those trig identities that you forgot after graduating from high school. That is why complex exponential notation is so widespread.

Complex Signals

A complete signal has a real part and an imaginary part.

$$f(t) = f_r(t) + jf_i(t)$$

The example in Fig. 4(a) showed two components of the NMR signal, which are $f_r(t)$ and $f_i(t)$.

You may wonder how can the NMR instrument give an 'imaginary' signal! When a signal is detected with a sine wave carrier synchronously, then we get the imaginary part signal, just as we get its real part with a cosine wave (or 90° phase shifted) carrier wave detection.

Complex Conjugate Signal

$$f^*(t) = f_r(t) - jf_i(t) \qquad \text{...(1.45)}$$

Real and imaginary parts can be found from the signal and its conjugate.

$$f_r(t) = \frac{1}{2}\left(f(t) + f^*(t)\right)$$

$$f_i(t) = \frac{1}{2j}\left(f(t) - f^*(t)\right) \qquad \text{...(1.46)}$$

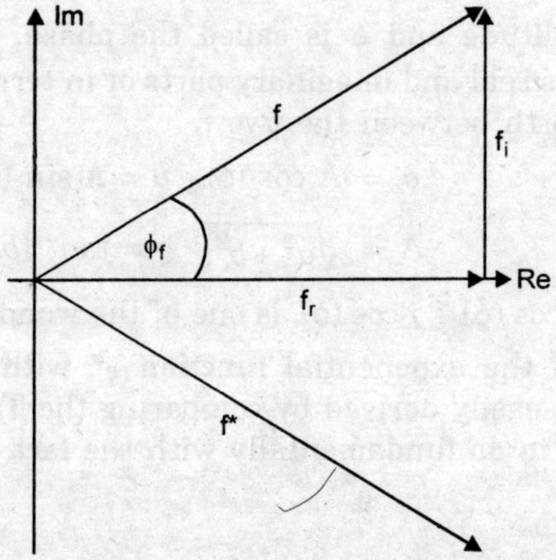

Squared magnitude: $$|f(t)|^2 = f(t).f^*(t) = f_r^2(t) + f_i^2(t) \qquad \text{...(1.47)}$$

Phase angle: $$\phi_f(t) = \tan^{-1}\frac{f_i(t)}{f_r(t)} \qquad \text{...(1.48)}$$

$$f(t) = \exp(j\omega_0 t)$$

Euler's identity:

$$\exp(j\omega_0 t) = \cos(\omega_0 t) + j\sin(\omega_0 t) \qquad \text{...(1.49)}$$

Real and imaginary part:

$$f_r(t) = \cos(\omega_0 t)$$

$$f_i(t) = \sin(\omega_0 t) \qquad \text{...(1.50)}$$

Magnitude: $$|f(t)| = 1 \quad ...(1.51)$$

Phase angle: $$\phi_f(t) = \omega_0 t \quad ...(1.52)$$

Complex conjugate signal:

$$f^*(t) = \exp(-j\omega_0 t) \quad ...(1.53)$$

Phasor of length 1 rotating counterclockwise (for $\omega_0 > 0$) at angular rate ω_0 is shown below.

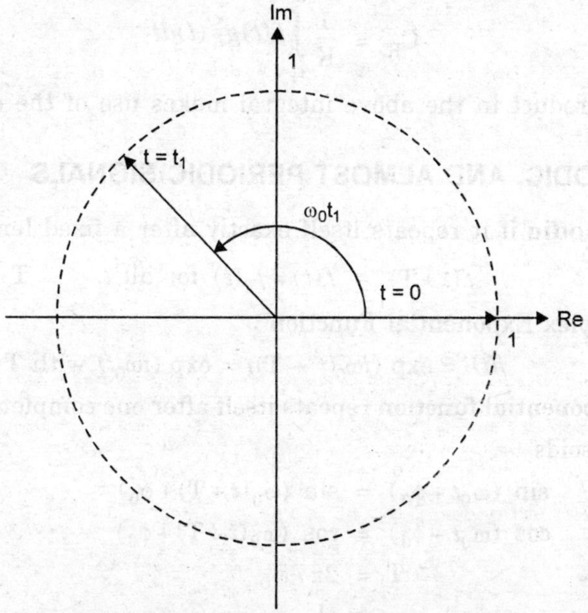

ORTHOGONALITY IN COMPLEX FUNCTIONS

So far we considered real variables for functions. But, just as a phasor of A.C. current can be considered as a complex variable, $e^{j\omega t} = \cos \omega t + j \sin \omega t$ is also a complex function. We can have functions of $e^{j\omega t}$ as $f(e^{j\omega t})$.

Two complex functions $f_1(t)$ and $f_2(t)$ are said to be 'orthogonal', if the integral is zero for them over the interval t_1 to t_2, say, as per :

$$\int_{t_1}^{t_2} f_1(t) f_2^*(t) dt = 0 \quad ...(1.54)$$

$$\int_{t_1}^{t_2} f_2(t) f_1^*(t) dt = 0 \quad ...(1.55)$$

In general, when we have a set of complex orthogonal functions $g_m(t)$,

$$\int g_m(t) g_n^*(t) dt = 0 \quad \text{when } m \neq n \quad ....(1.56)$$

But if $m = n$, $\int g_m(t) g_m^*(t) dt = K_m$. ...(1.57)

In this case, any given time function *f(t),* which itself is complex, can be expressed as a series in terms of the '*g*'-functions.

$$f(t) = C_0 + C_1 g_1(t) + C_2 g_2(t) + C_3 g_3(t) + \ldots \text{ etc.} \qquad \ldots(1.58)$$

Here we evaluate C_K coefficients in the above by

$$C_K = \frac{1}{K} \int_{t_1}^{t_2} f(t) g_K^*(t) dt \qquad \ldots(1.59)$$

Note that the product in the above integral makes use of the conjugate of *g*(*t*).

PERIODIC, APERIODIC, AND ALMOST PERIODIC SIGNALS

A signal is **periodic** if it repeats itself exactly after a fixed length of time.

$$f(t+T) = f(t) = f_T(t) \text{ for all } t, \qquad T : \text{period} \qquad \ldots(1.60)$$

Example: Complex Exponential Function

$$f(t) = \exp(j\omega_0(t + T)) = \exp(j\omega_0 t) \text{ with } T = 2\pi/\omega_0 \qquad \ldots(1.61)$$

The complex exponential function repeats itself after one complete rotation of the phasor.

Example: Sinusoids

$$\sin(\omega_0 t + \phi_0) = \sin(\omega_0(t+T) + \phi_0)$$
$$\cos(\omega_0 t + \phi_0) = \cos(\omega_0(t+T) + \phi_0)$$
$$T = 2\pi/\omega_0 \qquad \ldots(1.62)$$

Note:

$$\sin(\alpha) = \frac{1}{2j}(\exp(j\alpha) - \exp(-j\alpha))$$
$$\cos(\alpha) = 1/2\,(\exp(j\alpha) + \exp(-j\alpha)) \qquad \ldots(1.63)$$

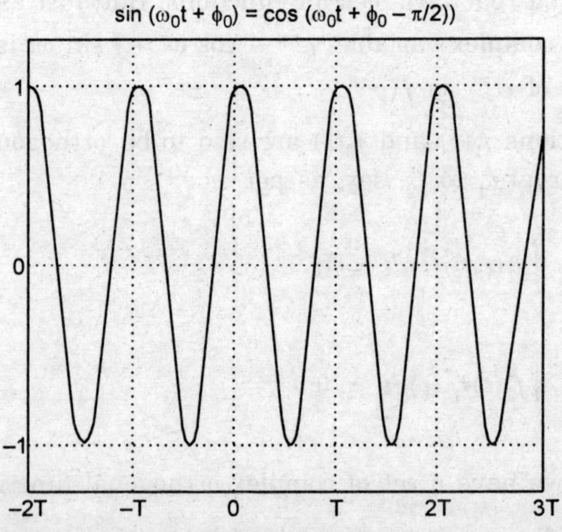

A signal is **nonperiodic** or **aperiodic** if there is no value of T such that *f*(*t* + T) = *f*(*t*).

SINGULARITY FUNCTIONS

Singularity functions have simple mathematical forms but they are either not finite everywhere or they do not have finite derivatives of all orders everywhere.

Definition of Unit Impulse Function

$$\int_a^b f(t)\delta(t-t_0)dt = \begin{cases} f(t_0) & a < t_0 < b \\ 0 & \text{elsewhere} \end{cases} \quad ...(1.64)$$

with $f(t)$ continuous at $t - t_0$, t_0 finite.

Area

$\delta(t)$ has unit area since for $f(t) = 1$: ...(1.65)

$$\int_a^b \delta(t-t_0)dt = 1, \qquad a < t_0 < b.$$

Amplitude

$$\delta(t-t_0) = \begin{cases} 0 & \text{for all } t \neq t_0 \\ \text{undefined} & \text{for } \quad t = t_0 \end{cases} \quad ...(1.66)$$

UNIT STEP FUNCTION AND RELATED FUNCTIONS

Unit Step Function

$$u(t) = \begin{cases} 0 & \text{for } t < 0 \\ 1 & \text{for } t > 0 \end{cases} \quad ...(1.67)$$

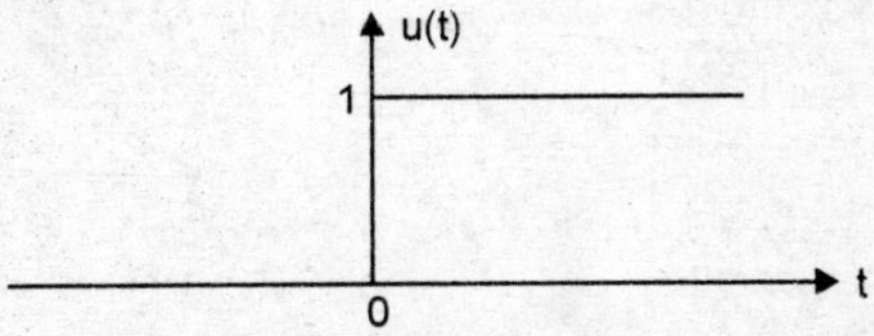

Gate Function

$$\text{rect}\ (t/\tau) = \begin{cases} 0 \text{ for } |t| > \tau/2 \\ 0 \text{ for } |t| < \tau/2 \end{cases}$$

$$= u\ (t+0.5\tau) - u(t-0.5\tau)$$

$$= u\ (t+0.5\tau).u(-t+0.5\tau) \quad ...(1.68)$$

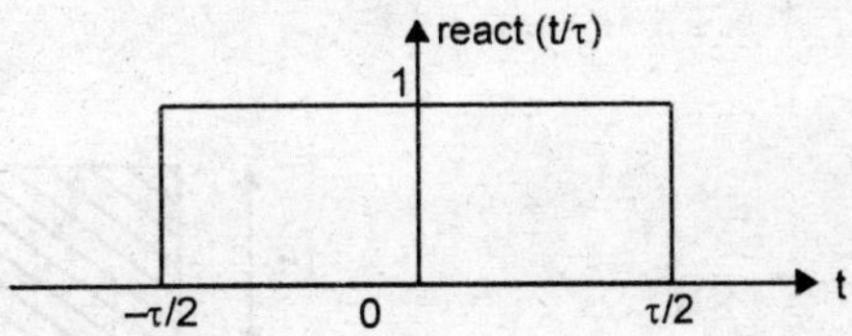

Signum Function

$$\text{sgn}(t) = \begin{cases} -1 & \text{for } \quad t < 0 \\ 0 & \text{for } \quad t = 0 \\ 1 & \text{for } \quad t > 0 \end{cases} \quad ...(1.69)$$

$$= 2u(t) - 1$$

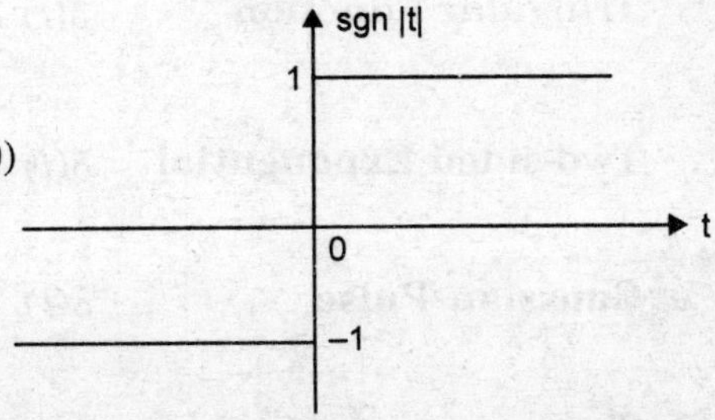

Triangular Function

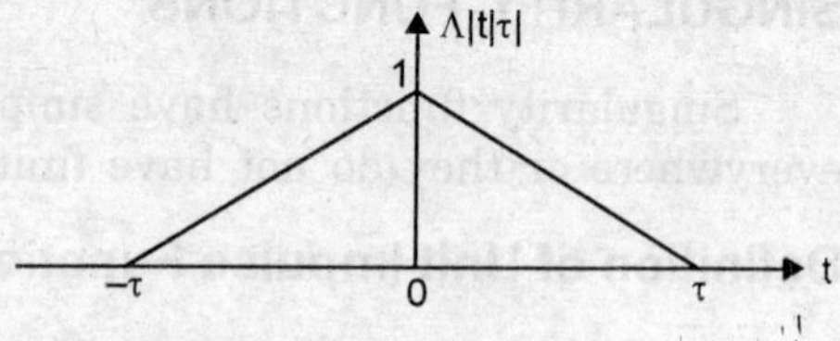

$$\Lambda(t/\tau) = \begin{cases} 1 - |t/\tau| & \text{for } |t| \leq \tau \\ 0 & \text{for } |t| > \tau \end{cases} \qquad \text{...(1.70)}$$

Graphical Representation

The area of the impulse is designated by a quantity in parenthesis beside the arrow and/ or by the height of the arrow. An arrow pointing down denotes a negative area.

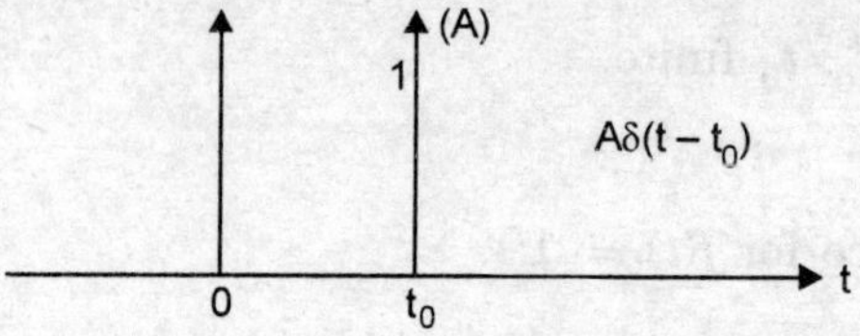

LIMIT OF A SEQUENCE OF WELL-BEHAVED FUNCTIONS

Gate function $$\delta(t) = \lim_{\tau \to 0} \frac{1}{\tau} \text{rect}\left(\frac{t}{\tau}\right)$$

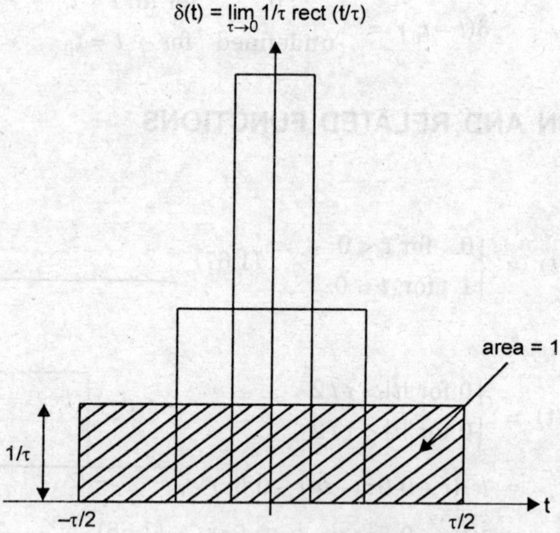

Tringular Function $$\delta(t) = \lim_{\tau \to 0} \frac{1}{\tau} \Lambda\left(\frac{t}{\tau}\right) \qquad \text{...(1.72)}$$

Two-sided Exponential $$\delta(t) = \lim_{\tau \to 0} \frac{1}{\tau} \exp(-2|t|/\tau) \qquad \text{...(1.73)}$$

Gaussian Pulse $$\delta(t) = \lim_{\tau \to 0} \frac{1}{\tau} \exp(-\pi(t/\tau)^2) \qquad \text{...(1.74)}$$

Sine-over-argument $$\delta(t) = \lim_{\tau \to 0} \frac{1}{\tau} \frac{\sin(\pi t/\tau)}{\pi t/\tau} \qquad \text{...(1.75)}$$

Even Symmetry $\delta(t) = \delta(-t)$, (same effect inside the integral) ...(1.76)

Time Scaling $\delta(at) = \dfrac{1}{|a|}\delta(t)$ ...(1.77)

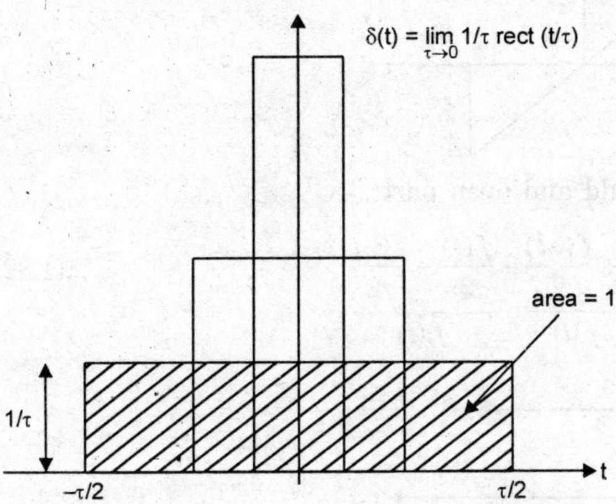

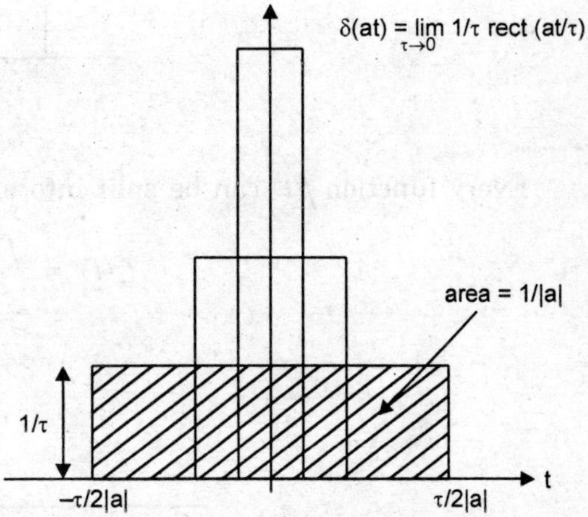

Multiplication by a Time Function

$$f(t).\delta(t-t_0) = f(t_0).\delta(t-t_0),\ f(t) \text{ continuous at } t_0. \qquad ...(1.78)$$

Relation to the Unit Step Function

$$\frac{d}{dt}\mathrm{A}.u(t-t_0) = \mathrm{A}.\delta(t-t_0) \qquad ...(179)$$

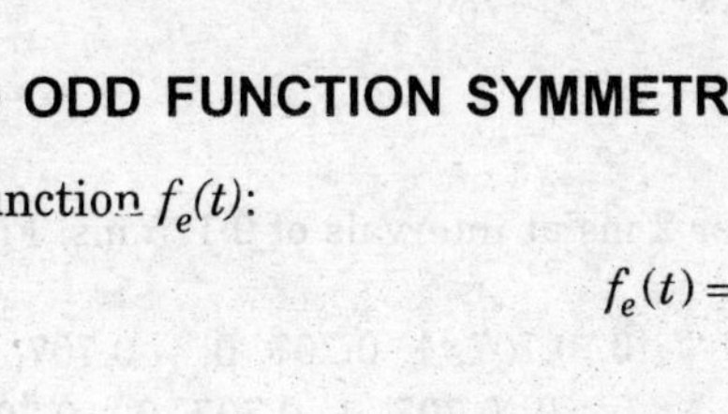

EVEN AND ODD FUNCTION SYMMETRY

Even function $f_e(t)$:

$$f_e(t) = -f_e(-t) \qquad ...(1.80)$$

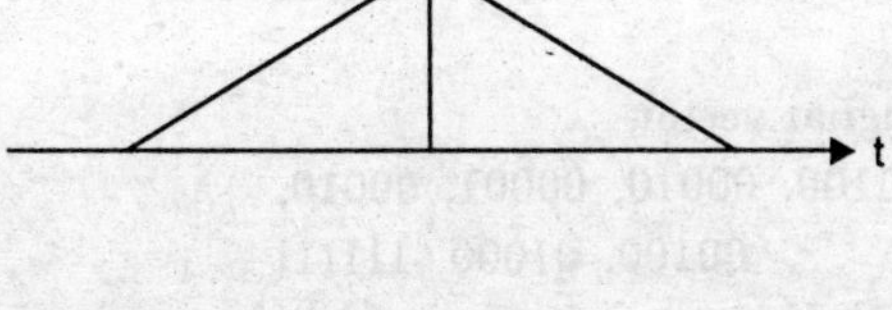

Odd function $f_0(t)$:

$$f_0(t) = -f_0(t) \qquad ...(1.81)$$

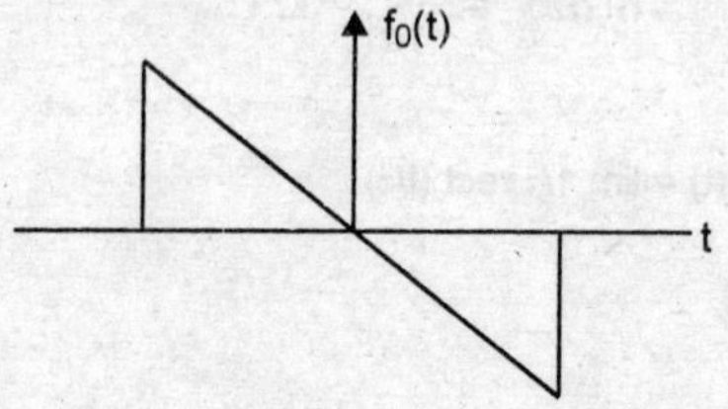

Every function *f(t)* can be split into an odd and even part:

$$f_e(t) = \underbrace{\frac{f(t)}{2} + \frac{f(-t)}{2}}_{f_e(t)} + \underbrace{\frac{f(t)}{2} - \frac{f(-t)}{2}}_{f_o(t)} \qquad ...(1.82)$$

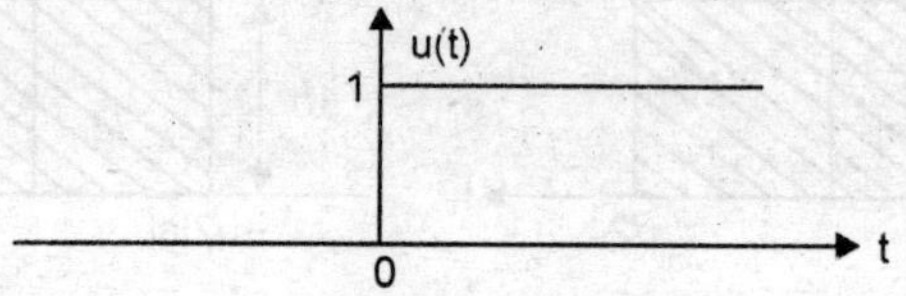

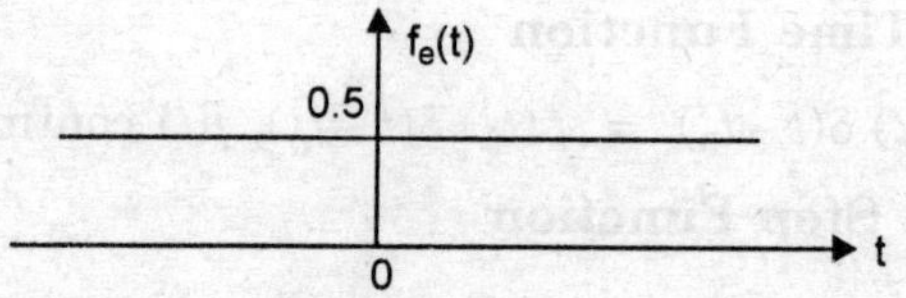

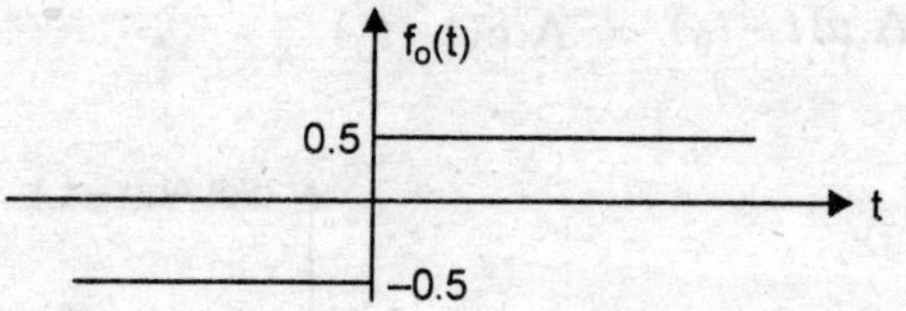

Example: Unit Step Function:

$$u(t) = \underbrace{\frac{u(t)}{2} + \frac{u(-t)}{2}}_{f_e(t)} + \underbrace{\frac{u(t)}{2} - \frac{u(t)}{2}}_{f_0(t)} \qquad ...(1.83)$$

EXERCISES

1. A sine wave signal of 1 kHz is measured for 2 ms at intervals of 0.125 ms. Find the signal vector, if amplitude is 1 volts.

   [0, 0.707, 1, 0.707, 0, – 0.707, –1, –0.707,0, 0, 0.707, 1, 0.707, 0, –0.707, –1, 0.707]

2. In a 5×7 matrix, the number 2 is represented by dot as shown.

   Write down the signal vector.

   [01100, 00010, 00001, 00010, 00100, 01000, 11111]

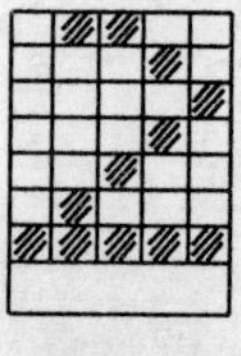

3. If a square wave is approximated by a sine wave (of the same period and amplitude), find the mean square error.

$$\frac{1}{\pi}\int_{o}^{\pi}(1-\sin\theta)^2 d\theta$$

4. Approximate the following signal vector of 6 points by a function $f(t) = a + bt$.

t =	−2	−1	0	2	3	5
Signal =	−1	0.7	2.3	5.6	7.4	10.7

[$f(t)$ = 2.3335 + 11.6713t.]

5. Find the first two sinusodial components of the signal vector's approximation to a Fourier series, using least square principle.

Q =	30°	60°	90°	120°	150°	180°	210°	240°	270°	300°	330°	360°
Value =	3.5	6.09	7.82	8.58	8.43	7.73	6.98	6.19	6.04	5.55	5.01	3.35

[12.545 −1.363 cos θ − 0.936 cos 2θ +1.97 sin θ + 0.235 sin 2θ]

6. Find the Legendre-Fourier series for the periodic function shown, for first 3 terms.

[0.5 + 3/4 $P_1(t)$ − 1/8 $P_2(t)$) + ...]

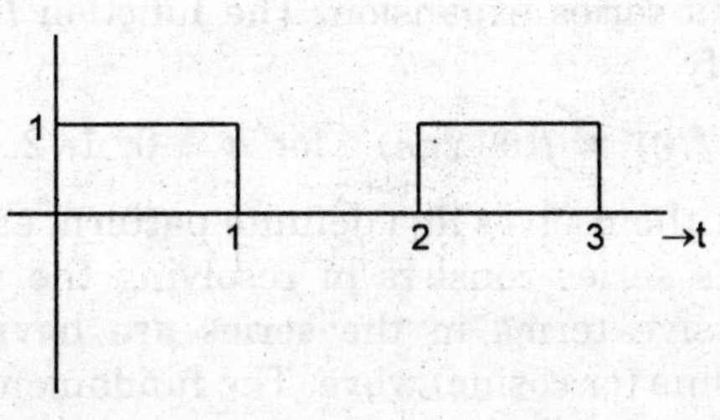

7. Show that $P_1(t) = t$ and $P_2(t) = \frac{3}{2}t^2 - \frac{1}{2}$ form an orthogonal set.

8. Show that (by integrating the product of the function and congugate) the complex function $e^{jm\theta} = f(\theta)$ is orthogonal.

$$\left[\textbf{Hint: } \int e^{jm\theta}e^{-jn\theta}d\theta = \int g_m(\theta)g_n^*(\theta)d\theta \text{ as per formula.}\right.$$

The limits are 0 to 2π. If $m \neq n$, answer is zero. If $m = n$, answer is 2π. Hence etc.]

9. Show that the following functions are orthogonal in the $\left(-\frac{1}{2}, \frac{1}{2}\right)$ interval.

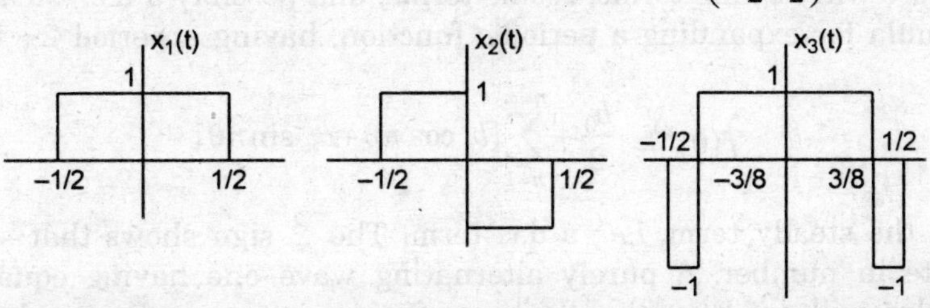

10. Expand the function $y(t)$ = sin $2\pi t$ terms of the above functions and find the error in the approximation. [$-2/\pi\ x_2(t)$ only; 0.0947]

2

FOURIER SERIES REPRESENTATION OF PERIODIC SIGNALS

FOURIER SERIES

We saw in chapter 1 how the sine-cosine series is a least squares approximation of a function $f(t)$. Any periodic wave which is non-sinusoidal can be resolved into a series of sinusoidal waves, by Fourier's series expansion. The function $f(\theta)$ which is periodic with a period 2π will therefore satisfy

$$f(\theta) = f(\theta + 2\pi n) \quad \text{for } n = 0, 1, 2... \qquad ...(2.1)$$

So, periodic waves repeat themselves in a definite pattern, each of the repeating patterns constituting a cycle. Fourier's series consists in resolving the wave into a sum of sine or cosine functions. The successive terms in the series are having frequencies an integral multiple of the *fundamental* sine (or cosine) wave. The fundamental wave is a sine (or cosine) wave having the same frequency as the given wave $f(\theta)$. For example, consider the square wave shown in Fig. 2.1 (*a*). The fundamental wave for this is shown in Fig. 2.1(*b*), below it. In this case, it is a sine wave. It is shown later that the square wave can be resolved into the following waves, by applying Fourier's series.

$$f(\theta) = \frac{4}{\pi}\left[\sin\theta + \frac{1}{2}\sin 3\theta + \frac{1}{5}\sin 5\theta + ...\right] \qquad ...(2.2)$$

The waves (*b*), (*c*), (*d*)... are the Fourier components of the given wave.

In this example of a square wave, we had only sine terms in the expansion. For a general complex wave there will be sine terms, cosine terms, and possibly a d.c. (steady terms) also. The general formula for expanding a periodic function, having a period 2π is :

$$f(\theta) = \frac{b_0}{2} + \sum_{n=1}^{n=\infty}[b_n \cos n\theta + a_n \sin n\theta] \qquad ...(2.3)$$

Here $b_0/2$ is the steady term, *i.e.*, a d.c. term. The Σ sign shows that sine and cosine terms are infinite in number. A purely alternating wave–one having equal positive and negative half cycles–will not have the d.c. term. Some waves may have only sine or cosine terms. The co-efficients a_n, b_n are calculated from

$$b_n = \frac{1}{\pi}\int_0^{2\pi} f(\theta)\cos n\theta \, d\theta \qquad \text{...(2.4)}$$

$$n = 0, 1, 2...$$

$$a_n = \frac{1}{\pi}\int_0^{2\pi} f(\theta)\sin n\theta \, d\theta \qquad \text{...(2.5)}$$

For the case of the square wave, $f(\theta) = 1$ between 0 and π; and -1 between π and 2π.

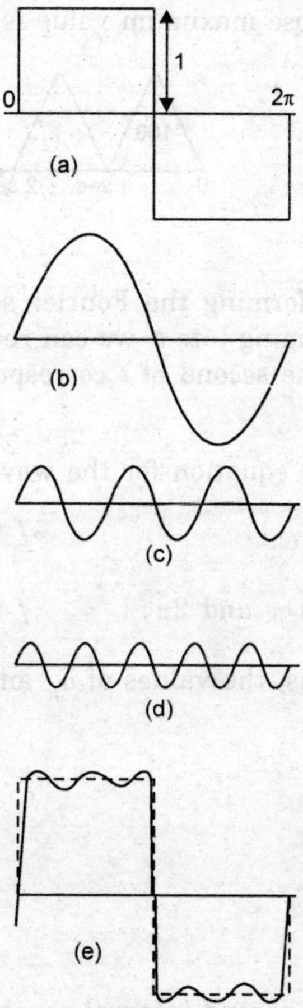

Fig. 1

(*a*) Given wave $f(\theta)$. One cycle corresponds to 2π. Amplitude is taken to be 1.

(*b*) The fundamental sine wave. Amplitude given by (2.2) as $4/\pi$.

(*c*) The wave $(4/\pi)\,\frac{1}{3}\sin 3\theta$, called the third *harmonic*. This is the second term in the expansion. Its frequency is 3 times the fundamental.

(*d*) The fifth harmonic, amplitude is 1/5 of the fundamental.

(*e*) When we add (*b*), (*c*) and (*d*), we get this wave, which nearly resembles the given wave in (*a*). If we added more terms in (2.2)—theoretically upto infinite terms—we get wave (*a*).

$$b_n = \frac{1}{\pi}\left\{\int_0^{\pi} 1\cos n\theta \, d\theta + \int_{\pi}^{2\pi} (-1)\cos n\theta \, d\theta\right\} = 0$$

$$a_n = \frac{1}{\pi}\left\{\int_0^{\pi} 1\sin n\theta \, d\theta + \int_{\pi}^{2\pi} (-1)\sin n\theta \, d\theta\right\}$$

$$= -\frac{2}{n\pi}[\cos n\pi - 1]$$

Hence cosine terms are absent in the series. For sine terms,

$$a_n = -\frac{2}{n\pi}[\cos n\pi - 1]$$

which is zero if n is even and $\frac{4}{n\pi}$ of n is odd. So the series becomes

$$\frac{4}{n}[\sin\theta + 1/3\sin 3\theta + 1/5\sin 5\theta + \ldots] \qquad \ldots(2.5)$$

Example 1. Find the Harmonics of a triangular wave shown whose period is 1 second and whose maximum value is 100 volts.

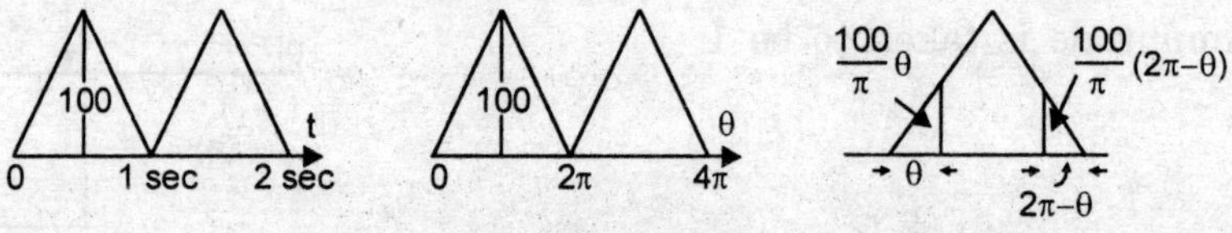

Fig. 2

In forming the Fourier series, let us consider a wave whose period is 2π. Then, by transforming into t, we can rewrite the series. Then let us take the wave shown in Fig. (b). Here, one second of t corresponds to 2π radians of θ.

$$t = \theta/2\pi$$

The equation for the wave must be written. Between θ and π,

$$f(\theta) = (100/\pi)\theta$$

Between π and 2π, $$f(\theta) = \frac{100}{\pi}(2\pi - \theta)$$

Thus, the values of a_n and b_n are to be found.

$$a_n = \frac{1}{\pi}\int_0^{2\pi} f(\theta)\sin n\theta\, d\theta$$

$$= \frac{100}{\pi^2}\left\{\int_0^{\pi}\theta\sin n\theta\, d\theta + \int_{\pi}^{2\pi}(2\pi-\theta)\sin n\theta\, d\theta\right\}$$

The second integral can be rewritten by putting $\alpha = 2\pi - \theta$ as

$$\int_{\pi}^{0}\alpha\sin n\alpha\, d\alpha = -\int_0^{\pi}\theta\sin n\theta\, d\theta$$

Thus, both first and second integrals are equal but opposite in sign. So, $a_n = 0$. Hence, no sine terms are present.

$$b_n = \frac{100}{\pi^2}\left(\int_0^{\pi}\theta\cos n\theta\, d\theta + \int_{\pi}^{2\pi}(2\pi-\theta)\cos n\theta\, d\theta\right)$$

The second integral, putting $\alpha = 2\pi-\theta$, becomes

$$-\int_{\pi}^{0} \alpha \cos n\alpha \, d\alpha = \int_{0}^{\pi} \theta \cos n\theta \, d\theta$$

Thus, both first and second integrals are equal. So,

$$b_n = \frac{200}{\pi^2}\int_0^{\pi} \theta \cos n\theta \, d\theta$$

$$= \frac{200}{\pi^2}\left[\frac{\theta \sin n\theta}{n} + \frac{\cos n\theta}{n^2}\right]_0^{\pi}$$

$$= \frac{200}{\pi^2 n^2}(\cos n\pi - 1)$$

which is equal to zero if n is even;

$$\text{to } -\frac{400}{\pi^2 n^2} \text{ if } n \text{ is odd.}$$

The average value, *i.e.*, the $b_0/2$ term is easily seen to be the average height of the triangle, which is 50. Hence

$$f(\theta) = 50 - \frac{400}{\pi^2}\left[\cos\theta + \frac{\cos 3\theta}{3^2} + \frac{\cos 5\theta}{5^2} + \ldots\right]$$

Replacing θ by the variable t,

$$f(\theta) = 50 - \frac{400}{\pi^2}\left[\cos 2\pi t + \frac{1}{3^2}\cos 6\pi t + \ldots\right]$$

CONVERGENCE OF FOURIER SERIES—DIRICHLET CONDITIONS

Dirichlet has formulated conditions for valid Fourier expansions. These conditions guarantee that the Fourier expansion of function *f(t)* will converge to *f(t)* at all points of continuity.

(1) *f(t)* must never become infinite in the defined time interval.

(2) *f(t)* must be single valued.

(3) *f(t)* must have at most a finite number of maxima and minima in this interval.

(4) *f(t)* must have, at the most, a finite number of discontinuities (including infinities) in the interval of definition.

It can be shown that at any point of discontinuity, say $x = a$, where a function is represented by a Fourier series, the Fourier series is $\frac{1}{2}[f(a+0)+f(a-0)]$.

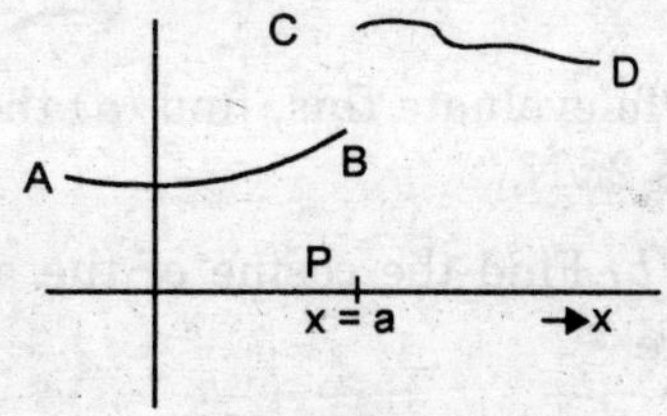

Fig. 2(a)

The function $f(x)$ is discontinuous at $x = a$. If $f(x)$ is expressed as Fourier series, its value at $x = a$ is $\frac{1}{2}(PB + PC)$.

Generally, since Fourier expansion is based on sinusoidal function, it is convenient to expand in the interval 0 to 2π, or $-\pi$ to $+\pi$.

EFFECT OF SYMMETRIES IN THE WAVE

The following simple rules enable us to find out which of the terms may be absent in the series, For example, in the previous example, the sine terms were absent.

The function of Fig. 3(*a*) is the same for both θ and $-\theta$. This is called *zero axis* symmetry. The function is such that

$$f(\theta) = +f(-\theta)$$

Fig. 3(*b*) is called an even function. For *even* functions, it may be shown that sine terms are absent.

This function is one possessing *zero-point* symmetry. Here

$$f(\theta) = -f(-\theta)$$

Hence is also called *odd* function. For odd functions, cosine functions are absent.

In Fig. 3(*c*), the wave is said to have half-wave symmetry *i.e.*, $f(\theta) = -f(\theta+\pi)$. In this case, the wave has no even harmonics.

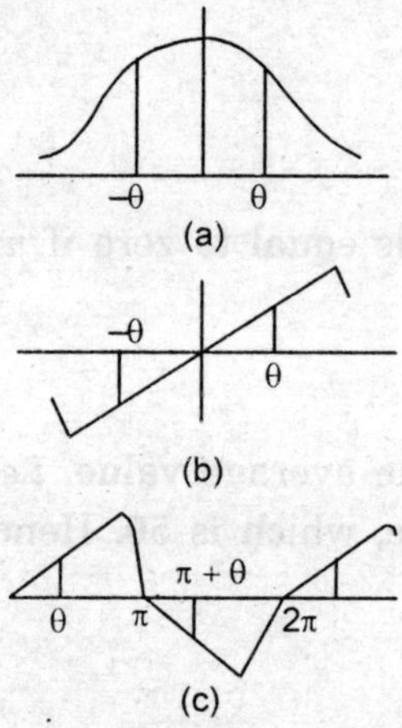

Fig. 3

NUMERICAL EVALUATION FOR A SIGNAL VECTOR

One period of the given wave is divided into N intervals and ordinates are measured at these intervals as shown.

$$b_0/2 = \text{average value} = \text{Area under the curve/Base.}$$

The area can be found using trapezoidal rule so that

$$b_0/2 = [(f_0 + f_N)/2 + f_1 + f_2 + f_3 + \ldots f_{N-1}] \div N$$

Next consider b_n.

$$b_n = (1/\pi)\int_0^{2\pi} f(\theta)\cos n\theta\, d\theta$$

$$= (1/\pi) \text{ Area under } f(\theta)\cos n\theta \text{ over one period.}$$

To evaluate this, find (*a*) the average ordinate in any interval which is $(f_i + f_{i+1})/2$ and $d\theta$ is $2\pi/N$.

(*b*) Find the cosine of the angle $n\theta$ at the middle of this interval. So $\cos n\theta_i$ is found where

$$\theta_i = n\left(i + \frac{1}{2}\right)2\pi/N.$$

(*c*) Multiply (*a*) and (*b*), sum up and then divide by π.

$$b_n = \frac{1}{N}\sum_{i=0}^{N-1}(f_i + f_{i+1})\cos\left[\left(i+\frac{1}{2}\right)2\pi n/N\right]$$

Similarly
$$a_n = \frac{1}{N}\sum_{i=0}^{N-1}(f_i + f_{i+1})\sin\left[\left(i+\frac{1}{2}\right)2\pi n/N\right] \quad ...(2.6)$$

The numerical method is to be employed when the irregular wave can not be expressed as a mathematical function.

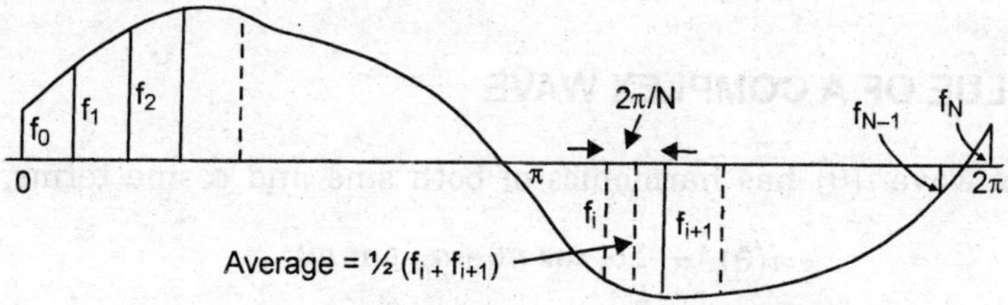

Fig. 4. Showing ordinates for irregular wave.

Example 2. Find the fundamental, 3rd harmonic and fifth harmonics of the following complex wave, with half wave symmetry, whose ordinates at 30° intervals starting from 0° are given in order as :

0, 1, 3, 4, 2, 1, 0, –1, –3, –4, –2, –1, 0 ....

The function can be roughly sketched. It will be found that it possesses half wave symmetry and is an *odd* function. Hence the series will have only sine terms of odd order. Equations (2.6) are applied. Here N = 12, as 2π/N = 30°.

For the fundamental, n = 1. The summation for the first 6 ordinates multiplied by 2 gives the total sum.

$$a_1 = 2\left[\sum_{0}^{5}(f_i + f_{i+1})\sin\left(i+\frac{1}{2}\right)30°\right] \div 12$$

$$= (1/6)[(0+1)\sin 15 + (1+3)\sin 45 + (3+4)\sin 75 + (4+2)\sin 105 + (2+1)\sin 135 + (1+0)\sin 165]$$

$$= 1/6[0.26 + 4 \times 0.707 + 7 \times 0.97 + 6 \times 0.97 + 3 \times 0.707 + 1 \times 0.26] = 3.026$$

$$a_3 = 1/6\left[\sum_{}^{5}(f_i + f_{i+1})\sin\left(i+\frac{1}{2}\right)3\times 30°\right]$$

$$= 1/6\,[1\sin 45 + 4\sin 135 + 7\sin 225 + 6\sin 315 + 3\sin 405 + 1\sin 495]$$

$$= 1/6[0.707 + 4 \times 0.707 - 7 \times 0.707 - 0.707 \times 6 + 3 \times 0.707 + 0.707 \times 1]$$

$$= -0.471$$

$$a_5 = 1/6\Sigma(f_i + f_{i+1})\sin\left(i+\frac{1}{2}\right)5\times30°$$

$$= 1/6[1 \sin 75 + 4 \sin 225 + 7 \sin 375 + 6 \sin 165$$
$$+ 3 \sin 315 + 1 \sin 105]$$
$$= 1/6\ [0.97 - 4 \times 0.707 + 7 \times 0.26 + 6 \times 0.26 - 3 \times 0.707 + 0.97]$$
$$= -0.37/6 = -0.06.$$

Hence, $$f(\theta) = 3.026 \sin\theta - 0.471 \sin 3\theta - 0.06 \sin 5\theta + \ldots$$

THE R.M.S. VALUE OF A COMPLEX WAVE

If a complex wave $i(\theta)$ has harmonics of both sine and cosine terms,

$$i(\theta) = \Sigma b_n \cos n\theta + a_n \sin n\theta$$
$$= \Sigma\sqrt{a_n^2 + b_n^2}\cos(n\theta + \varphi_n) \ldots (\tan\varphi = b_n / a_n)$$
$$= I_1\cos(\theta + \varphi_1) + I_2\cos(2\theta + \varphi_2) + \ldots$$

where I_1, I_2, I_3... are the maximum values of the harmonics and $I_n = [a_n^2 + b_n^2]^{1/2}$. The power produced in a resistor R by each harmonic current is $I_n^2R/2$; so, the total power will be $\Sigma I_n^2\ R/2$.

$$\therefore\quad I_{RMS} \text{ of a complex wave} = \sqrt{\Sigma I_n^2/2} \qquad \ldots(2.7)$$

If there is a d.c. component also, it becomes $\sqrt{I_{dc}^2 + \Sigma I_n^2/2}$.

Example 3. A voltage given by

$$v = 100 \sin\omega t + 50 \sin(3\omega t + 30°) + 20 \sin(\omega t - 90°)$$

is applied to an R-L circuit of impedance $(3 + j4)$ to the fundamental ω. Find the current and the power absorbed. Find also the overall p.f.

The current due to each of the harmonics is found in the usual way.

The current due to the fundamental component

$$= \frac{100}{3 + j4} = 20\angle -53°8'$$

The impedance for 3rd harmonic is $3+j\,(3 \times 4) = 3+j\,12$ and so, the third harmonic current is $50/(3+j12) = 4.042\ \angle -76°$.

The impedance due to 5th harmonic $= 3 + j\,(5 \times 4) = 3 + j\,20$, so that 5th harmonic current is $20/(3 + j20) = 1\ \angle -81°30'$.

Hence $$i = 20 \sin(\omega t - 53°8') + 4.042 \sin(3\omega t + 30 - 76°)$$
$$+ 1 \sin(5\omega t - 90 - 81°30')$$

The R.M.S. current can be found as

$$\sqrt{(20/\sqrt{2})^2 + (4.04/\sqrt{2})^2 + (1/\sqrt{2})^2} = 14.44\,\text{A}.$$

The power = $R\ I_{rms}^2 = 3 \times 14.44^2 = 627$ W.

The power factor = 627 / (RMS Volts × RMS Current)

$= 627 \div [\sqrt{(100^2 + 50^2 + 20^2)/2} \times 14.44]$

= 0.53.

RMS Voltage = $\sqrt{100 + 50^2 + 20^2 / 2} = 80$ V.

Excample. Determine the first 3 harmonics of the Fourier series for the values

x	0°	30°	60°	90°	120°	150°	180°	210°	240°	270°	300°	330°
y	–3.00	2.07	5.29	9.75	10.20	4.68	3.00	6.12	6.30	2.25	–3.79	–6.87

To tabulate the a's, tabulate the values as given below :–

x	y	$\cos\theta$	$y\cos\theta$	$\cos 2\theta$	$y\cos 2\theta$	$\cos 3\theta$	$y\cos 3\theta$
0°	–3.00	1	–3.00	1	–3.00	1	–3.0
30°	2.07	0.87	1.70	0.5	1.04	0	0
60°	5.29	0.5	2.65	–0.5	–2.65	–1	5.29
90°	9.57	0	0	–1	–9.57	0	0
120°	10.20	–0.5	–5.10	–0.5	–5.10	1	10.20
150°	4.68	–0.87	–4.07	0.5	2.34	0	0
180°	3.00	–1	–3.00	1	3.00	–1	–3.00
210°	6.12	–0.87	–5.32	0.5	3.06	0	0
240°	6.30	–0.5	–3.15	0.5	–3.15	1	6.30
270°	2.25	0	0	–1	–2.25	0	0
300°	–5.79	0.5	–1.89	0.5	1.89	–1	3.79
330°	–6.87	0.87	–5.98	0.5	–3.44	0	0
	35.82		–27.16		–17.83		9.00

$$a_0 = \frac{1}{6}(35.82) = 5.97$$

$$a_1 = \frac{1}{6}(-27.16) = 4.50$$

$$a_2 = \frac{1}{6}(-17.83) = 2.97$$

$$a_3 = \frac{1}{6}(9.00) = 1.50$$

To get the b's, the values are tabulated as below :

x	y	$\sin\theta$	$y\sin\theta$	$\sin 2\theta$	$y\sin 2\theta$	$\sin 3\theta$	$y\sin 3\theta$
0°	–3.00	0	0	0	0	0	0
30°	2.07	0.50	1.04	0.87	1.70	1	2.07
60°	5.29	0.87	4.60	0.87	4.60	0	0
90°	9.57	1.00	9.57	0	0	–1	–9.57
120°	10.20	0.87	8.87	–0.87	–8.87	0	0
150°	4.68	0.50	2.34	–0.87	–4.07	1	4.68
180°	3.00	0	0	0	0	0	0
210°	6.12	–0.50	–3.06	0.87	5.32	–1	–6.12
240°	6.30	–0.87	–5.48	0.87	5.48	0	0
270°	2.25	–1	–2.25	0	0	1	2.25
300°	–3.79	–0.87	3.30	–0.87	3.30	0	0
330°	–6.87	–0.5	3.44	–0.87	5.98	–1	6.87
			22.37		13.44		0.8

$$b_1 = \frac{1}{6}(22.37) = 3.73$$

$$b_2 = \frac{1}{6}(13.44) = 2.26.$$

$$b_3 = \frac{1}{6}(0.18) = 0.03$$

Hence $$y = 2.96 - 4.50\cos x + 3.73\sin x - 2.97\cos 2x$$
$$+2.26 \sin 2x + 1.50 \cos 3x + 0.03 \sin 3x$$

THE COMPLEX FORM OF FOURIER SERIES

The function $f(\theta)$ can be rewritten, after substituting the exponential forms for *sine* and *cosine* functions as

$$f(\theta) = \frac{1}{2}b_0 + \Sigma\left\{b_n.\frac{1}{2}(e^{jn\theta} + e^{-jn\theta}) - ja_n.\frac{1}{2}(e^{jn\theta} - e^{-jn\theta})\right\}$$

$$= \frac{1}{2}b_0 + \Sigma\left[\frac{1}{2}(b_n - ja_n)e^{jn\theta} + (b_n + ja_n)e^{-jn\theta}\right]$$

Let us find the value of the co-efficient $C_n = \frac{1}{2}(b_n - ja_n)$.

Then, from equation (2.4) giving b_n and a_n, we get

$$C_n = \frac{1}{2\pi}\int_0^{2\pi} f(\theta)(\cos n\theta - j\sin n\theta)\,d\theta$$

$$= \frac{1}{2\pi}\int_0^{2\pi} f(\theta)e^{-jn\theta}$$

Similarly, $$C_{-n} = \frac{1}{2}(b_n + ja_n) = \frac{1}{2\pi}\int_0^{2\pi} f(\theta)\, e^{jn\theta}\, d\theta \quad \text{(for } n = 0\text{)}$$

The factor $\frac{1}{2}b_0 = \frac{1}{2\pi}\int_0^{2\pi} f(\theta)\,d(\theta) = \frac{1}{2\pi}\int f(\theta)e^{j0\theta}\, d\theta = C_0$

Hence $$f(\theta) = C_0 + \Sigma\, C_n e^{jn\theta} + C_{-n}e^{-jn\theta}$$

$$= \sum_{-\infty}^{\infty} C_n e^{jn\theta} \quad ...(2.8)$$

where $$C_n = \left(\frac{1}{2}\pi\right)\int_0^{2\pi} f(\theta)e^{-jn\theta} d\theta \quad ...(2.9)$$

Note the n takes all values right from $-\infty$ to $+\infty$, inclucive of zero. The above is the complex from of Fourier's series.

Example 5. For a saw tooth wave of Fig. 5, find the complex form of the series.

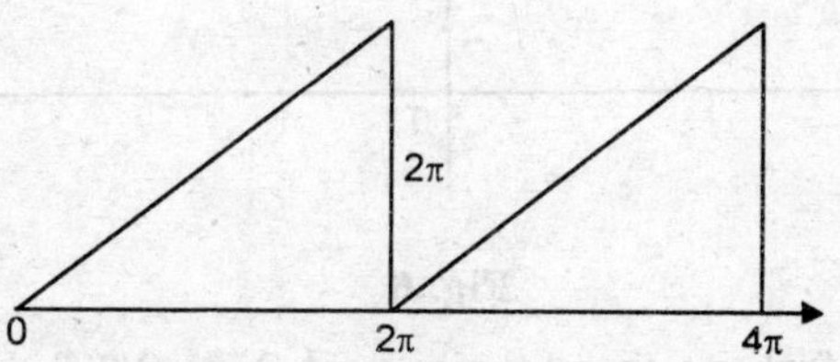

Fig. 5

$$f(\theta) = (\theta) \quad (0 < \theta < 2\pi)$$

$$C_n = \left(\frac{1}{2}\pi\right)\int_0^{2\pi} \theta\, e^{-jn\theta} d\theta$$

$$= j/n \quad \text{(for all } n \text{ except 0)}$$

$$= \pi \quad \text{(for } n = 0\text{)}$$

Hence the series is, from (2.7),

$$f(\theta) = \pi + \sum_{-\infty}^{\infty}\left(\frac{j}{n}\right)e^{jn\theta}$$

$$= -j\frac{e^{-j3\theta}}{3} - j\frac{e^{-j2\theta}}{2} - je^{-j\theta} + \pi + je^{j\theta} + \frac{je^{j2\theta}}{2} + \frac{je^{j3\theta}}{3} + ...$$

Clubbing the corresponding negative and positive terms,

$$f(\theta) = \pi - 2\left(\sin\theta + \frac{1}{2}\sin 2\theta + \frac{1}{3}\sin 3\theta + \ldots\right)$$

COMPLEX-FOURIER SPECTRUM

If we were to plot the 'spectrum', *i.e.*, the frequency versus real/imaginary parts, we get, for the previous problem only imaginary part (since sine terms only are there); there is a zero frequency value of π, of course.

Since the spectrum includes negative frequencies also, the imaginary part has both negative and positive frequencies at $\sin\theta$, $\sin 2\theta$ and $\sin 3\theta$....

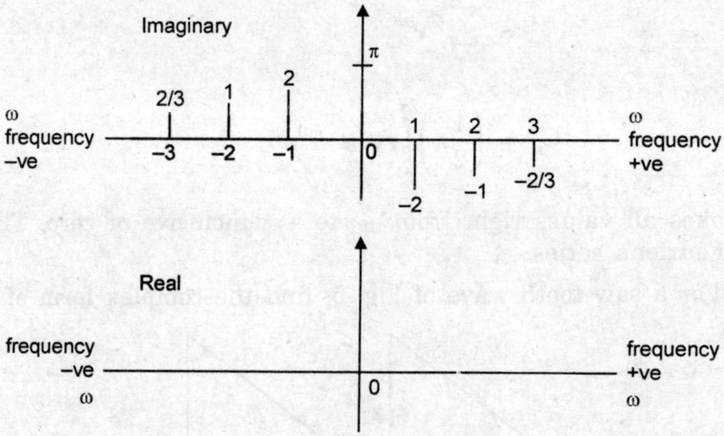

Fig. 6

Values are –2, –1, –2/3 for +ve frequencies and 2, 1, 2/3 for –ve frequencies.

PROBLEM SOLVING IN FOURIER SERIES

The following half range expansions in the interval $0 < x < l$ will be found useful :–

(*i*) $1 = \frac{4}{\pi}\left(\sin\frac{\pi x}{l} + \frac{1}{3}\sin\frac{3\pi x}{l} + \frac{1}{5}\sin\frac{5\pi x}{l}\ldots\right).$

(*ii*) $x = \frac{2l}{\pi}\left(\sin\frac{\pi x}{l} - \frac{1}{2}\sin\frac{2\pi x}{l} + \frac{1}{3}\sin\frac{3\pi x}{l}\ldots\right).$

(*iii*) $x = \frac{1}{2} - \frac{4l}{\pi^2}\left(\cos\frac{\pi x}{l} + \frac{1}{3^2}\cos\frac{3\pi x}{l} + \frac{1}{5^2}\cos\frac{5\pi x}{l}\ldots\right).$

(*iv*) $x^2 = \frac{l^2}{3} - \frac{4l^2}{\pi^2}\left(\cos\frac{\pi x}{l} - \frac{1}{2^2}\cos\frac{2\pi x}{l} + \frac{1}{3^2}\cos\frac{3\pi x}{l}\ldots\right).$

(v) $$x^2 = \frac{2l^2}{\pi^3}\left\{\left((\pi^2-4)\sin\frac{\pi x}{l}-\frac{\pi^2}{2}\sin\frac{2\pi x}{l}\right)\right\}+\left(\frac{\pi^2}{3}-\frac{4}{3^2}\right)\sin\frac{3\pi x}{l}-\frac{n^2}{4}\sin\frac{4\pi x}{l}$$

$$+\left(\frac{\pi^2}{5}-\frac{4}{5^3}\right)\sin\frac{5\pi x}{l}-\frac{\pi^2}{6}\sin\frac{6\pi x}{l}...\right\}$$

How the above series can be combined to give Fourier expansions in half range is illustrated in the following examples.

Example 1. Find a sine and a cosine series for the function $f(x) = 3x - 2$; $0 < x < 4$.

We have in the interval $0 < x < 4$.

$$x = \frac{8}{\pi}\left(\sin\frac{\pi x}{4}-\frac{1}{2}\sin\frac{2\pi x}{4}+\frac{1}{3}\sin\frac{3\pi x}{4}...\right)$$

$$2 = \frac{8}{\pi}\left(\sin\frac{\pi x}{4}-\frac{1}{3}\sin\frac{3\pi x}{4}+\frac{1}{5}\sin\frac{5\pi x}{4}...\right)$$

$$\therefore 3x-2$$

$$= \frac{24}{\pi}\left(\sin\frac{\pi x}{4}-\frac{1}{2}\sin\frac{2\pi x}{4}+\frac{1}{3}\sin\frac{3\pi x}{4}...\right)$$

$$-\frac{8}{\pi}\left(\sin\frac{\pi x}{4}+\frac{1}{3}\sin\frac{3\pi x}{4}...\right)$$

$$= \frac{8}{\pi}\left(2\sin\frac{\pi x}{4}-\frac{3}{2}\sin\frac{2\pi x}{4}+\frac{2}{3}\sin\frac{3\pi x}{4}\right.$$

$$\left.-\frac{3}{4}\sin\frac{4\pi x}{4}+\frac{2}{5}\sin\frac{5\pi x}{4}...\right)$$

The above is the sine series.

In the interval $0 < x < 4$, for finding the cosine series:

$$x = 2-\frac{16}{\pi^2}\left(\cos\frac{\pi x}{4}+\frac{1}{3^2}\cos\frac{3\pi x}{4}+\frac{1}{5^2}\cos\frac{5\pi x}{4}...\right)$$

$$\therefore \quad 3x-2 = 4-\frac{48}{\pi^2}\left(\cos\frac{\pi x}{4}+\frac{1}{3^2}\cos\frac{3\pi x}{4}+\frac{1}{5^2}\cos\frac{5\pi x}{4}...\right)$$

Example 2. Express in the interval $0 < x < l$ as a half range cosine and sine series the function $f(x) = x(l - x)$.

In the interval $0 < x < l$, we have

$$x = \frac{l}{2}-\frac{4l}{\pi^2}\left(\cos\frac{\pi x}{l}+\frac{1}{3^2}\cos\frac{3\pi x}{l}+\frac{1}{5^2}\cos\frac{5\pi x}{l}...\right)$$

$$x^2 = \frac{l^2}{3} - \frac{4l^2}{\pi^2}\left(\cos\frac{\pi x}{l} - \frac{1}{2^2}\cos\frac{2\pi x}{l} + \frac{1}{3^2}\cos\frac{3\pi x}{l}\ldots\right)$$

$$\therefore \quad x(l-x) = lx - x^2$$

$$= \frac{l^2}{2} - \frac{4l^2}{\pi^2}\left(\cos\frac{\pi x}{l} + \frac{1}{3^2}\cos\frac{3\pi x}{l} + \frac{1}{5^2}\cos\frac{5\pi x}{l}\ldots\right),$$

$$+ \frac{l^2}{3} - \frac{4l^2}{\pi^2}\left(\cos\frac{\pi x}{l} - \frac{1}{2^2}\cos\frac{2\pi x}{l} + \frac{1}{3^2}\cos\frac{3\pi x}{l}\ldots\right)$$

$$= \frac{l^2}{6} - \frac{4l^2}{\pi^2}\left(\frac{1}{2^2}\cos\frac{2\pi x}{l} + \frac{1}{4^2}\cos\frac{4\pi x}{l}\frac{1}{6^2}\cos\frac{6\pi x}{l}\ldots\right)$$

In the interval $0 < x < l$, we have

$$x = \frac{2l}{\pi}\left(\sin\frac{\pi x}{l} - \frac{1}{2}\sin\frac{2\pi x}{l} + \frac{1}{3}\sin\frac{3\pi x}{l}\ldots\right)$$

$$x^2 = \frac{2l^2}{\pi^3}\left\{(\pi^2-4)\sin\frac{\pi x}{l} - \frac{\pi^2}{2}\sin\frac{2\pi x}{l} + \left(\frac{\pi^2}{3} - \frac{4}{3^3}\right)\sin\frac{3\pi x}{l} - \frac{\pi^2}{4}\sin\frac{4\pi x}{l}\ldots\right\}$$

$$\therefore \quad x(l-x) = lx - x^2$$

$$= \sin\frac{\pi x}{l}\left\{\frac{2l^2}{\pi} - \frac{2l^2}{\pi^3}(x^2-4)\right\} + \sin\frac{2\pi x}{l}\left\{-\frac{l^2}{n\pi} + \frac{l^2}{\pi}\right\}$$

Example 3. Find a Fourier series with period 3 to represent $f(x) = 2x - x^2$ in the range (0, 3).

Hence $\quad 2l = 3, \therefore l = \frac{3}{2}$

$$f(x) = \frac{a_0}{2} + \sum_{n=1}^{\infty}\left(a_n\cos\frac{n\pi x}{l} + b_n\sin\frac{n\pi x}{l}\right).$$

In this case $\quad f(x) = \frac{a_0}{2} + \sum_{n=1}^{\infty}\left(a_n\cos\frac{2n\pi x}{3} + b_n\sin\frac{2n\pi x}{3}\right)$

where $\quad a_n = \frac{1}{l}\int_0^{2l} f(x)\cos\frac{n\pi x}{l}\,dx$

$$= \frac{2}{3}\int_0^3 (2x - x^2)\cos\frac{2n\pi x}{3}\,dx = -\frac{9}{n^2\pi^2}$$

$$a_0 = \frac{2}{3}\int_0^3 (2x - x^2)dx = 0$$

$$b_n = \frac{2}{3}\int_0^3 (2x - x^2)\sin\frac{2n\pi x}{3}dx = \frac{3}{n\pi}$$

$$\therefore \quad 2x - x^2 = -\frac{9}{\pi^2}\sum_{n=1}^{\infty}\frac{1}{n^2}\cos\frac{2n\pi x}{3} + \frac{3}{\pi}\sum_{n=1}^{\infty}\frac{1}{n}\sin\frac{2n\pi x}{3}.$$

PROPERTIES OF FOURIER SERIES AND APPLICATIONS

These properties of odd and even functions can be used to shorten the computation when we have to find the Fourier series of either an even or odd function for the interval $-\pi < x < \pi$.

If $f(x)$ be expanded as a Fourier series of the form $\frac{a_0}{2} + \sum_{n=1}^{\infty}(a_n \cos nx + b_n \sin nx)$,

we have
$$a_n = \frac{1}{\pi}\int_{-\pi}^{\pi} f(x)\cos nx\,dx \qquad (n = 0, 1, 2, \ldots)$$

$$b_n = \frac{1}{\pi}\int_{-\pi}^{\pi} f(x)\cos nx\,dx \qquad (n = 0, 1, 2, \ldots)$$

Case (*i*) **$f(x)$ is an odd function,** then $f(x)$ cos nx is also an odd function.

$$\therefore \quad \int_{-\pi}^{\pi} f(x)\cos nx\,dx = 0$$

Hence
$$a_n = 0$$

$f(x)$ sin nx is an even function.

$$\therefore \quad \int_{-\pi}^{\pi} f(x)\sin nx\,dx = 2\int_0^{\pi} f(x)\sin nx\,dx.$$

Hence
$$b_n = \frac{2}{\pi}\int_0^{\pi} f(x)\sin nx\,dx$$

Case (*ii*) **If $f(x)$ is an even function,** then $f(x)$ sin nx is an odd function and hence

$$\int_{-\pi}^{\pi} f(x)\sin nx = 0$$

$$\therefore \qquad b_n = 0$$

$f(x) \cos nx$ is an even function.

$$\therefore \qquad \int_{-\pi}^{\pi} f(x) \cos nx\,dx = 2\int_0^{\pi} f(x) \cos nx\,dx$$

$$\therefore \qquad a_n = \frac{2}{\pi}\int_0^{\pi} f(x) \cos nx\,dx.$$

Hence we get the results that

(*i*) If $f(x)$ is an even function, $f(x)$ can be expanded as a series of the form

$$\frac{a_0}{2} + \sum_{n=1}^{\infty} a_n \cos nx$$

in the interval $(-\pi < x < \pi)$ where

$$a_n = \frac{2}{\pi}\int_0^{\pi} f(x) \sin nx\,dx. \qquad (n = 0, 1, 2, ...);$$

(*ii*) If $f(x)$ is an odd function, $f(x)$ can be expanded as a series of the form

$$\sum_{n=1}^{\infty} b_n \sin nx$$

in the interval $(-\pi < x < \pi)$ where

$$b_n = \frac{2}{\pi}\int_0^{\pi} f(x) \sin nx\,dx$$

Example 1. Express $f(x) = x$ $(-\pi < x < \pi)$ as a Fourier series with period 2π.

$f(x) = x$ is an odd function.

Hence in the expansion, the cosine terms are absent.

$$\therefore \qquad x = \Sigma b_n \sin nx$$

where

$$b_n = \frac{2}{\pi}\int_0^{\pi} x \sin nx\,dx$$

$$= \frac{2}{\pi}\left[\frac{-x \cos nx}{n}\right]_0^{\pi} + \int_0^{\pi} \frac{\cos nx}{n}\,dx$$

$$= -\frac{2}{\pi}\cos n\pi = -\frac{2}{\pi}(-1)^n = \frac{(-1)^{n+1}2}{n}$$

$$\therefore \qquad x = 2\left(\sin x - \frac{1}{2}\sin 2x + \frac{1}{3}\sin 3x - ...\right).$$

Example 2. If $f(x) = -x \text{ in } -\pi < x < 0.$

$= x \text{ in } 0 \le x < \pi$

expand $f(x)$ as a Fourier series in the interval $-\pi$ to π.

Deduce that $\dfrac{\pi^2}{8} = 1+\dfrac{1}{3^2}+\dfrac{1}{5^2}+\dfrac{1}{7^2}+\dots$

We easily see that $f(x)$ is an even function by drawing the graph of the function and noting that it is symmetrical with respect to the y axis.

$$\therefore \quad f(x) = \frac{a_o}{2}+\sum_{n=1}^{\infty} a_n \cos nx$$

where $a_n = \dfrac{2}{\pi}\displaystyle\int_0^{\pi} f(x)\cos nx\, dx (n = 0, 1, 2\dots)$

$$a_0 = \frac{2}{\pi}\int_0^{\pi} f(x)\,dx = \frac{2}{\pi}\int_0^{\pi} x\,dx = \frac{2}{\pi}\left[\frac{x^2}{2}\right]_0^{\pi} = \pi$$

$$a_n = \frac{2}{\pi}\int_0^{\pi} x\cos nx\,dx.$$

$$= \frac{2}{\pi}\left\{\left[\frac{x\sin nx}{n}\right]_0^{\pi} - \int_0^{\pi}\frac{\sin nx}{n}dx\right\}$$

$$= \frac{2}{n^2\pi}\left[\cos nx\right]_0^{\pi} = \frac{2}{n^2\pi}(\cos n\pi - 1)$$

$$= \frac{2}{n^2\pi}\left\{(-1)^n - 1\right\}$$

When n is odd, $a_n = \dfrac{4}{n^2\pi}$

When n is even, $a_n = 0$

$$\therefore \quad f(x) = \frac{\pi}{2}-\frac{4}{\pi}\left[\frac{\cos x}{1^2}+\frac{\cos 3x}{3^2}+\frac{\cos 5x}{5^2}+\dots\right]$$

When $x = 0, f(x) = 0.$

$$\therefore \quad 0 = \frac{\pi}{2}-\frac{4}{\pi}\left(\frac{1}{1^2}+\frac{1}{3^2}+\frac{1}{5^2}+\dots\right).$$

$$\therefore \quad \frac{1}{1^2}+\frac{1}{3^2}+\frac{1}{5^2}+\dots = \frac{\pi^2}{8}.$$

HALF RANGE FOURIER SERIES

It is often convenient to obtain a Fourier expansion of a function to hold for a range which is half the period of the Fourier series, that is to expand $f(x)$ in the range $(0, \pi)$ in a Fourier series of period 2π. In the half range $f(x)$ can be expanded as a series containing cosines alone or sines alone.

The following identities are very useful in this connection :

(i) $\int_0^\pi \cos mx\, dx = 0$ if m is an integer.

(ii) $\int_0^\pi \cos mx \cos nx\, dx = 0$ if $m \neq n$ and m and n are integers.

(iii) $\int_0^\pi \sin mx \sin nx\, dx = 0$ if $m \neq n$ and m and n are integers.

(iv) $\int_0^\pi \cos mx \cos nx\, dx = \int_0^\pi \cos^2 mx\, dx$ if $m = n = \frac{\pi}{2}$.

(v) $\int_0^\pi \sin mx \sin nx\, dx = \int_0^\pi \sin^2 mx\, dx$ if $m = n = \frac{\pi}{2}$.

DEVELOPMENT IN COSINE SERIES

Let $f(x)$ be expanded as a series containing cosines only and let

$$f(x) = \frac{a_o}{2} + \sum_{n=1}^{\infty} a_n \cos nx \qquad ...(1)$$

If we integrate both sides of (1) between limits 0 and π, then

$$\int_0^\pi f(x)\, dx = \int_0^\pi \frac{a_0}{2} dx + \sum_{n=1}^{\infty} a_n \int_0^\pi \cos nx\, dx = \frac{a_0 \pi}{2}$$

$$\therefore \qquad a_0 = \frac{2}{\pi}\int_0^\pi f(x)\, dx$$

If we multiply both sides of the equation (1) by cos nx and integrate between 0 and π, then

$\int_0^\pi f(x) \cos nx\, dx = a_n \frac{\pi}{2}$, since all the terms except the term containing a_n vanish.

$$\therefore \qquad a_n = \frac{2}{\pi}\int_0^\pi f(x) \cos nx\, dx.$$

DEVELOPMENT IN SINE SERIES

Let $f(x)$ be expanded as a series containing sines only and let

$$f(x) = \sum_{n=1}^{\infty} b_n \sin nx.$$

Multiply both sides of the above equation by sin nx and integrate from 0 to π.

Then $\int_0^{\pi} f(x) \sin nx\, dx = b_n \frac{\pi}{2}$ since all the terms except the term containing b_n vanish.

$$\therefore \qquad b_n = \frac{2}{\pi}\int_0^{\pi} f(x) \sin nx\, dx$$

Example 1. Find a sine series for $f(x) = c$ in the range 0 to π.

Let
$$f(x) = \sum_{n=1}^{\infty} b_n \sin nx$$

where $b_n = \frac{2}{\pi}\int_0^{\pi} f(x) \sin nx\, dx$

$$= \frac{2}{\pi}\int_0^{\pi} c \sin nx\, dx = -\frac{2c}{n\pi}\left[\frac{\cos nx}{n}\right]_0^{\pi}$$

$$= \frac{2c}{n\pi}(1 - \cos n\pi)$$

$$= \frac{2c}{n\pi}\left[1-(-1)^n\right]$$

When n is even, $\qquad b_n = 0$

When n is odd, $\qquad b_n = \frac{4c}{n\pi}$

Hence
$$c = \frac{4c}{\pi}\left(\sin x + \frac{1}{3}\sin 3x + \frac{1}{5}\sin 5x \ldots\right).$$

Example 2. If
$$f(x) = x \text{ when } 0 < x < \frac{\pi}{2}$$
$$= \pi - x \text{ when } x > \frac{\pi}{2}$$

expand $f(x)$ as a sine series in the interval $(0, \pi)$.

Let $$f(x) = \sum_{n=1}^{\infty} b_n \sin nx$$

where $$b_n = \frac{2}{\pi}\int_0^{\pi} f(x)\sin nx\, dx$$

$$= \frac{2}{\pi}\int_0^{\pi/2} x\sin nx\, dx + \frac{2}{\pi}\int_0^{\pi/2}(\pi - x)\sin nx\, dx$$

$$= \frac{2}{\pi}\left\{\left[\frac{-x\cos nx}{n}\right]_0^{\pi/2} + \int_0^{\pi/2}\frac{\cos nx}{n}dx\right\}$$

$$+\frac{2}{\pi}\left\{\left[\frac{-(\pi - x)\cos nx}{n}\right]_{\pi/2}^{\pi} - \int_{\pi/2}^{\pi}\frac{\cos nx}{n}dx\right\}$$

$$= \frac{2}{n^2\pi}[\sin nx]_0^{\pi/2} - \frac{2}{\pi}[\sin nx]_{\pi/2}^{\pi}$$

$$= \frac{2}{n^2\pi}\sin\frac{n\pi}{2} + \frac{2}{n^2\pi}\sin\frac{n\pi}{2}$$

$$= \frac{4}{n^2\pi}\sin\frac{n\pi}{2}.$$

Example 3. Find a cosine series in the range 0 to π for

$$f(x) = x \text{ for } 0 < x < \frac{\pi}{2}$$

$$= (\pi - x)) \text{ for } \frac{\pi}{2} < x < \pi.$$

Let $$f(x) = \frac{a_0}{2} + \sum_{n=1}^{\infty} a_n \cos nx$$

where $a_n = \frac{2}{\pi}\int_0^{\pi} f(x)\cos nx\, dx.$

$\therefore$ $$a_0 = \frac{2}{\pi}\int_0^{\pi/2} x\,dx + \frac{2}{\pi}\int_{\pi/2}^{\pi}(\pi - x)dx$$

$$= \frac{2}{\pi}\left[\frac{x^2}{2}\right]_0^{\pi/2} - \frac{2}{\pi}\left[\frac{(\pi - x)^2}{2}\right]_{\pi/2}^{\pi} = \frac{\pi}{2}$$

$$a_n = \frac{2}{\pi}\int_0^{\pi/2} x\cos nx\,dx + \frac{2}{\pi}\int_{\pi/2}^{\pi}(\pi - x)\cos nx\,dx$$

$$= \frac{2}{\pi}\left\{\left[\frac{(x\sin nx)}{n}\right]_0^{\pi/2} - \frac{1}{n}\int_0^{\pi}\sin nx\,dx\right\}$$

$$+\frac{2}{\pi}\left\{\left[\frac{(\pi - x)\sin nx}{n}\right]_{\pi/2}^{\pi} + \frac{1}{n}\int_{\pi/2}^{\pi}\sin nx\,dx\right\}$$

$$= \frac{2}{\pi}\left\{\frac{\pi}{2\pi}\sin\frac{n\pi}{2} - \frac{1}{n^2}[\cos nx]_{\pi/2}^{\pi}\right\}$$

$$+\frac{2}{\pi}\left\{-\frac{\pi}{2n}\sin\frac{n\pi}{2} - \frac{1}{n^2}[\cos nx]_{\pi/2}^{\pi}\right\}$$

$$= \frac{2}{\pi}\left\{-\frac{1}{n^2} + \frac{\cos\frac{n\pi}{2}}{n^2} - \frac{1}{n^2}\cos n\pi + \frac{1}{n^2}\cos\frac{n\pi}{2}\right\}$$

$$= \frac{2}{\pi}\left\{\frac{-1-(-1)^n + 2\cos\frac{n\pi}{2}}{n^2}\right\}$$

When n is odd, $a_n = 0.$

When n is even and is of the form $4p$,

$$a_n = 0$$

When n is even and is of the form $4p + 2$,

$$a_n = -\frac{8}{n^2\pi}$$

$$\therefore \quad a_1 = a_3 = a_5 \ldots = 0$$

$$a_4 = a_8 = a_{12} \ldots = 0$$

$$a_2 = -\frac{8}{2^2\pi} = -\frac{2}{1^2\pi}$$

$$a_6 = -\frac{8}{6^2\pi} = -\frac{2}{3^2\pi}$$

$$a_{10} = -\frac{8}{10^2\pi} = -\frac{2}{5^2\pi}$$

$$\therefore \quad f(x) = \frac{\pi}{4} - \frac{2}{\pi}\left(\frac{\cos 2x}{1^2} + \frac{\cos 6x}{3^2} + \frac{\cos 10x}{5^2} + \ldots\ldots\right).$$

Spectrum of Periodic Signals

We begin with the plots of Fourier spectra using some examples.

1. *Determine the fundamental frequency of the following signals, and plot their sinusoidal spectra (both magnitude and phase) and their exponential spectra.*

(a) $x(t) = 2 + 3\cos(0.2t) + \cos(0.25 + \pi/2) + 4\cos(0.3t - \pi)$

(b) $x(t) = 1 + 10\cos(2\pi(60)t + \pi/8) + 2\cos(2\pi(300)t - \pi/4)$.

Solution. *(a)* $$x(t) = 2 + 3\cos(0.2t) + \cos(0.25t + \pi/2) + 4\cos(0.3t - \pi)$$

Frequencies are 0.2 rad/sec, 0.25 rad/sec, 0.3 rad/sec

The greatest common divisor is 0.05 rad/sec (all the frequencies are divided by 0.05 in an integer number of ways)

$$\Rightarrow \qquad \omega_0 = 0.05 \text{ rad/sec}$$

Sinusoidal spectrum

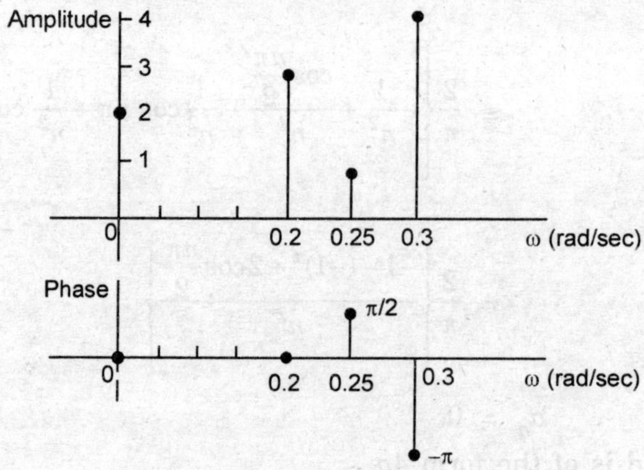

Exponential spectrum

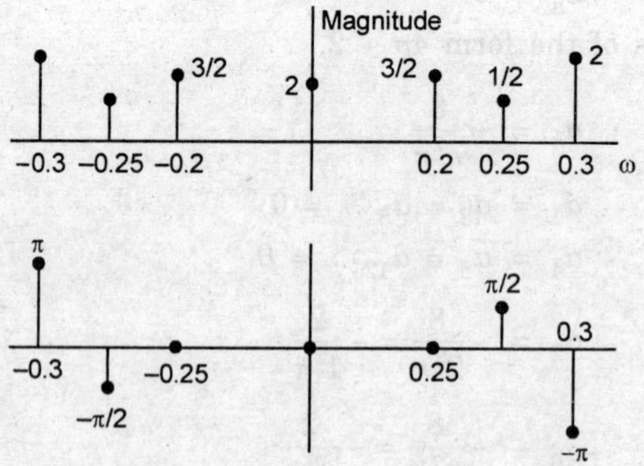

(b) $$x(t) = 1 + 10\cos(0.2\pi(60)t + \pi/8) + 2\cos(2\pi(300)t - \pi/4)$$

Greater common divisor is $2\pi(60)$ rad/sec

or 60Hz $\Rightarrow$ $$f_0 = 60 \text{ Hz}$$

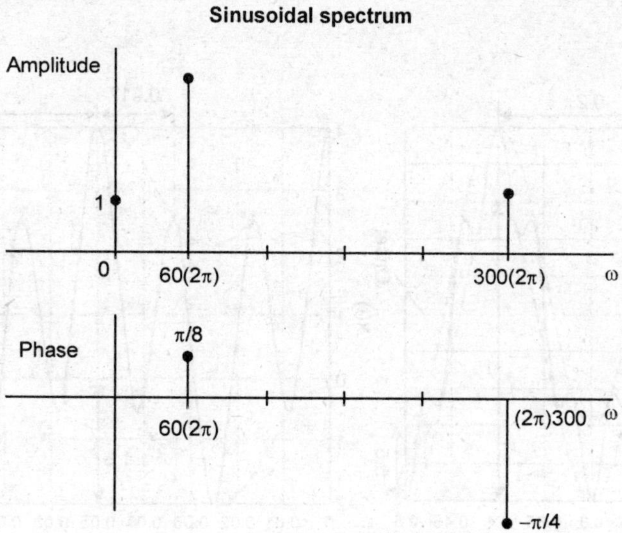

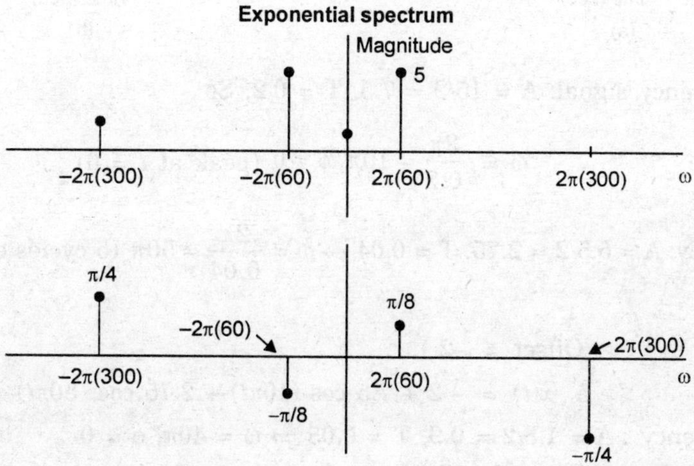

2. *Find an expression for x(t) and plot the spectrum for each graph.*

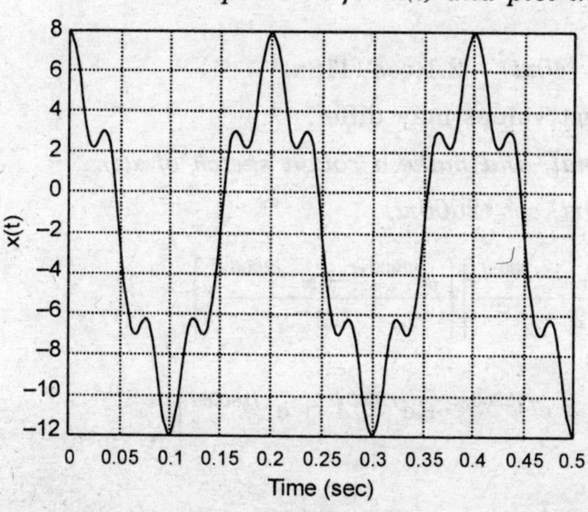

(a)

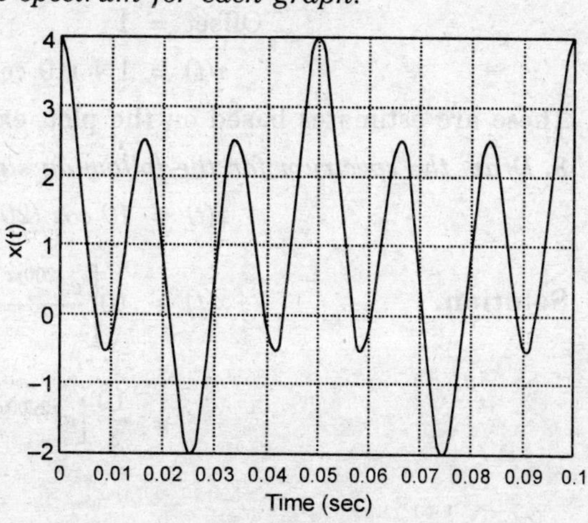

(b)

Solution.

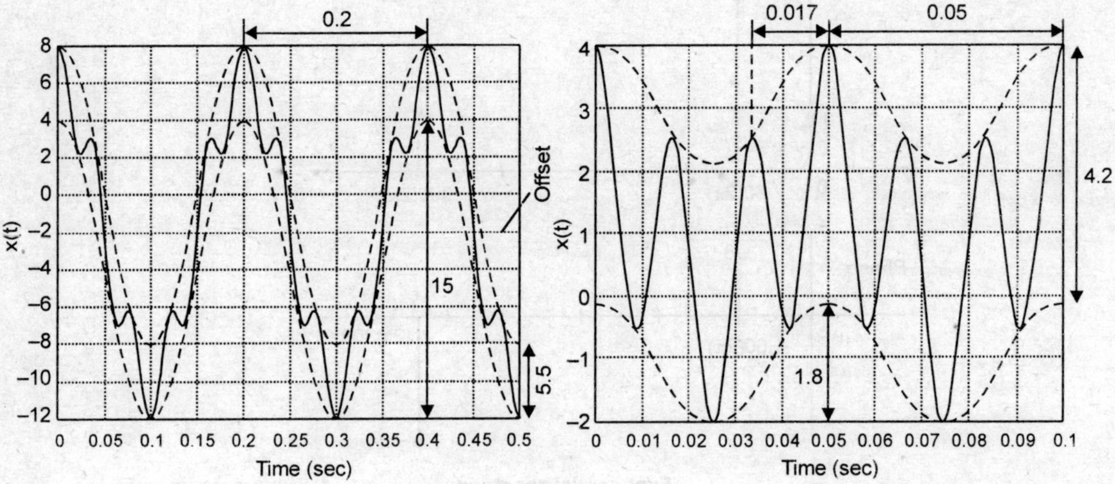

(a) (b)

(*a*) Low frequency signal: A = 15/2 = 7.5, T = 0.2, So

$$\omega = \frac{2\pi}{0.2} = 10\pi, \phi = 0 \text{ (peak at } t = 0)$$

High frequency: A = 5.5/2 = 2.75, T = 0.04 → $\omega = \dfrac{2\pi}{0.04} = 50\pi$ (5 cycles of high frequency in 0.2 sec), $\phi = 0$

$$\text{Offset} = -2$$

$$x(t) = -2 + 7.5 \cos (10\pi t) + 2.75 \cos (50\pi t)$$

(*b*) Low frequency : A = 1.8/2 = 0.9, T = 0.05 ⇒ $\omega = 40\pi$, $\phi = 0$

High frequency: A = 4.2/2 = 2.1, T = 0.017 (or 3 cycles in 0.05 sec), $\phi = 0$, $\omega = 118\pi$

$$\text{Offset} = 1$$

$$x(t) = 1 + 0.9 \cos (40\pi t) + 2.1 \cos (118\pi t)$$

These are estimates based on the plot, exact values may differ.

3. *Draw the spectrum for the following signal, and make a rough sketch of x(t).*

$$x(t) = 10 \cos (200\pi t) \cos (2000\pi t)$$

Solution.

$$x(t) = 10\left[\frac{e^{j200\pi t} + e^{-j200\pi t}}{2}\right]\left[\frac{e^{j2000\pi t} + e^{-j2000\pi t}}{2}\right]$$

$$= \frac{10}{4}\left[e^{j2200\pi t} + e^{-j2200\pi t} + e^{j1800\pi t} + e^{-j1800\pi t}\right]$$

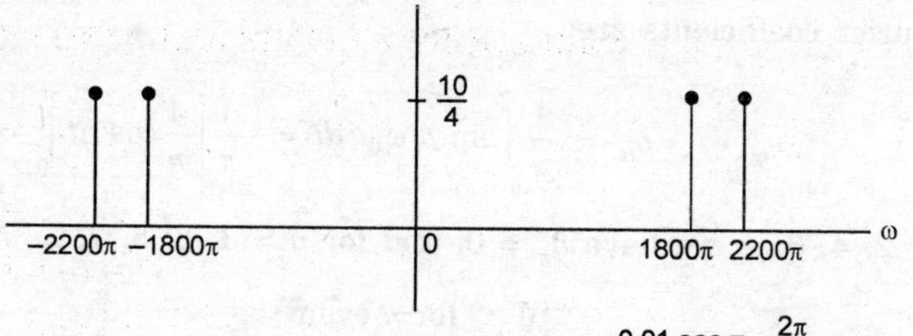

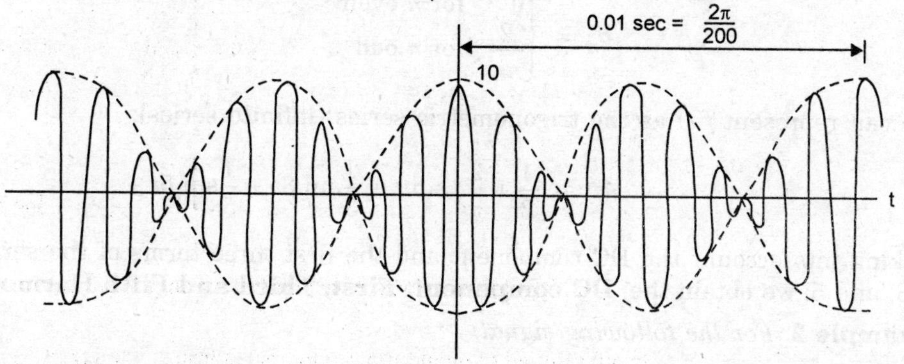

Waveform-based Problems

Example 1. *Represent the following using the **Trigonometric Fourier Series** representation.*

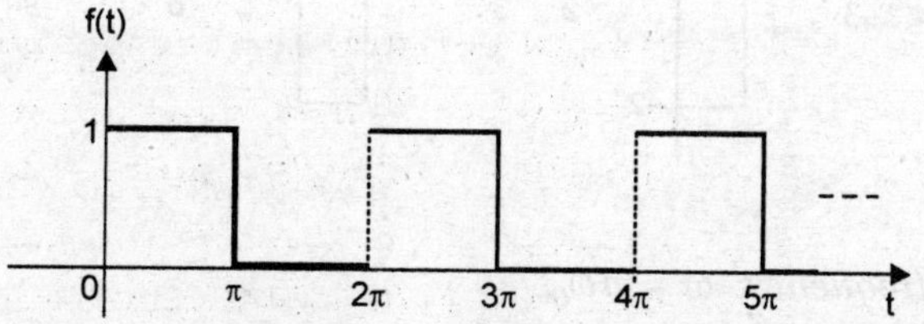

$$f(t) = \begin{cases} 1 & 0 < t < \pi \\ 0 & \pi < t < 2\pi \end{cases}$$

For the signal the period is:

$$T_0 = 2\pi, \text{ and } \omega_0 = \frac{2\pi}{T_0} = 1$$

The average or **DC** component for $f(t)$ is given by:

$$a_0 = \frac{1}{2\pi}\int_0^{2\pi} f(t)\,dt = \frac{1}{2\pi}\int_0^{\pi} dt = \frac{1}{2\pi}[t]_0^{\pi} = \frac{1}{2}$$

Note that $f(t) = 0$ in the time interval π to 2π.

The a_n Fourier coefficients are:

$$a_n = \frac{1}{\pi}\int_0^{\pi} \cos n\,\omega_0 t\,dt = \frac{1}{n\pi}[\sin nt]_0^{\pi} = \frac{\sin n\pi}{n\pi} = 0$$

Remember that $\omega_0 = 1$ in the example.

The b_n Fourier coefficients are:

$$b_n = \frac{1}{\pi}\int_0^{\pi} \sin n\,\omega_0 t\, dt = -\frac{1}{\pi}\left[\frac{1}{n}\cos nt\right]_0^{\pi} = -\frac{[\cos n\pi - 1]}{n\pi}$$

For $n = 0, 2, 4, 6, \ldots$ we have $b_n = 0$, and for $n = 1, 3, 5, 7, \ldots$, $b_n = 2/n\pi$, then

$$b_n = \begin{cases} 0 & \text{for } n \text{ even} \\ \dfrac{2}{n\pi} & \text{for } n \text{ odd} \end{cases}$$

and we can represent $f(t)$ as the trigonometric series (infinite series):

$$f(t) = \frac{1}{2} + \frac{2}{\pi}\left[\sin t + \frac{1}{3}\sin 3t + \frac{1}{5}\sin 5t + \ldots\right]$$

Taking into account the DC component and the first three terms of the series, this is, $n = 1, 3$, and 5; we obtain the, **DC component, First, Third and Fifth Harmonic of $f(t)$.**

Example 2. *For the following signal:*

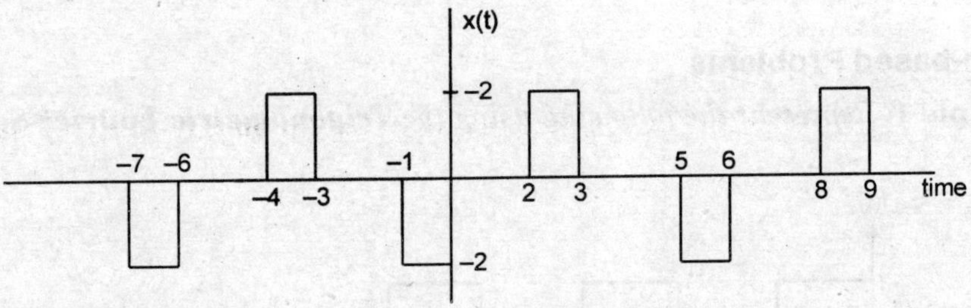

(a) Find the Fourier series

(b) Plot ∴he spectra verus frequency, $\omega = n\omega_0$.

Solution.

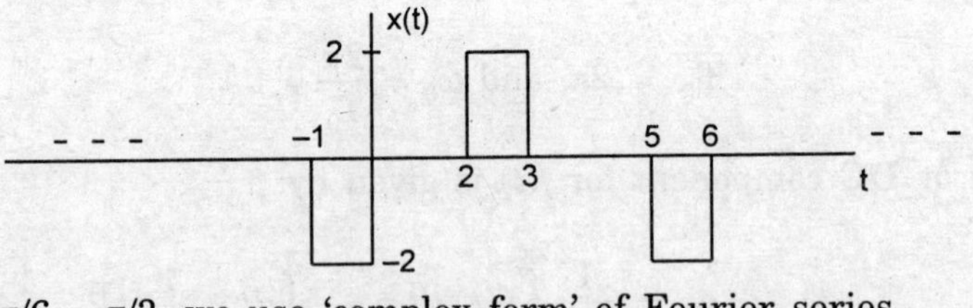

T = 6, $\omega_0 = 2\pi/6 = \pi/3$, we use 'complex form' of Fourier series.

$$x(t) = \sum_{n=-\infty}^{\infty} C_n e^{jn\omega_0 t}$$

$$C_n = \frac{1}{6}\int_{-1}^{5} x(t)e^{-jn\omega_0 t}dt$$

$$= \frac{1}{6}\int_{-1}^{0} -2e^{-jn\omega_0 t}dt + \frac{1}{6}\int_{2}^{3} 2e^{-jn\omega_0 t}dt$$

$$= \frac{-1}{3}\frac{-1}{jn\omega_0}e^{-jn\omega_0 t}\Big|_{-1}^{0} + \frac{1}{3}\frac{-1}{jn\omega_0}e^{-jn\omega_0 t^3}\Big|_{2}^{3}, n \neq 0$$

$$= \frac{1}{jn\pi}(1-e^{jn\pi/3}) - \frac{1}{jn\pi}\left(e^{-jn\pi} - e^{-jn2\pi/3}\right), n \neq 0$$

$$= \frac{1}{jn\pi}\left[e^{jn\pi/6}\left(e^{-jn\pi/6} - e^{jn\pi/6}\right) - e^{-jn5\pi/6}\left(e^{-jn\pi/6} - e^{jn\pi/6}\right)\right]$$

$$= \frac{2}{n\pi}\left[-e^{jn\pi/6}\sin\left(\frac{n\pi}{6}\right) + e^{-jn\frac{5\pi}{6}}\sin\left(\frac{\pi n}{6}\right)\right]$$

$$= \frac{2}{n\pi}\sin\left(\frac{\pi n}{6}\right)\left[-e^{jn3\pi/6} + e^{-jn3\pi/6}\right]e^{-2\pi/6nj}$$

$$\boxed{C_n = \frac{-4j}{n\pi}\sin\left(\frac{\pi n}{6}\right)\sin\left(\frac{n\pi}{2}\right)e^{-jn\pi/3}}$$

if $n = 0$

$$C_0 = \frac{1}{6}\int_{-1}^{0} -2dt + \frac{1}{6}\int_{2}^{3} 2dt = 0$$

$$C_1 = -0.5513 - 0.318j = 0.64e^{-j2.6}, C_{-1} = \overline{C_1}$$

$$C_2 = 0$$

$$C_3 = 0 - j\,0.42 = 0.42e^{-j1.6}, C_{-3} = \overline{C_3}$$

$$C_4 = 0$$

$$C_5 = 0.1103 - 0.0637j = 0.127e^{-j0.52}, C_{-5} = \overline{C_5}$$

Since the values are complex, the spectrum is shown for magnitude and phase separately as below:

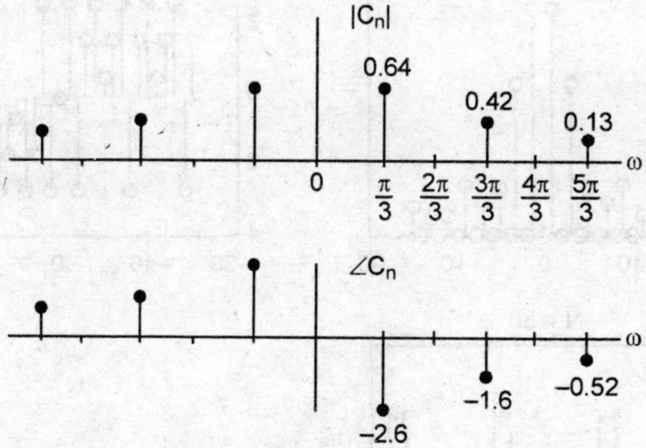

Spectrum

Some of the work can be done using MATLAB:

```
» n = 1:15;
» cn = -4*j./n/pi.*sin(pi*n/6).*sin(n*pi/2).*exp(-j*n*pi/3);
» n = -15: —1;
```

```
» c_n = -4 * j./n / pi.*sin(pi * n / 6).*sin(n * pi / 2).*exp(-j * n * pi / 3);
» cn = [c_n 0 cn];
» n = -15:15;
» subplot (221), stem (n, abs (cn))
» title ('|c_n|')
» subplot (222), stem (n, angle (cn))
» title ('angle (c_n) in rad')
```

To plot the Fourier series to check your answers:

```
T = 6;
ω0 = 2*pi/T;
t = -1.5*T:T/1000:1.5*T;
N = input ('Number of harmonics');
c0 = 0;
x = c0*ones (1, length (t)); % dc component
for n = 1 : N,
    cn = -4*j/n/pi*sin(pi*n/6)*sin(n*pi/2)*exp(-j*n*pi/3);
    c_n = conj(cn);
    x = x + cn*exp(j*n*ω0*t) + c_n*exp(-j*n*ω0*t);
end
plot (t, x)
title (['N = ', num2str(N)])
```

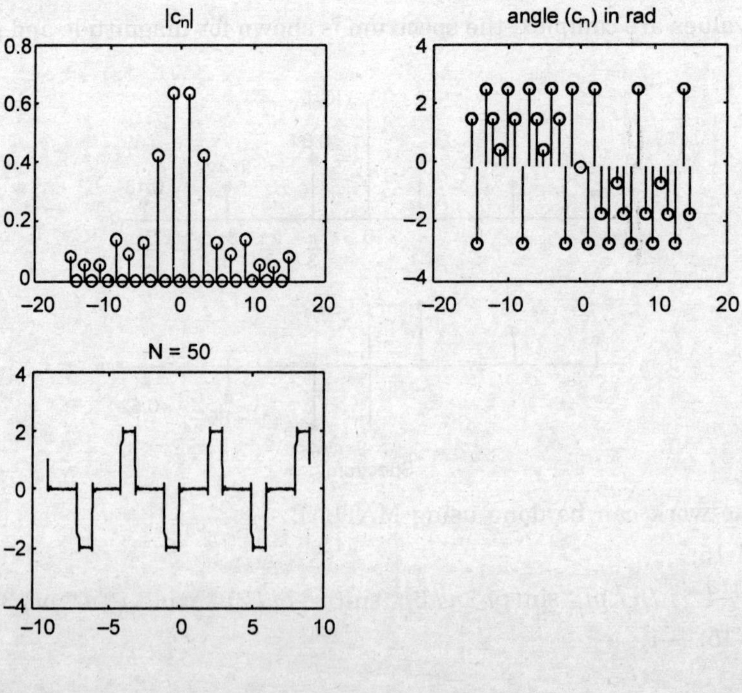

Example 3. *Repeat problem 1 for the following signal:*

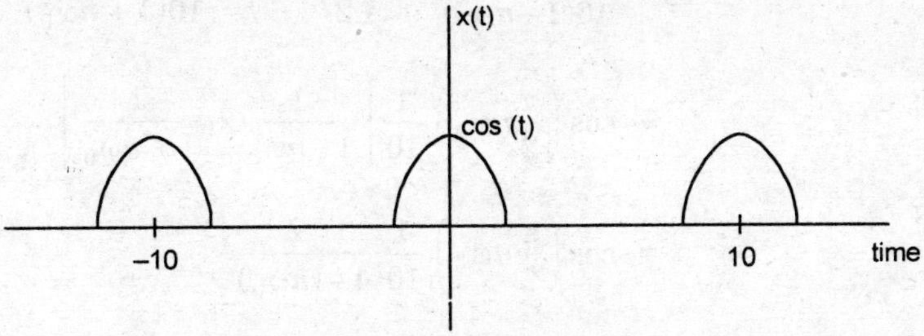

Solution.

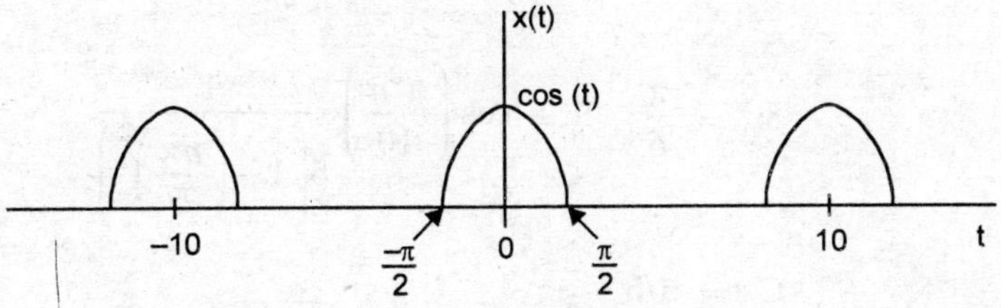

$T = 10 \Rightarrow \omega_0 = 2\pi/10 = \pi/5$

$$x(t) = \sum_{n=-\infty}^{\infty} C_n\, e^{jn\omega_0 t}$$

$$C_n = \frac{1}{T}\int_T x(t)\, e^{-jn\omega_0 t} dt$$

$$= \frac{1}{10}\int_{-\pi/2}^{\pi/2} \cos(t)\, e^{-jn\omega_0 t} dt$$

$$= \frac{1}{20}\int_{-\pi/2}^{\pi/2} \left(e^{jt} + e^{-jt}\right) e^{-jn\omega_0 t} dt$$

$$= \frac{1}{20}\int_{-\pi/2}^{\pi/2} e^{jt(1-n\omega_0)} + e^{-jt(1+n\omega_0)} dt$$

$$= \frac{1}{20}\,\frac{1}{j(1-n\omega_0)}\left[e^{j\pi/2(1-n\omega_0)} - e^{-j\pi/2(1-n\omega_0)}\right]$$

$$+\frac{-1}{20j(1+n\omega_0)}\left[e^{-\pi/2j(1+n\omega_0)} - e^{j\pi/2(1+n\omega_0)}\right]$$

$$= \frac{1}{20j}\,\frac{1}{(1-n\omega_0)}\left[je^{-j\pi/2n\omega_0} + je^{+j\pi/2n\omega_0}\right]$$

$$-\frac{1}{20j(1+n\omega_0)}\left[je^{-j\pi/2n\omega_0} - je^{j\pi/2n\omega_0}\right]$$

$$= \frac{1}{10(1-n\omega_0)}\cos\left(\frac{\pi}{2}n\omega_0\right) + \frac{1}{10(1+n\omega_0)}\cos\left(\frac{\pi}{2}n\omega_0\right)$$

$$= \cos\left(\frac{\pi}{2}n\omega_0\right)\frac{1}{10}\left[\frac{1}{1-n\omega_0} + \frac{1}{1+n\omega_0}\right]$$

$$= \cos\left(\frac{\pi}{2}n\omega_0\right)\frac{1}{10}\frac{2}{1-(n\omega_0)^2}$$

$$C_n = \cos\left(\frac{\pi}{2}n\omega_0\right)\frac{1}{5(1-(n\omega_0)^2)}$$

for $\omega_0 = \frac{\pi}{5};\ C_n = \cos\left(\frac{\pi^2 n}{10}\right)\frac{1}{5\left(1-\left(\frac{n\pi}{5}\right)^2\right)}$

$$C_0 = 1/5$$

$$C_1 = \cos\left(\frac{\pi^2}{10}\right)\frac{1}{5\left(1-\left(\frac{\pi}{5}\right)^2\right)} = 0.1822$$

$$C_2 = 0.1355$$

$$C_3 = 0.0771; \qquad C_{-n} = \overline{C}_n = C_n$$

$$C_4 = 0.026$$

$$C_5 = -0.005$$

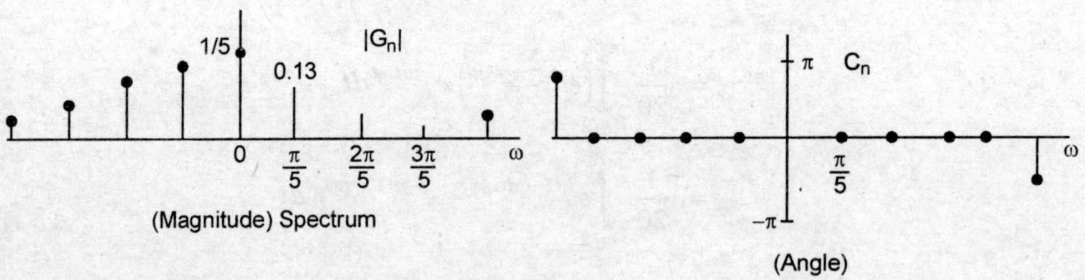

Some of the work can be done using MATLAB:

» n = –10 : 10;

» cn = cos (pi/2*n*w_0)/5./(1 – (n*w_0).^2) ;

» subplot (221), stem (n, abs (cn))

» title ('|c_n|')

» subplot (222), stem (n, angle (cn))

» title ('angle (c_n) in rad')

will give first plots shown below. To check your answer, you can plot the truncated series and see if it converges correctly.

```
T = 10;
w0 = 2*pi/T;
t = -1.5 : T/1000 : 1.5*T;
N = input ('Number of harmonics');
c0 = 1/5;
x = c0*ones (1, length (t)); % dc component
for n = 1 : N,
    cn = cos (pi/2*n*w0)/5/(1 - (n*w0)^2);
    c_n = cn;
    x = x + cn*exp (j*n*w0*t) + c_n*exp (-j*n*w0*t);
end
plot (t, x)
title (['N = ', num2str (N)])
```

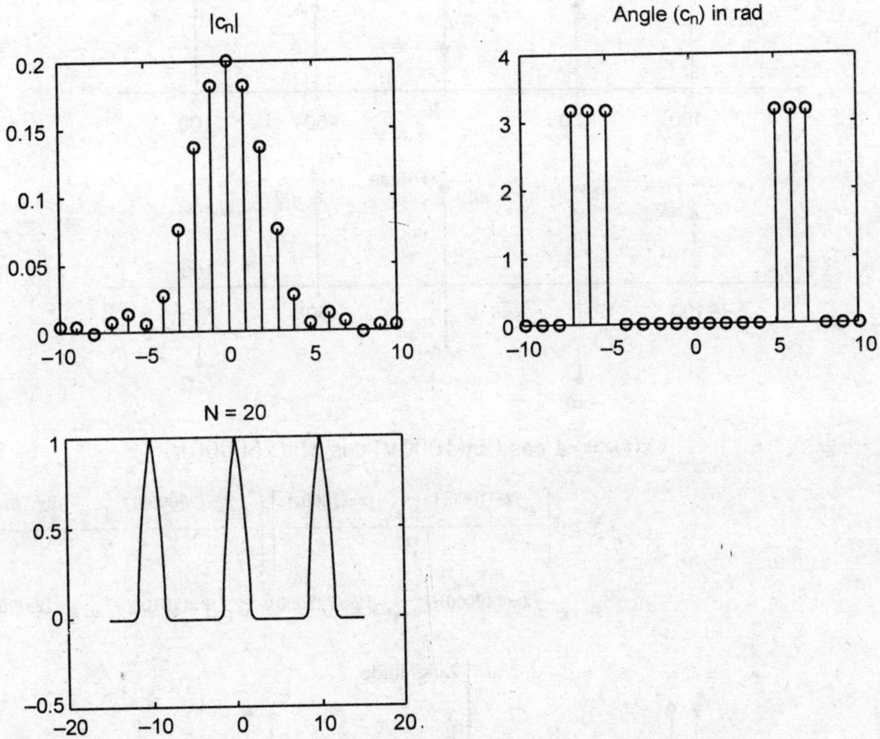

Example 4. *Compute the Fourier series for the following signals:*

(a) $x(t) = 2 + 4\cos(50t + \pi/2) + 12\cos(100t - \pi/3)$

(b) $x(t) = 4\cos(2\pi(1000)t)\cos(2\pi 750000t)$

(c)

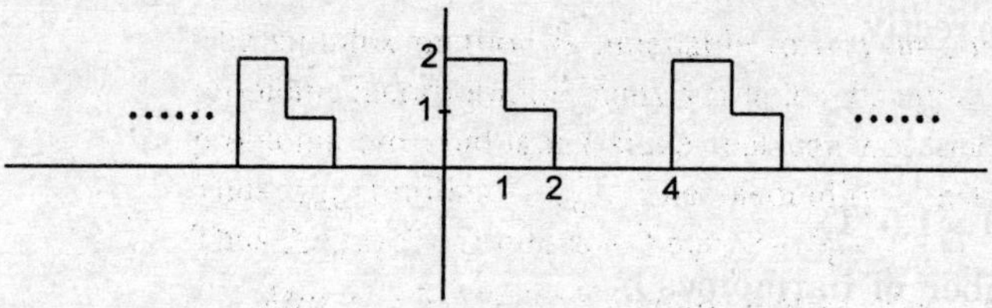

(d)

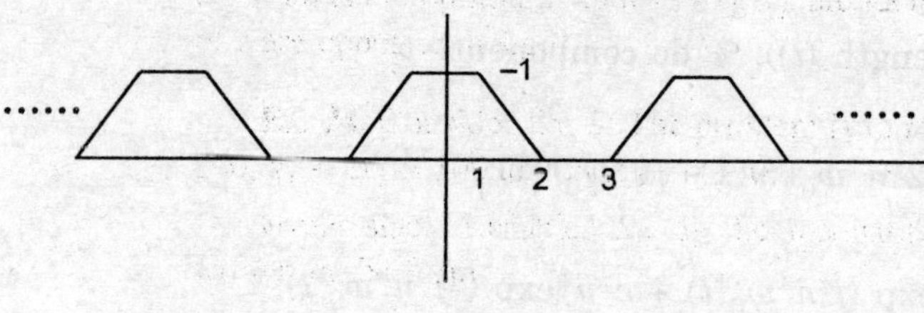

Solution. (*a*)

$$x(t) = 2 + 4 \cos (50t + \pi/2) + 12 \cos (100t - \pi/3)$$

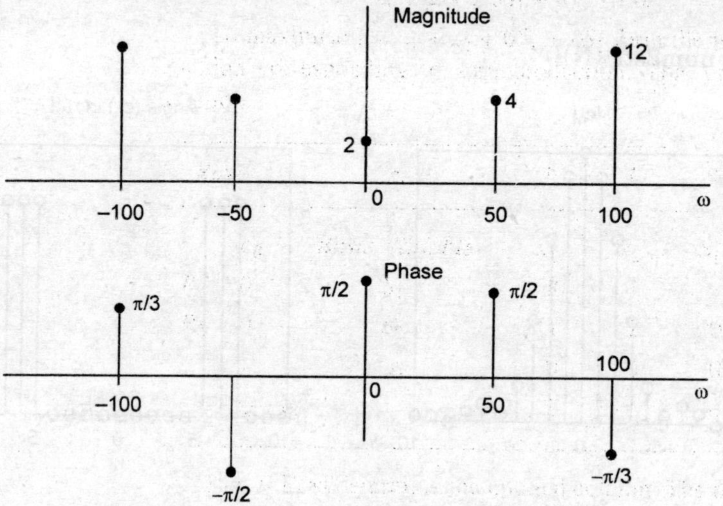

(*b*)

$$x(t) = 4 \cos (2\pi(1000)t) \cos (2\pi(750000)t)$$

$$= 4\left[\frac{e^{j2\pi(1000t)} + e^{-j2\pi(1000t)}}{2}\right]\left[\frac{e^{j2\pi(750000t)} + e^{-j2\pi(75000t)}}{2}\right]$$

$$= e^{-j2\pi(749000)t} + e^{-j2\pi(749000)t} + e^{j2\pi(751000t)} + e^{-j2\pi(751000t)}$$

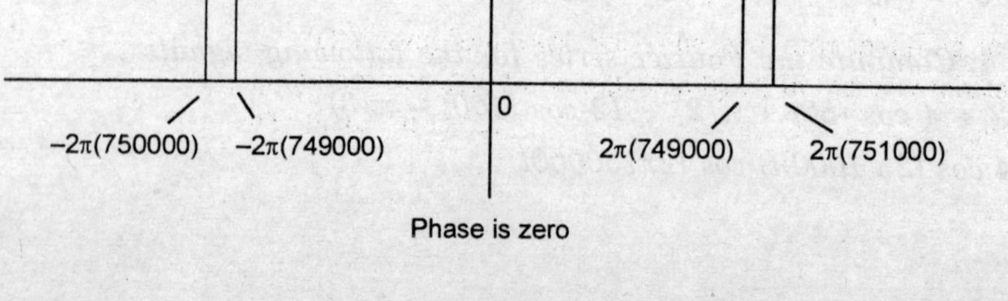

Phase is zero

ω_0 is greatest common factor,

$$\omega_0 = 2\pi(1000)$$

(c)

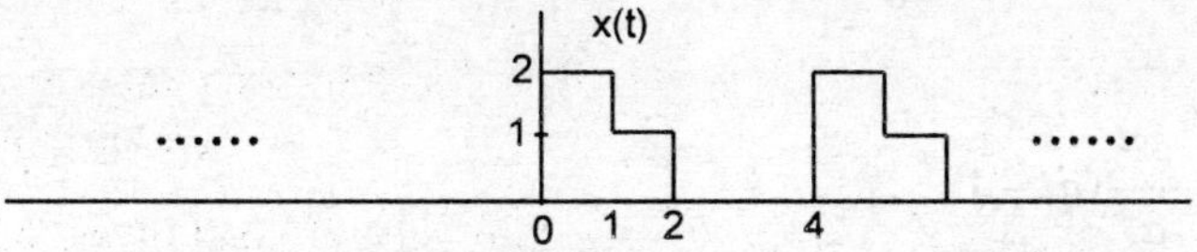

T = 4, so $\omega_0 = 2\pi/4 = \pi/2$

$$C_0 = \frac{1}{4}\int_0^2 x(t)dt = \frac{3}{4}$$

$$C_n = \frac{1}{4}\int_0^2 x(t)\, e^{-jn\pi/2t}dt$$

$$= \frac{1}{4}\int_0^1 2e^{-jn\pi/2t}dt + \frac{1}{4}\int_1^2 e^{-jn\pi/2t}dt$$

$$= \frac{1}{2}\,\frac{1}{-jn\pi/2}\,e^{-jn\pi/2t}\Big|_0^1 + \frac{1}{4}\,\frac{1}{-jn\pi/2}\,e^{-jn\pi/2t}\Big|_1^2$$

$$= \frac{1}{-jn\pi}\left[e^{-jn\pi/2} - 1\right] + \frac{1}{-j2n\pi}\left[e^{-jn\pi} - e^{-jn\pi/2}\right]$$

This can be simplified

$$e^{-jn\pi} = (-1)^n$$

$$C_n = \frac{1}{jn\pi}\left[1 - \frac{1}{2}(-1)^n - \frac{1}{2}e^{-jn\pi/2}\right]$$

$$C_0 = 3/4$$

$$x(t) = \sum_{n=-\infty}^{\infty} C_n e^{jn\pi t/2}$$

(d)

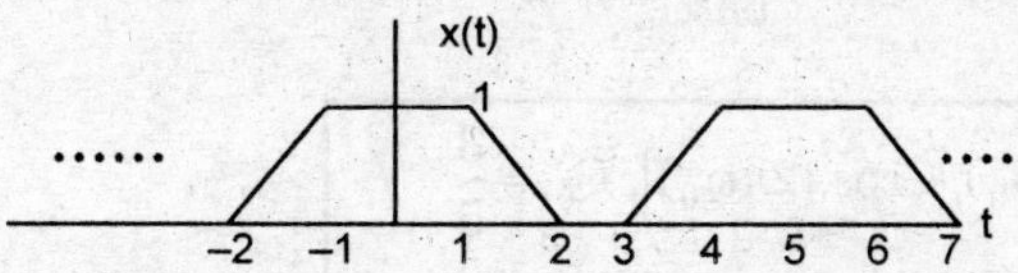

T = 5 $\omega_0 = \frac{2\pi}{5}$

$$C_0 = \frac{1}{5}\int_{-2}^{2} x(t)dt = \frac{3}{5}$$

$$C_n = \frac{1}{5}\int_{-2}^{-1}(t+2)\,e^{-j\omega_0 nt}dt + \frac{1}{5}\int_{-1}^{1}e^{-j\omega_0 nt}dt + \frac{1}{5}\int_{1}^{2}(-t+2)\,e^{-jn\omega_0 t}dt$$

Recall $\int te^{at}dt = \frac{e^{at}}{a^2}(at-1)$

$$C_n = \frac{1}{5}\left[\frac{e^{-jn\omega_0 t}}{(-jn\omega_0)^2}(-jn\omega_0 t-1) - \frac{2}{jn\omega_0}e^{-jn\omega_0 t}\Big|_{-2}^{-1}\right.$$

$$\left. + \frac{-1}{j\omega}e^{-jn\omega_0 t}\Big|_{-1}^{1} + \frac{-e^{-jn\omega_0 t}}{(-jn\omega_0)^2}(-jn\omega_0 t-1)\frac{-2}{jn\omega_0}e^{-jn\omega_0 t}\Big|_{1}^{2}\right]$$

$$C_n = \frac{1}{5}\left[\frac{e^{jn\omega_0}}{-n^2\omega_0}(jn\omega_0-1) - \frac{2}{jn\omega_0}e^{jn\omega_0} + \frac{e^{jn\omega_0^2}}{\omega_0^2 n^2}(jn\omega_0 2-1) + \frac{2}{jn\omega_0}e^{j\omega_0 n^2} - \frac{1}{j\omega_0 n}e^{-jn\omega_0}\right.$$

$$+ \frac{1}{j\omega_0 n}e^{jn\omega_0} + \frac{e^{jn\omega_0^2}}{n^2\omega_0^2}(-jn\omega_0 2-1) - \frac{2}{jn\omega_0}e^{-jn\omega_0^2}$$

$$\left. - \frac{e^{-jn\omega_0}}{n^2\omega_0^2}(-jn\omega_0-1) + \frac{2}{jn\omega_0}e^{-jn\omega_0}\right]$$

$$= \frac{1}{5}e^{jn\omega_0}\left(\frac{-j}{n\omega_0} + \frac{1}{n^2\omega_0^2} + \frac{2j}{n\omega_0} - \frac{j}{\omega_0 n}\right) + \frac{1}{5}e^{-jn\omega_0}\left(\frac{j}{n\omega_0} + \frac{j}{n\omega_0} + \frac{1}{n^2\omega_0^2} - \frac{2j}{n\omega_0}\right)$$

$$+ \frac{1}{5}e^{j2n\omega_0}\left(\frac{j2}{\omega_0^2 n^2} - \frac{1}{\omega_0^2 n^2} - \frac{2j}{n\omega_0}\right) + \frac{1}{5}e^{-j2n\omega_0}\left(\frac{-j2}{n\omega_0} - \frac{1}{n^2\omega_0^2} + \frac{2j}{n\omega_0}\right)$$

$$= \frac{1}{5n^2\omega_0^2}\left(e^{jn\omega_0} + e^{-jn\omega_0 t}\right) - \frac{1}{5n^2\omega_0^2}\left(e^{j2n\omega_0} + e^{-j2n\omega_0}\right)$$

$$C_n = \frac{2}{5n^2\omega_0^2}\left[\cos(n\omega_0) - \cos(2n\omega_0)\right],\ C_0 = \frac{3}{5}$$

$$x(t) = \sum_{n=-\infty}^{\infty} C_n e^{jn2\pi/5t}$$

EXERCISES

1. If the function $y = x$ in the range 0 to π is expanded as a sine series, show that it is equal to

$$2\left(\frac{\sin x}{1} - \frac{\sin 2x}{2} + \frac{\sin 3x}{3} \ldots\right).$$

2. Expand $\frac{\pi x}{8}(\pi - x)$ in a sine series valid when

$$0 \le x \le \pi.$$

3. Find a sine series for

$$\begin{aligned} f(x) &= x;\ 0 < x < \frac{\pi}{2} \\ &= 0;\ \frac{\pi}{2} < x < \pi \end{aligned}$$

4. Show that as a sine series in the range 0 to π

$$\begin{aligned} f(x) &= 0;\ 0 < x < \frac{\pi}{2} \\ &= c;\ \frac{\pi}{2} < x < \pi \end{aligned}$$

can be represented as

$$f(x) = \frac{2c}{\pi}\left[\frac{\sin x}{1} - \frac{2 \sin 2x}{2} + \frac{\sin 3x}{3} + \frac{\sin 5x}{5} - \frac{2 \sin 6x}{6} \ldots\right]$$

but as a cosine series in the same range it is

$$f(x) = \frac{2c}{\pi}\left(\frac{\pi}{4} - \frac{\cos x}{1} + \frac{\cos 3x}{3} - \frac{\cos 5x}{5} + \ldots\right).$$

5. If

$$\begin{aligned} f(x) &= \frac{\pi x}{4} \quad \left(0 < x < \frac{\pi}{2}\right) \\ &= \frac{\pi}{4}(\pi - x) \quad \left(\frac{\pi}{2} < x < \pi\right) \end{aligned}$$

Prove that for the range $0 < x < \pi$

$$f(x) = \frac{\pi^2}{16} - \frac{1}{2}\left(\frac{\cos 2x}{1^2} + \frac{\cos 6x}{3^2} + \ldots\right).$$

Find the expansion in series of sines for the same function.

6. Show that as a cosine series in the half range 0 to π,

$$\sin x = \frac{4}{\pi}\left(\frac{1}{2} - \frac{\cos 2x}{1.3} - \frac{\cos 4x}{3.5} - \frac{\cos 6x}{5.7} \ldots\right)$$

whilst as a sine series in the same range,

$$\cos x = \frac{4}{\pi}\left(\frac{2\sin 2x}{2^2-1}+\frac{4\sin 4x}{4^2-1}+\frac{6\sin 6x}{6^2-1}+\ldots\right).$$

7. Show that in the range $(0, \pi)$, the sine series for $\pi x - x^2$ is $\frac{8}{\pi}\left(\sin x+\frac{1}{3^3}\sin 3x+\frac{1}{5^3}\sin 5x+\ldots\right)$.

8. Find a Fourier cosine series corresponding to the function $f(x) = x$, defined in the interval $(0, \pi)$.

9. Find the Fourier sine series and the Fourier cosine series corresponding to the function,

$$f(x) = \pi - x \text{ when } 0 < x < \pi$$

defined in the interval 0 to π.

10. Show that, when $0 < x < \pi$.

$$f(x) = \sin 2x + \frac{1}{2}\sin 4x + \frac{1}{4}\sin 8x + \frac{1}{5}\sin 10x\ldots$$

$$= \frac{2}{\sqrt{3}}\left(\cos x - \frac{1}{5}\cos 4x + \frac{1}{7}\cos 7x - \frac{1}{11}\cos 11x\ldots\right).$$

where $$f(x) = \frac{\pi}{3} \quad \left(0 < x < \frac{\pi}{3}\right)$$

11. Determine the first three harmonics of the Fourier series for the values.

x	30°	60°	90°	120°	150°	180°	210°	240°	270°	300°	330°	360°
y	2.34	3.01	3.68	4.15	3.69	2.20	0.83	0.51	0.88	1.09	1.19	1.64

[**Ans.** 2.1; –0.28, –.018 1.61, –0.5, + 0.2]

12. The values of y, a periodic function of x, are given below for twelve equidistant values of x covering the whole period. Express y in a Fourier series as far as the second harmonic if the first value is for $x = 30°$.

3.5, 6.09, 7.82, 5.58, 8.43, 7.73,

6.98, 6.19, 6.04, 5.55 5.01, 3.35,

[**Ans.** 6.27 –1.99 cos x – 0.67 cos $2x$ + 0.75 sin x – 0.69 sin $2x$]

13. The following 12 values of $f(\theta)$ correspond to equidistant values of the angle $\theta°$ in the range (0° to 360°) :

θ	0°	30°	60°	90°	120°	150°	180°	210°	240°	270°	300°	330°
$f(\theta)$	10.5	20.2	26.4	29.3	27.0	21.5	12.8	1.6	–11.2	–18.0	–15.8	–3.5

Determine approximately, the Fourier expansion for $f(\theta)$ in terms of θ, up to the terms involving sin 3θ and cos 3θ.

[**Ans** 8.4 + 23.1 sin θ –1.7 cos θ + 1.1 sin θ = 20 + 3.1 cos 2θ –0.6 sin 3θ + 0.45 cos 3θ]

14. Find the Fourier series for the following signal. Also, sketch the approximation if a large number of terms are kept in the series (say N = 30).

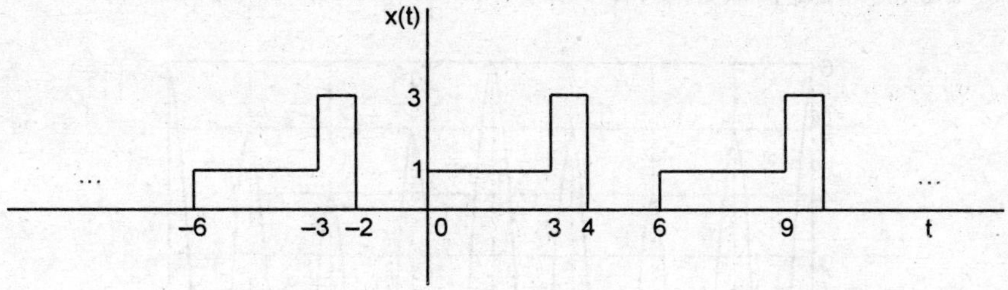

Ans. $$x(t) = \sum_{m=-\infty}^{\infty} C_n\, e^{jn\omega_0 t}; \qquad T = 6,\ \omega_0 = \frac{2\pi}{6} = \frac{\pi}{3}$$

$$C_n = \frac{1}{6}\int_0^6 x(t)\, e^{-jn\omega_0 t}\, dt$$

$$= \frac{1}{6}\int_0^3 e^{-jn\frac{\pi}{3}t}\, dt + \frac{1}{6}\, 3\int_3^4 e^{-jn\frac{\pi}{3}t}\, dt$$

$$= \frac{1}{6}\,\frac{-1}{j\frac{\pi}{3}n}\, e^{-jn\frac{\pi}{3}t}\Bigg|_0^3 + \frac{1}{2}\,\frac{-1}{jn\frac{\pi}{3}}\, e^{-jn\frac{\pi}{3}t}\Bigg|_3^4, n \neq 0$$

$$C_n = \frac{-1}{j2\pi n}\left(e^{-jn\pi} - 1\right) - \frac{3}{j2\pi n}\left(e^{-jn4\frac{\pi}{3}} - e^{-jn\pi}\right), n \neq 0$$

if $n = 0$, $$C_0 = \frac{1}{6}\int_0^6 x(t)\, dt = 1$$

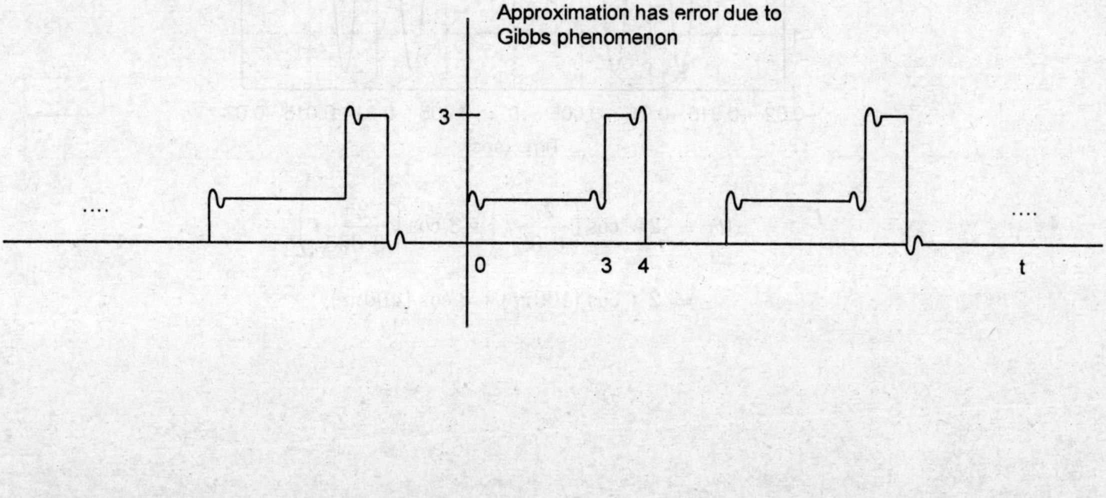

15. For the following plot, find the Fourier series and plot its spectra.

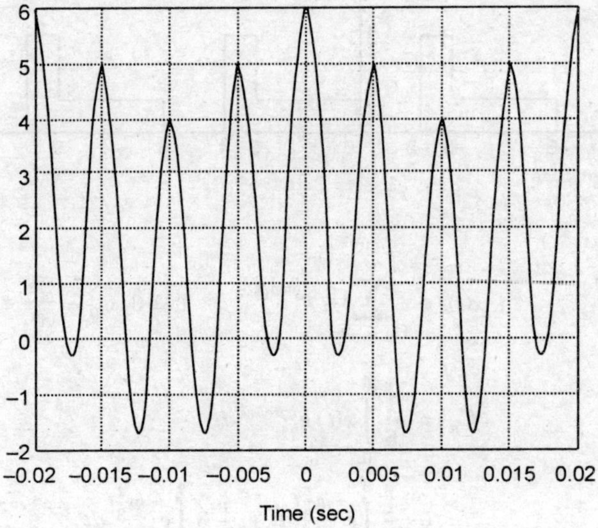

Ans.

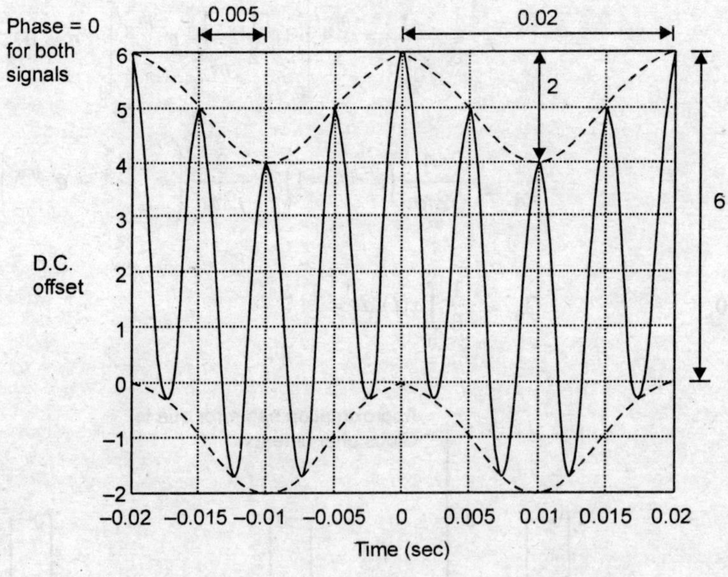

$$x(t) = 2 + \cos\left(\frac{2\pi}{0.02}t\right) + 3\cos\left(\frac{2\pi}{0.005}t\right)$$

$$= 2 + \cos(100\pi t) + 3\cos(400\pi t)$$

16. Time signals and their corresponding spectra are shown below. However, they are in random order. Match them up.

1. .... 2. .... 3. .... 4. .... 5. ....

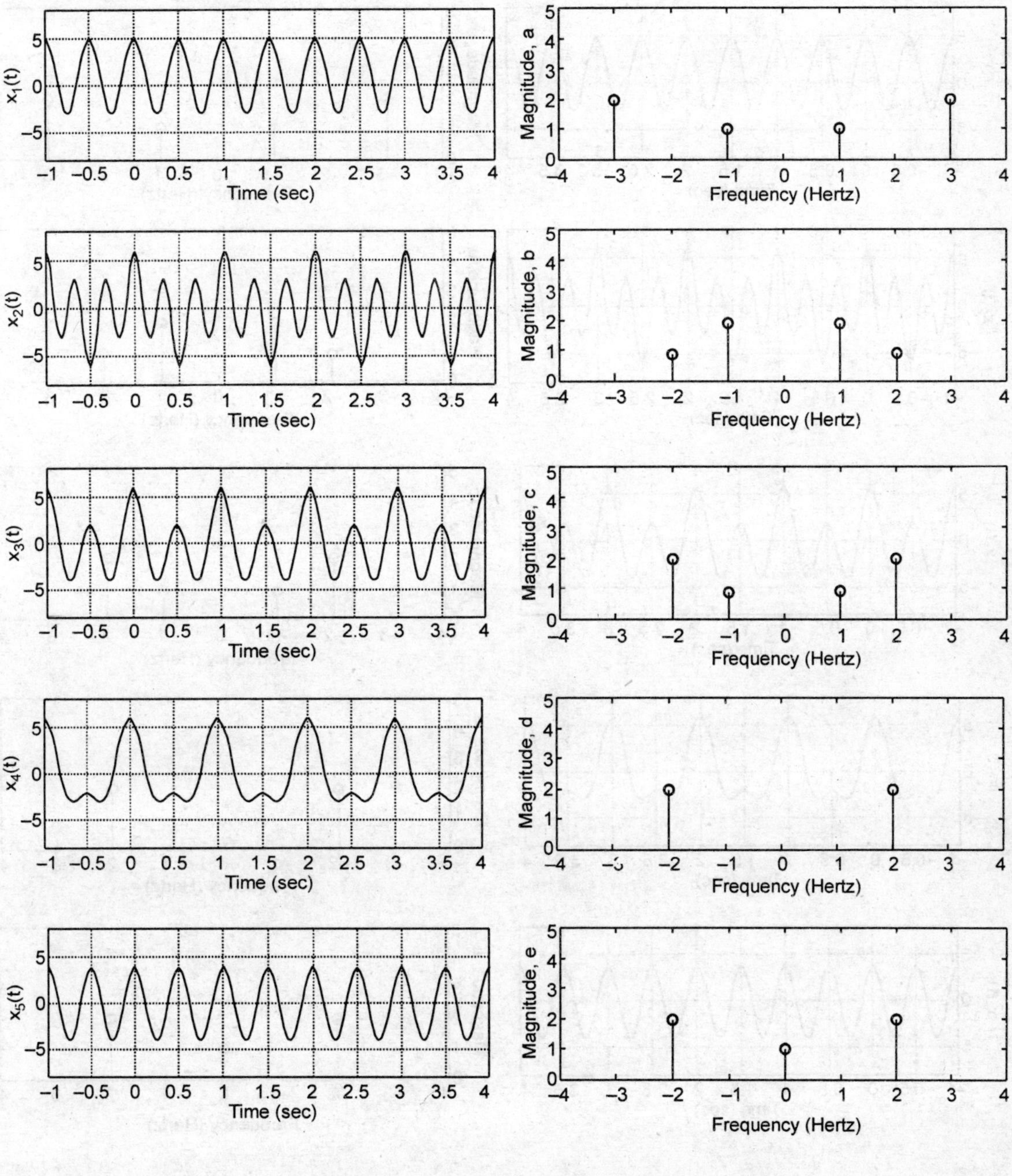

Ans. Time signals and their corresponding spectra are shown below. However, they are in random order. Match them up.

1. (*e*) 2. (*a*) 3. (*c*) 4. (*b*) 5. (*d*)

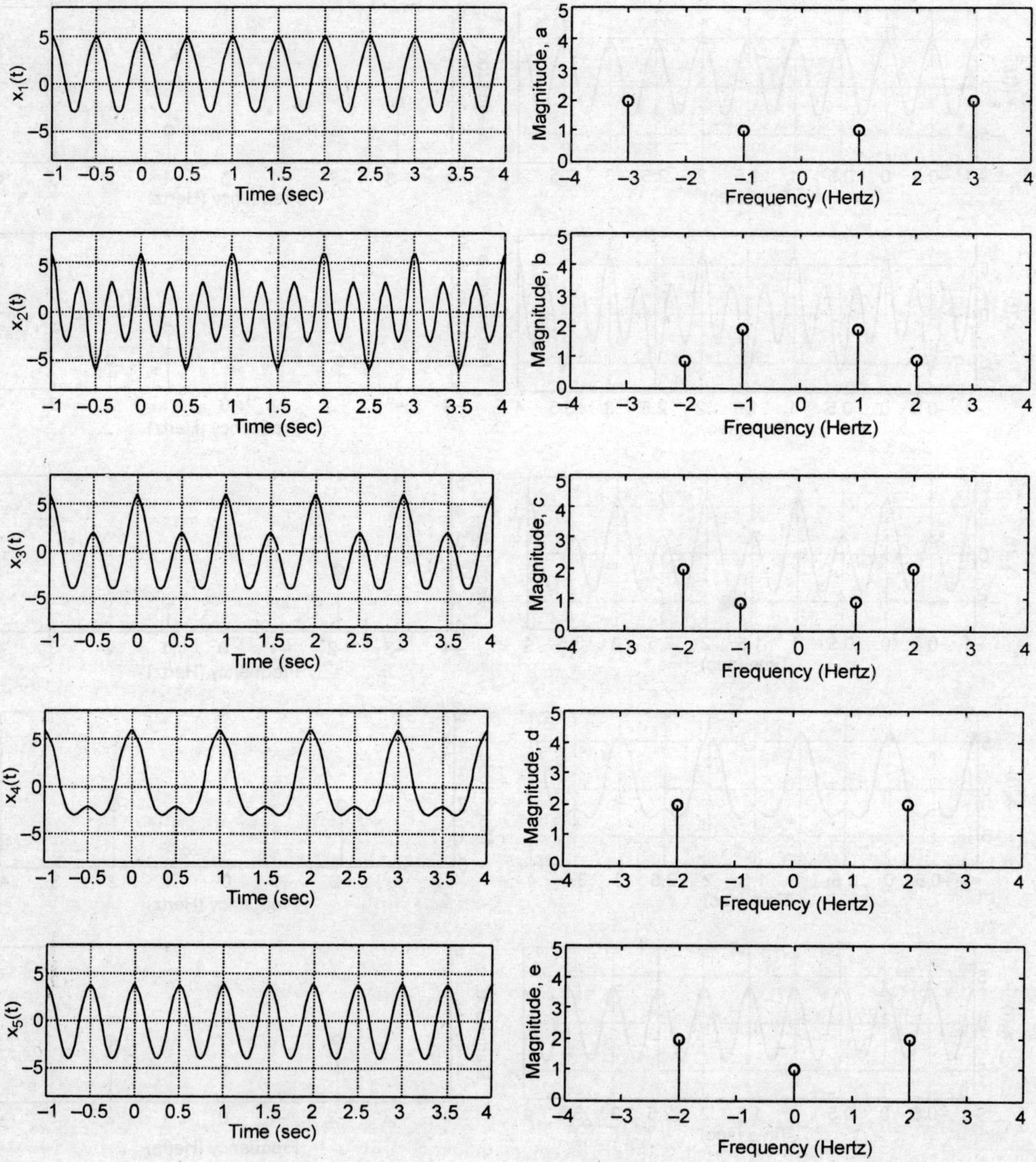

3

THE FOURIER TRANSFORM AND ITS APPLICATIONS

FOURIER SERIES FOR NON-PERIODIC SIGNALS

For periodic signals, such as a square wave, we saw that it can be represented as a series of sine and cosine waves, from the fundamental and all the harmonics.

$$f(t) = \sum a_n \sin nt + \sum b_n \cos nt \quad ...(3.1)$$

For a square wave, which is continuous and also periodic, with a frequency, say 50 Hz, we will have n values from 50 Hz, and in multiples of 50 Hz, as

50, 100, 150, 200, 250, 300, ...

We have found the series for a square wave and noted that the harmonics are lesser and lesser in amplitude, being

$1, 0, \frac{1}{3}, 0, \frac{1}{5}, 0, ...$

for the above frequencies. Note that the even harmonics are zero.

Consider a non-periodic signal. It is also a signal which exists for a certain span of time. For instance, Fig. 1.4 gives the NMR instrument signal. This is a signal which lasts for 2.5 seconds and comprises a lot of frequency components in it. This was got by the instrument from the chemical in a combined RF and magnetic field, after exciting the chemical with a powerful RF pulse. It is an one-time signal, and is not a periodic signal.

For such a signal, we can consider the period to be infinite.

For the periodic signal, of period T, the basic or fundamental frequency is $\frac{1}{T}$, and $f_0 = \frac{1}{T}$.

All Fourier series components one in *'multiples'* of this f_0. Thus, we have seen 50, 100, 150, 200, 250, 300, ... in the above example. Note that intermediate frequencies do not exist, (such as 75 or 80 or whatever).

In the non-periodic signal, since $f_0 \to 0$, or infinitesimally small, we will have *all* frequencies in the Fourier components. The difference between the several components diminishes and so we will have a continuous 'spectrum' of frequency components possible for a non-periodic signal.

In that case, the summation in (3.1) becomes an integration.

$$f(t) = \int_0^{\infty} \{X_r(f) \cos 2\pi ft + X_i(f) \sin 2\pi ft\}\, df \qquad ...(3.2)$$

Note that the b_n and a_n have been written as $X_r(f)$, $X_i(f)$, where f varies from 0 to ∞ continuously instead of in steps as for Fourier series. 'r' and 'i' are *real* and *imaginary* parts of Fourier components, meaning cosine and sine component amplitudes.

Note that the summation is for all frequencies and hence df is in the integration.

The above equation is usually written in terms of the complex notation for sinusoids,

$$e^{j\omega t} = e^{j2\pi ft} = (\cos \omega t + j \sin \omega t) \qquad ...(3.3)$$

Also, the integral should be actually from $-\infty$ to $+\infty$, because $e^{j\omega t}$ refers to positive values of frequencies, while $e^{-j\omega t}$ refers to negative values. And we know that

$$\frac{(e^{j\omega t} + e^{-j\omega t})}{2} = \cos \omega t$$

$$\frac{(e^{+j\omega t} - e^{-j\omega t})}{2j} = \sin \omega t \qquad ...(3.4)$$

Therefore (3.2) is re-written as

$$\boxed{x(t) = \frac{1}{2\pi}\int_{-\infty}^{\infty} X(\omega)\, e^{j\omega t}\, d\omega} \quad \textbf{(Inverse Fourier Transform)} \qquad ...(3.5)$$

We have changed from f to ω, the angular frequency in the above, putting $\omega = 2\pi f$ and $d\omega = 2\pi df$ in (3.2) and changing from sine-cosine form into complex exponential form using exp $(j\omega t)$. The division by 2π is also noteworthy.

If the function is $f(t)$ in lieu of $x(t)$,

$$f(t) = \frac{1}{2\pi}\int_{-\infty}^{\infty} F(\omega)\, e^{j\omega t}\, d\omega \qquad ...(3.6)$$

What we saw in (3.5) is really the inverse Fourier transform. The relation that finds the frequencies of a given signal is the Fourier transform, $F(\omega)$. From

$$f(t) = \frac{1}{2\pi}\int_{-\infty}^{\infty} F(\omega)\, e^{j\omega t}\, d\omega, \qquad ...(3.6)$$

let us find the value of $F(\omega)$ now. Form the integral below

$$\boxed{F(\omega) = \int_{-\infty}^{\infty} f(t)\, e^{-j\omega t}\, dt} \quad \textbf{(Fourier Transform)} \qquad ...(3.7)$$

Take the right hand side of (3.6). Substitute (3.7) in it

$$\frac{1}{2\pi}\int_{-\infty}^{\infty} F(\omega)\, e^{j\omega t}\, d\omega = \frac{1}{2\pi}\int_{-\infty}^{\infty} e^{j\omega t}\, d\omega \int_{-\infty}^{\infty} f(x)\, e^{-j\omega t}\, dx$$

Note we have change from t to x in making this substitution, because we already have t in $e^{j\omega t}$.

Interchanging the order of integration,

$$I = \frac{1}{2\pi}\int_{-\infty}^{\infty} F(\omega)\, e^{j\omega t}\, d\omega = \frac{1}{2\pi}\int_{-\infty}^{\infty} f(x)\, dx \int_{-\infty}^{\infty} e^{j\omega(t-x)}\, d\omega$$

The second integral is the $\delta(t - x)$ function.

$$\int_{-\infty}^{\infty} e^{-j\omega(t-x)}\, d\omega = \delta(t - x).$$

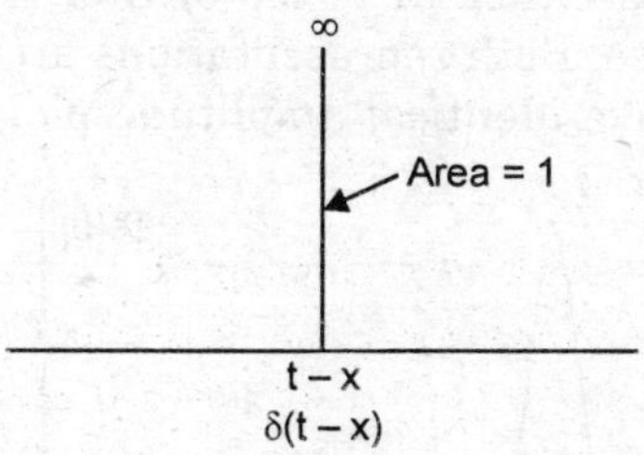

Fig. 1

So, $$I = \int_{-\infty}^{\infty} f(x)\, \delta(t-x)\, dx$$

$$= f(t) \qquad \text{(provided } f(t) \text{ is 'continuous')}$$

because the product $f(x)\ \delta(t - x)$ becomes zero for all x except when $x = t$ and then product is just unity.

Thus the inter-relationship between (3.6) and (3.7) is proved.

FOURIER TRANSFORM PLOTS

Any signal can be written as a sum of scaled sinusoids, as was expressed in Eq. (3.1). That equation is usually written using complex exponential notation :

$$x(t) = \int_{-\infty}^{\infty} \hat{x}(f)\, e^{j\omega t}\, d\omega \qquad ...(3.8)$$

The complex exponential notation, remember, is just a shorthand for sinusoids and co-sinusoids, but it is mathematically more convenient. The $\hat{x}(\omega)$ are the Fourier transform coefficients for each frequency component ω. These coefficients are complex numbers and can be expressed either in terms of their real (cosine) and imaginary (sine) parts or in terms of their amplitude and phase.

A second equation tells us how to compute the Fourier transform coefficients, $\hat{x}(\omega)$, from the input signal :

$$\hat{x}(\omega) = \mathcal{F}\{x(t)\} = \frac{1}{2\pi}\int_{-\infty}^{\infty} x(t)\, e^{-j\omega t}\, dt \qquad ...(3.9)$$

These two equations are inverses of one another. Eq. (3.9) is used to compute the Fourier transform coefficients from the input signal, and then Eq. (3.8) is used to reconstruct the input signal from the Fourier coefficients.

The equations for the Fourier transform are rather complex not merely because they involve complex exponent. The best way to get an intuition for the frequencies domain is to look at a few examples. Figure 2 plots sinusoidal signals of two different frequencies, along with their Fourier transform amplitudes. A sinusoidal signal contains only one frequency component, hence the frequency plots contain impulses. Both sinusoids are modulated between plus and minus one, so the impulses in the frequency plots have unit amplitude. The only difference between the two sinusoids is that one has 4 cycles per second and the other has 8 cycles per second. Hence the impulses in the frequency plots are located at 4 Hz and 8 Hz, respectively.

Figure 3 shows the Fourier transforms of a sinusoid and a cosinusoid. We can express the Fourier transform coefficients either in terms of their real and imaginary parts, or in terms of their amplitude and phase. Both representations are plotted in the figure. Sines and cosines of the same frequency have identical amplitude plots, but the phases are different.

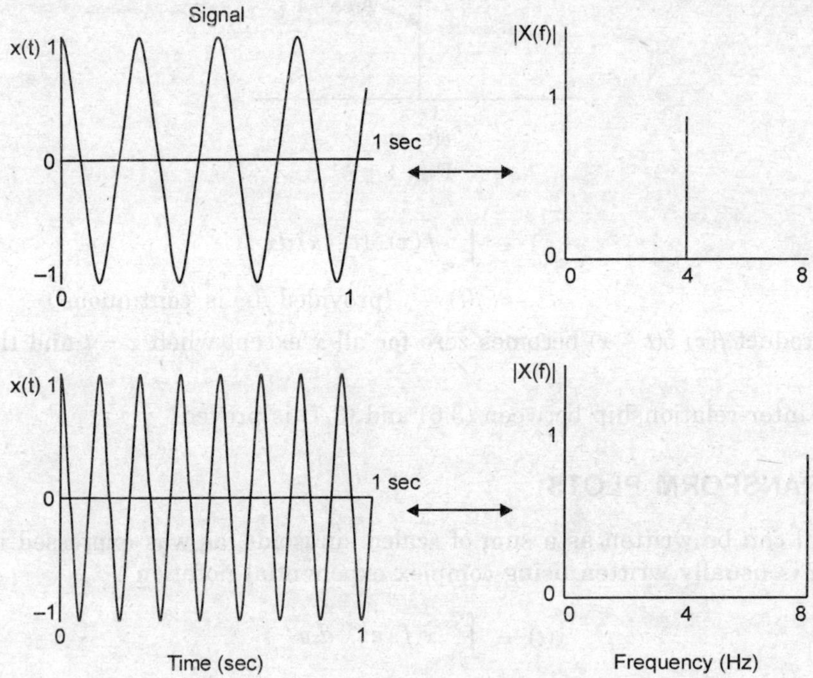

Fig. 2. Fourier transforms of sinusoidal signals of two different frequencies.

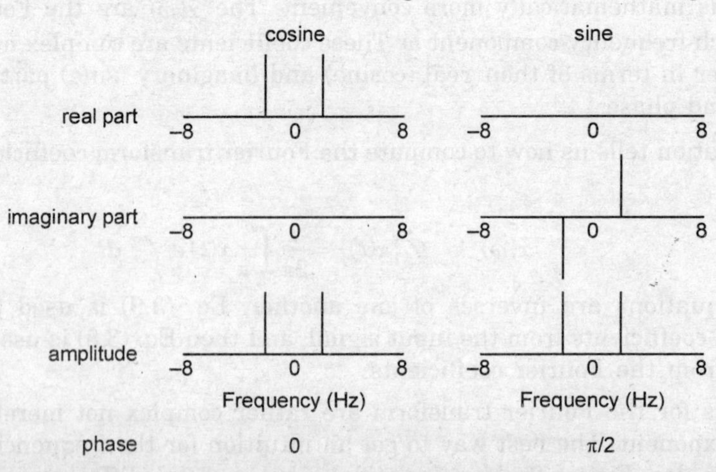

Fig. 3. Fourier transforms of sine and cosine signals. The amplitudes and the same, but the phases are different.

Do not be put off by the negative frequencies in the plots. The equations for the Fourier transform and its inverse include both positive and negative frequencies. This is really just a mathematical convenience. The information in the negative frequencies is redundant with that in the positive frequencies. Since cos $(-f) = \cos(f)$, the negative frequency components in the real part of the frequency domain will always be the same as the corresponding positive frequency components. Since sin $(-f) = -\sin(f)$ the negative frequency components in the imaginary part of the frequency domain will always be minus one times the corresponding positive frequency components. Often, people plot only the positive frequency components, as was done in Fig. 2, since the negative frequency components provide no additional information. Sometimes, people plot only the amplitude. In this case, however, there is information missing.

There are a few facts about the Fourier transform that often come in handy. The first of the properties is that the Fourier transform is itself a linear system, which one can check one by making sure that Eq. (3.9) obeys both homogeneity and additivity. This is important because it makes it easy for us to write the Fourier transforms of lots of things. For example, the Fourier transform of the sum of two signals is the sum of the two Fourier transforms:

$$\mathcal{F}\{x(t)+y(t)\} = \mathcal{F}\{x(t)\}+\mathcal{F}\{y(t)\} = \hat{x}(\omega)+\hat{y}(\omega), \qquad \text{...(3.10)}$$

where $\mathcal{F}\{.\}$ has been used as a shorthand notation for "the Fourier transform of". The linearity of the Fourier transform was one of the tricks that made it easy to write the transforms of both sides of Eq. 3.6.

A second fact, known as the convolution property of the Fourier transform, is that the Fourier transform of a convolution equals the product of the two Fourier transforms :

$$\mathcal{F}\{h(t)*x(t)\} = \mathcal{F}\{h(t)\}\,\mathcal{F}\{x(t)\} = \hat{h}(\omega)\hat{x}(\omega) \qquad \text{...(3.11)}$$

This property was also used to write the Fourier transform of Eq. (3.6). Indeed this property is central to much of the discussion. Above it was emphasized that for a shift-invariant linear system (*i.e.*, convolution), the system's response are always given by shifting and scaling the frequency components of the input signal. This fact is expressed mathematically by the convolution property above, where $\hat{x}(\omega)$ are the frequency components of the input and $\hat{h}(\omega)$ is the frequency response, the (complex-valued) scale factors that shift and scale each frequency component.

1. The Fourier Transform of an Impulse Function

The Fourier transform of a unit impulse function $\delta(t)$ is given by

$$\mathcal{F}[\delta(t)] = \int_{-\infty}^{\infty} \delta(t)\, e^{-j\omega t}\, dt \qquad \text{...(3.12)}$$

From the sampling property of an impulse function, it is evident that the integral on the right-hand side of Eq. (3.12) is unity. Hence

$$\mathcal{F}[\delta(t)] = 1 \qquad \text{...(3.13)}$$

Thus the Fourier transform of a unit impulse function is unity.

It is therefore evident that an impulse function has a uniform spectral density over the entire frequency interval. In other words, an impulse function contains all frequency components with the same relative amplitudes.

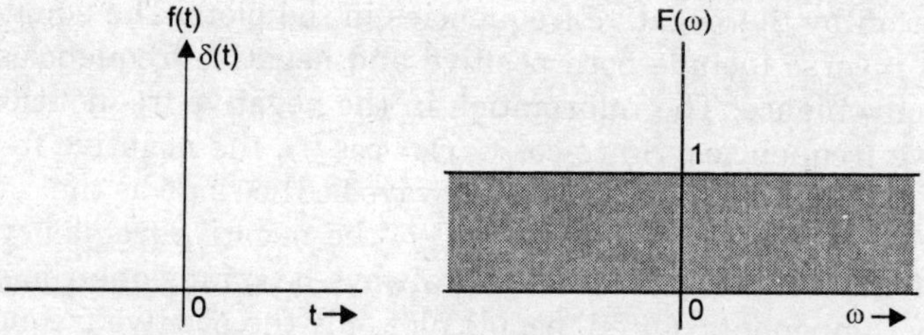

Fig. 4. An impulse function and its transform.

2. A Constant—What is its F.T.?

Let $f(t) = k$, a constant.

A constant is nothing but a zero frequency component. So, we only have a zero frequency in the F.T. spectrum. There is only one line at $f = 0$. Its amplitude is $2\pi k$.

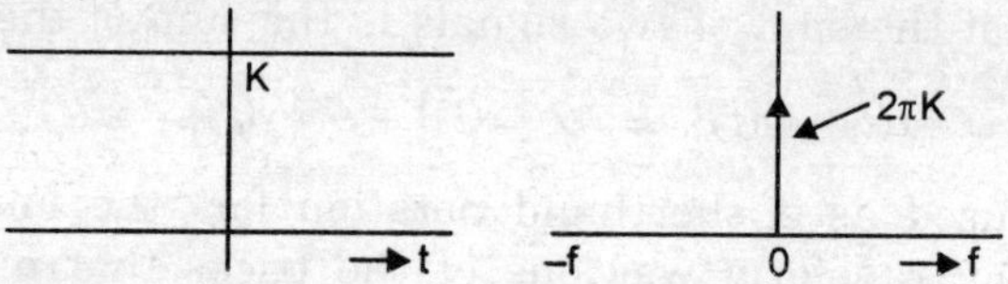

Fig. 5. Constant term and its Fourier transform.

Note that 2π comes in, that is because of ω to t relation.

3. Transform of sgn (*t*)

The signum function abbreviated as sgn (t) is defined by

$$\text{sgn}\,(t) = \begin{cases} 1 & t > 0 \\ -1 & t > 0 \end{cases} \quad \text{...(3.14a)}$$

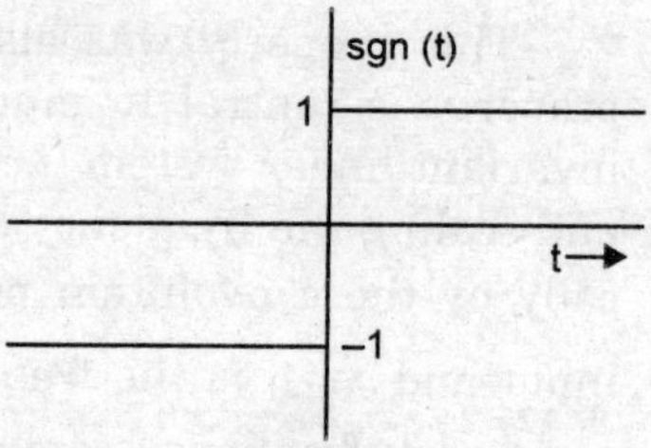

Fig. 6

It can be seen easily that

$$\text{sgn}\,(t) = 2u(t) - 1 \quad \text{...(3.14b)}$$

The Fourier transform of sgn (t) can be easily got when we observe that

$$\text{sgn}\,(t) = \lim_{a \to 0}\left[e^{-at}u(t) - e^{at}u(-t)\right] \quad \text{...(3.15)}$$

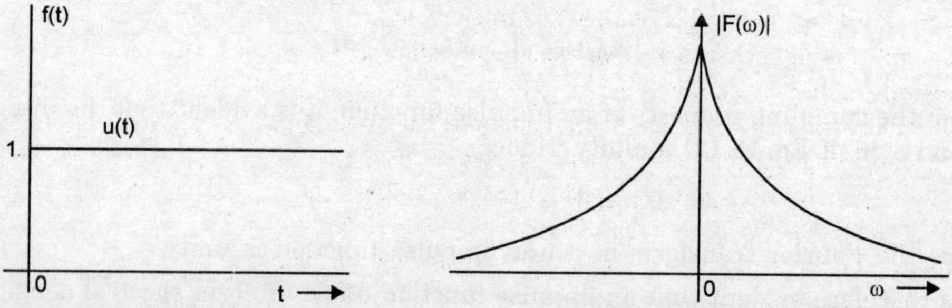

Fig. 7. The unit step function and its spectral density function.

$$\mathscr{F}[\operatorname{sgn}(t)] = \lim_{a \to 0}\left[\int_0^{\infty} e^{-at}e^{-j\omega t}dt - \int_{-\infty}^{0} e^{at}e^{-j\omega t}dt\right]$$

$$= \lim_{a \to 0}\left[\frac{-2j\omega}{a^2+\omega^2}\right] = \frac{2}{j\omega} \quad \text{...(3.16)}$$

4. Transform of Unit Step Function *u*(*t*)

From Eq. (3.14b) it follows that

$$u(t) = \frac{1}{2}[1+\operatorname{sgn}(t)]$$

Hence $$\mathscr{F}[u(t)] = \frac{1}{2}[\mathscr{F}[1]+\mathscr{F}[\operatorname{sgn}(t)]]$$

Using Eqs. (3.13b) and (3.16), we obtain

$$\mathscr{F}[u(t)] = \pi\delta(\omega)+\frac{1}{j\omega} \quad \text{...(3.17)}$$

The spectral density function contains an impulse at $\omega = 0$ (Fig. 7). Thus the function $u(t)$ contains a large d-c component as expected. In addition, it has other frequency components. The function $u(t)$ appears to be a pure d-c signal, and hence frequency components other than $\omega = 0$ may appear rather strange. However, the function $u(t)$ is not a true d-c signal. It is zero for $t < 0$, and there is an abrupt discontinuity at $t = 0$, giving rise to other frequency components. For a true d-c signal, $f(t)$ is constant for the entire interval $(-\infty, \infty)$. We have already seen (Eq. 3.14) that such a signal indeed has no frequency components other than d-c ($\omega = 0$).

5. Eternal Sinusoidal Signals cos ω_0t and sin ω_0t

We shall now consider the sinusoidal signals $\cos \omega_0 t$ and $\sin \omega_0 t$ in the entire interval $(-\infty, \infty)$. These signals do not satisfy the condition of absolute integrability; yet their Fourier transforms exist and can be found by a limiting process analogous to that used for a constant $f(t) = A$. We shall first assume these functions to exist in the interval $-\tau/2$ to $\tau/2$ and zero outside this interval. In the limit, τ will be made infinity. The procedure is now demonstrated.

$$\mathscr{F}(\cos \omega_0 t) = \lim_{\tau \to \infty}\int_{-\tau/2}^{\tau/2} \cos \omega_0 t\, e^{-j\omega t}\, dt$$

$$= \lim_{\tau \to \infty}\frac{\tau}{2}\left\{\frac{\sin[(\omega-\omega_0)\tau/2]}{(\omega-\omega_0)\tau/2}+\frac{\sin[(\omega+\omega_0)\tau/2]}{(\omega+\omega_0)\tau/2}\right\}$$

$$= \lim_{\tau \to \infty}\left\{\frac{\tau}{2}\operatorname{Sa}\left[\frac{\tau(\omega-\omega_0)}{2}\right]+\frac{\tau}{2}\operatorname{Sa}\left[\frac{\tau(\omega+\omega_0)}{2}\right]\right\} \quad \text{...(3.18)}$$

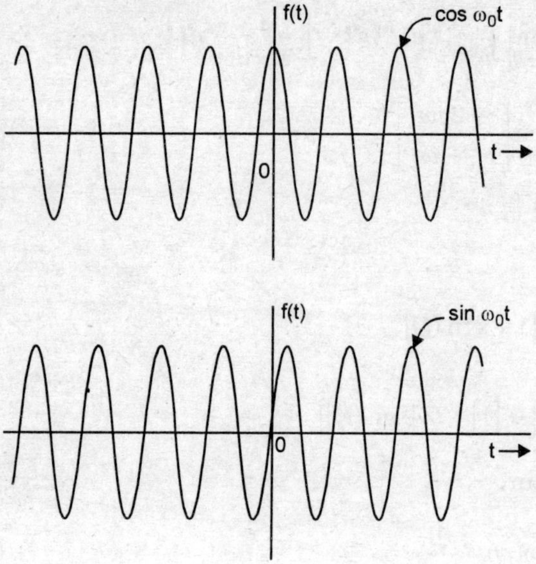

Fig. 8

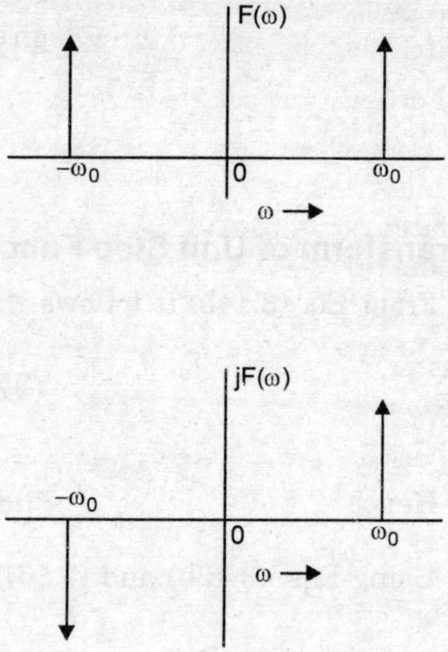

Fig. 9

In the limit the sampling function Sa ( ), becomes an impulse function (π amplitude), and we have

$$\mathscr{F}(\cos\,\omega_0 t) \;=\; \pi\big[\delta(\omega-\omega_0)+\delta(\omega+\omega_0)\big] \qquad \text{...(3.19)}$$

Similarly, it can be shown that

$$\mathscr{F}(\sin\,\omega_0 t) \;=\; j\pi\big[\delta(\omega+\omega_0)-\delta(\omega-\omega_0)\big] \qquad \text{...(3.20)}$$

Therefore the Fourier spectrum for these functions consists of two impulses at ω_0 and $-\ \omega_0$, respectively.

6. Transform of a Sinusoidal Burst Signal

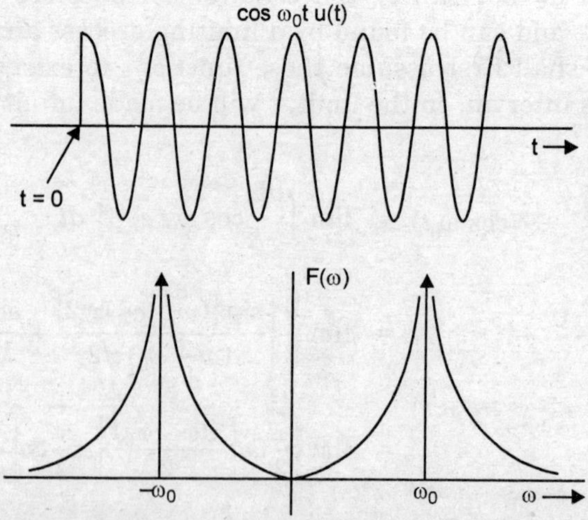

Fig. 10. F.T. spectrum of a signal cos $\omega_0 t.u(t)$.

A sudden onset of a cosine signal or sine signal is usual when we gate a signal through a switch for output.

For a cosine signal, when we give a burst at $t = 0$, since there is a sudden change at $t = 0$ by the step function $u(t)$, the transform shows smear on either side of ω_0, as shown.

Like this, whenever sine/cosine signals do not exist for all time, as in bursts of sine oscillations, there will be a smear (which is an exponential curve) around the frequency of the sine/cos signal.

THE FOURIER TRANSFORM OF PERIODIC FUNCTIONS

We have developed a Fourier transform as a limiting case of the Fourier series by letting the period of a periodic function become infinite. We shall now proceed in the opposite direction and show that the Fourier series is just a limiting case of the Fourier transform. This point of view is very useful, since it permits a unified treatment of both the periodic and the non-periodic functions.

Strictly speaking, the Fourier transform of a periodic function does not exist, since it fails to satisfy the condition of absolute integrability. For any periodic function $f(t)$:

$$\int_{\infty}^{-\infty} |f(t)|\, dt = \infty$$

But the transform does exist in the limit. We have already found the Fourier transform of cos $\omega_0 t$, sin $\omega_0 t$ in the limit.

Example : A number of impulses of unit strength and separated by T seconds as shown in Fig. 11. Find the Fourier transform.

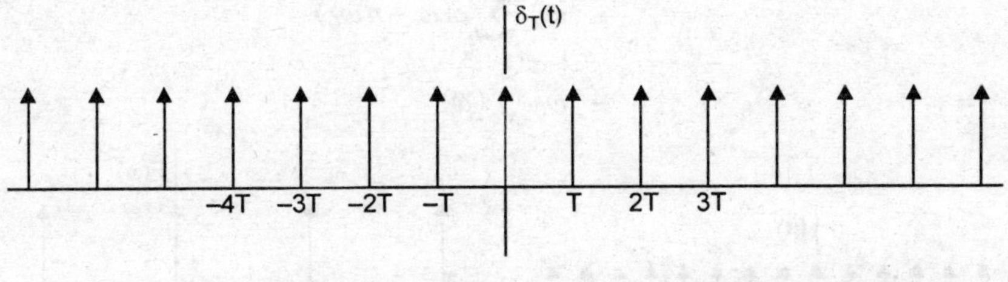

Fig. 11. The sequence of a uniform equidistant impulse function.

This function is very important in sampling theory, and hence it is convenient t_0 denote it by a special symbol $\delta_T(t)$. Thus

$$\delta_T(t) = \delta(t) + \delta(t - T) + \delta(t - 2T) + ... + \delta(t - nT) + ...$$
$$+ \delta(t + T) + \delta(t + 2T) + ... + \delta(t + nT) + ...$$

$$= \sum_{n=-\infty}^{\infty} \delta(t - nT) \qquad ...(3.21)$$

This is obviously a periodic function with period T. We shall first find the Fourier series for this function.

$$\delta_T(t) = \sum_{n=-\infty}^{\infty} F_n e^{jn\omega_0 t}$$

where
$$F_n = \frac{1}{T}\int_{-T/2}^{T/2} \delta_T(t)\, e^{-jn\omega_0 t}\, dt$$

Function $\delta_T(t)$ in the interval (–T/2, T/2) is simply $\delta(t)$. Hence

$$F_n = \frac{1}{T}\int_{-T/2}^{T/2} \delta(t)\, e^{-jn\omega_0 t}\, dt \qquad ...(3.22)$$

From the sampling property of an impulse function as expressed in (Eq. 3.12) the above equation reduces to

$$F_n = \frac{1}{T}$$

Consequently, F_n is a constant, 1/T. It therefore follows that the impulse train function of period T contains components of frequencies $\omega = 0, \pm\omega_0, \pm 2\omega_0, ..., \pm n\omega_0, ...,$ etc., $(\omega_0 = 2\pi/T)$ in the same amount.

$$\delta_T(t) = \frac{1}{T}\sum_{n=-\infty}^{\infty} e^{jn\omega_0} \qquad ...(3.23)$$

To find the Fourier transform of $\delta_T(t)$, we use Eq. (3.21). Since in this case $F_n = 1/T$, it is evident that

$$\mathscr{F}[\delta_T(t)] = 2\pi \sum_{n=-\infty}^{\infty} \frac{1}{T}\delta(\omega - n\omega_0)$$

$$= \frac{2\pi}{T}\sum_{n=-\infty}^{\infty} \delta(\omega - n\omega_0)$$

$$= \omega_0 \sum_{n=-\infty}^{\infty} \delta(\omega - n\omega_0)$$

$$= \omega_0 \delta_{\omega_0}(\omega) \qquad ...(3.24)$$

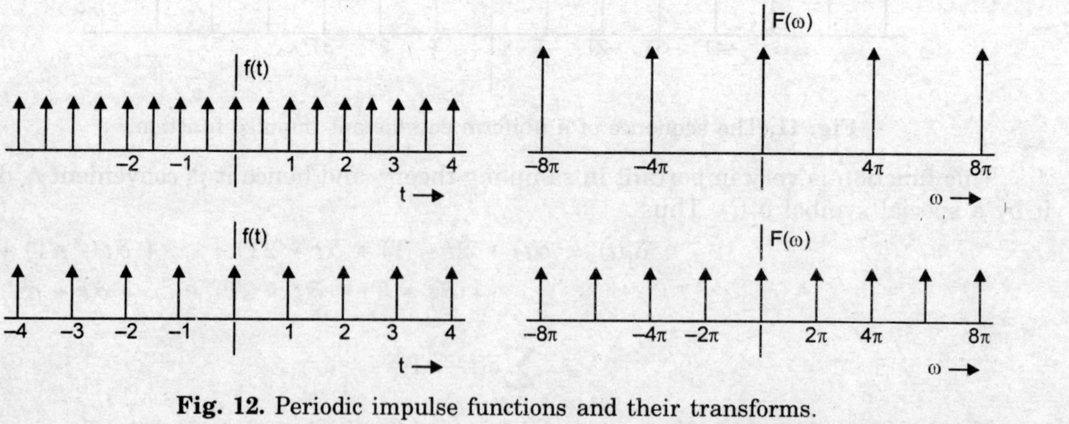

Fig. 12. Periodic impulse functions and their transforms.

Relation (3.24) is very significant. It states that the Fourier transform of a unit impulse train of period T is also a train of impulses of strength ω_0 and separated by ω_0 radians $(\omega_0 = 2\pi / T)$. Therefore the impulse train function is its own transform. The sequence of impulses with periods $T = \frac{1}{2}$ and T = 1 second and their respective transforms are shown in Fig. 12.

FOURIER TRANSFORM OF DISCRETE SIGNALS

In digital signal processing, we take samples of signals and use them. These are 'discrete' signals. Instead of the continuous Fourier transform, we use its discrete version: Here $x(t)$ or $f(t)$ becomes a set or sequence of numbers $x[n]$ or $f[n]$.

$$F(\omega) = \int_{-\infty}^{\infty} f(t)\, e^{-j\omega t} dt \;\;\rightarrow \sum_{k=0}^{N-1} f[n] \exp(-j2\pi\, kn/N) = F\left(\frac{n}{NT}\right) \qquad ...(3.25)$$

Note that the ωt term, which is $(2\pi f)t$ becomes $2\pi\, nk/N$.

The n denotes the frequency number, N is the total number of samples taken, k denotes the summation index, and T is the time between samples. The limits are $k = 0$ to $N - 1$ for N samples. We will illustrate this with an example.

Example 1. *Compute the Discrete Time Fourier Transform (DIFT) of the following signal:*

$$x[n] = \left[\frac{1}{4}\;\frac{1}{4}\;\frac{1}{4}\;\frac{1}{4}\right]$$

Note that there are only 4 samples.

$$X[n] = \left[\frac{1}{4}\;\frac{1}{4}\;\frac{1}{4}\;\frac{1}{4}\right]$$

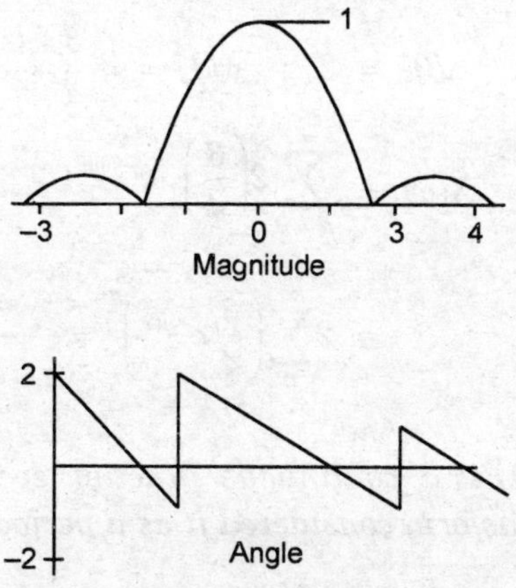

Fig. Example 1.

$$X[\omega] = \sum_{-\infty}^{\infty} \frac{1}{4}\left(1 + e^{-j\omega} + e^{-2j\omega} + e^{-j3\omega}\right)$$

$$= \frac{1}{4} e^{-j3\omega/2}\left[e^{j3\omega/2} + e^{j\omega/2} + e^{-j\omega/2} + e^{-j3\omega/2}\right]$$

$$= \frac{1}{2} e^{-j3\omega/2}\left[\cos\frac{3\omega}{2} + \cos\frac{\omega}{2}\right].$$

The magnitude and phase angle have sketch as above.

Example 2. $x[n] = [1\ -2\ 1]$

Here there are just 3 samples.

$$X(\omega) = 1 - 2e^{-j\omega} + e^{-j2\omega} = e^{-j\omega}[e^{j\omega} - 2 + e^{-j\omega}]$$

$$= e^{j\omega}\, 2(\cos\omega - 1) = e^{-j(\omega \pm \pi)}\, 2(1 - \cos(\omega)).$$

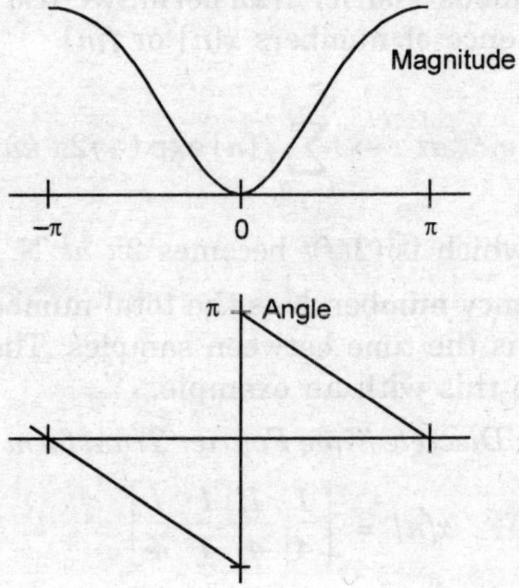

Example 3. *If* $$x[n] = 2\left(\frac{3}{4}\right)^n u(n).$$

u(n) is unit step signal. $x[0] = 2.$ $x[1] = 2.\frac{3}{4}.1$ *etc.*

$$X(\omega) = \sum_{n=0}^{\infty} 2\left(\frac{3}{4}\right)^n e^{-j\omega n}$$

$$= 2\sum_{0}^{\infty}\left(\frac{3}{4}e^{-j\omega}\right)^n = \frac{2}{1-\frac{3}{4}e^{-j\omega}}$$

Note that the figure (Ex. 1) is a 'continuous' function even though the signal is discrete. That is because the Fourier transform considered it as a periodic signal, repeating itself in the same four value pattern.

Example 4. *Find the response of the system having the F.T. as*

$$H(j\omega) = e^{-j\omega/2}\left(1 - \cos\frac{\omega}{2}\right)$$

for an input signal in discrete form as

$$x[n] = 2 + 2\cos\frac{\pi n}{4} + \cos\left(\frac{2\pi n}{3} + \frac{\pi}{2}\right).$$

Solution. $H(j0) = 0$ by substitution $\omega = 0$ in $e^{-j\omega/2}\left(1 - \cos\frac{\omega}{2}\right)$

$$H(\pi/4) = e^{-j\pi/8}\left(1 - \cos\frac{\pi}{8}\right) = 0.076\, e^{-j\pi/8} = 0.076 \angle -\frac{\pi}{8}$$

$$H\left(\frac{2\pi}{3}\right) = e^{-j2\pi/6}\left(1 - \cos\frac{2\pi}{6}\right) = 0.5\, e^{-j\pi/3} = 0.5 \angle -\pi/3.$$

$$y[n] = 0 + 0.152 \cos\left(\frac{\pi n}{4} - \frac{\pi}{8}\right) + \frac{1}{2}\cos\left(\frac{2\pi n}{3} + \frac{\pi}{6}\right),$$

Which is obtained by multiplying by the frequency response for each frequency, $\left(\text{D.C.}, \frac{\pi}{4} \text{ and } \frac{2\pi}{3}\right)$.

DISCRETE TIME FOURIER TRANSFORM (DTFT) AND DISCRETE FOURIER TRANSFORM (DFT)

Both pertain only to samples of signals taken; hence both are 'Discrete time'. But the first one means that the signal samples are repeating in this same pattern all the time (from $-\infty < t < \infty$). The second calculates the Fourier transform coefficient directly by substitution.

For example, take just a two-sample function

$$x[n] = \underset{n=0}{[1} \qquad \underset{n=1}{-1]}$$

DTFT

$$X(\omega) = \sum_{n=-\infty}^{\infty} x[n] \exp(-j\omega n)$$

$$= e^{0} - e^{-j\omega} = e^{-j\omega/2}\left[e^{j\omega/2} - e^{-j\omega/2}\right]$$

$$= e^{-j\omega/2} \,.\, 2j \,.\, \sin(\omega/2)$$

$$= e^{-j\omega/2}\, e^{\pm j\pi/2}\, 2\sin\omega/2.$$

This is a continuous function of ω.

DFT

Here we just calculate that F.T. values also discretely for different ω.

$$X_k = \sum_{n=0}^{1} x[n] \exp(-2j\, k\pi/2.n)$$

$$= 1 - e^{-j\pi k} \quad \text{for just } k = 0 \text{ and } k = 1 \text{ (using } x_0 = x_1 = 1)$$

$$\therefore \quad X_0 = 1 - e^{0} = 0$$

$$X_1 = 1 - e^{-j\pi} = 2.$$

Thus, the two DFT values are [0, 2].

FOURIER TRANSFORMS FOR CERTAIN MATHEMATICAL FUNCTIONS

1. A single rectangular pulse has the F.T.

$$\tau.\frac{\sin \omega\tau/2}{\omega\tau/2} = \tau.Sa\left(\frac{\omega\tau}{2}\right)$$

$Sa(\ ) \equiv$ Sampling Function

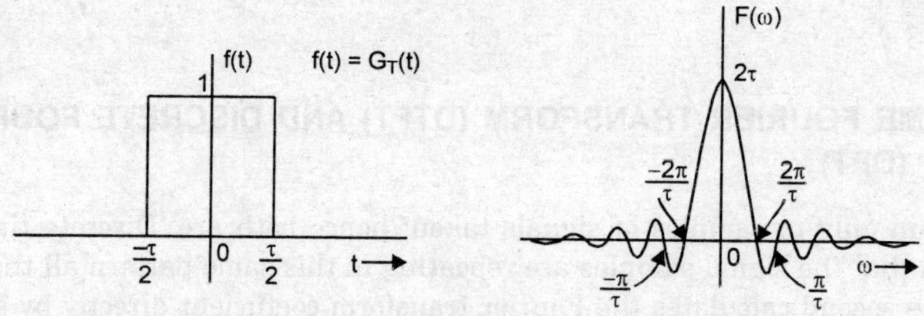

Fig. 13

2. Triangular pulse gate

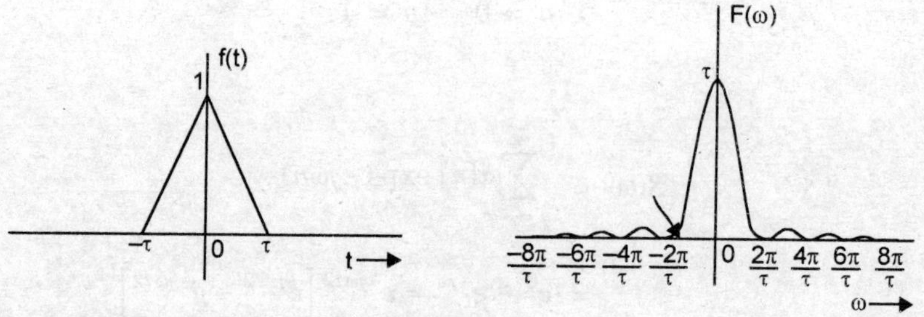

Fig. 14

3. Gaussian function. The F.T. is also Gaussian.

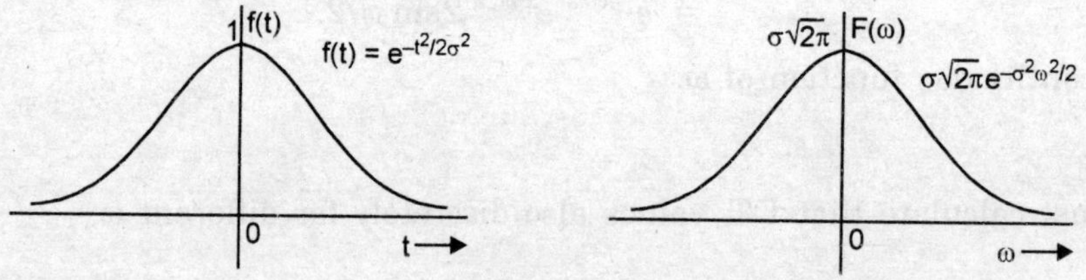

Fig. 15

4. Double sided exponential decay.

Note : Phase is zero.

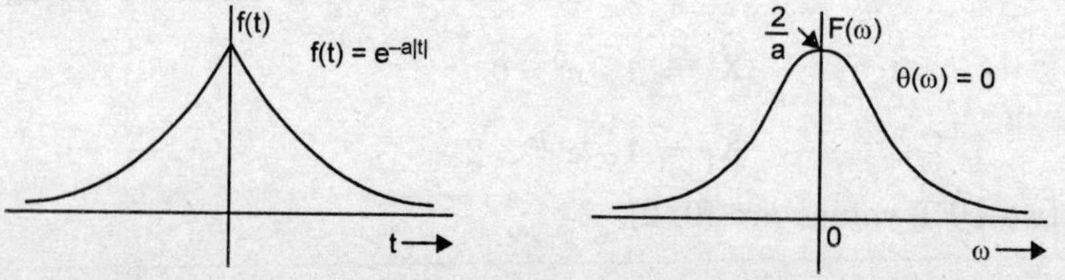

Fig. 16

5. Decaying oscillation

Note peak at $\sqrt{a^2 - \omega_0^2}$ and smear on either side. The smear is narrow if decay is fast (α large).

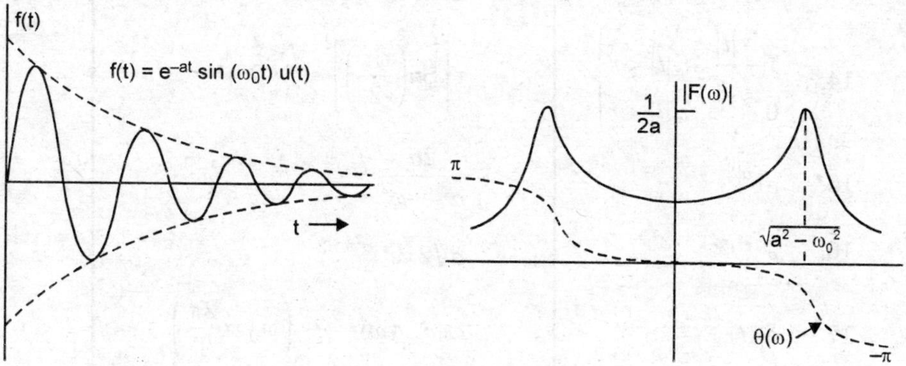

Fig. 17

SOME FOURIER TRANSFORM PAIR

Fourier Transforms

	$f(t)$	$F(\omega)$
1.	$e^{-at}\,u(t)$	$\dfrac{1}{a + j\omega}$
2.	$te^{-at}\,u(t)$	$\dfrac{1}{(a + j\omega)^2}$
3.	$\lvert t \rvert$	$\dfrac{-2}{\omega^2}$
4.	$\delta(t)$	1
5.	1	$2\pi\delta(\omega)$
6.	$u(t)$	$\pi\delta(\omega) + \dfrac{1}{j\omega}$
7.	$\cos\omega_0 t\; u(t)$	$\dfrac{\pi}{2}[\delta(\omega - \omega_0) + \delta(\omega + \omega_0)] + \dfrac{j\omega}{\omega_0^2 - \omega^2}$
8.	$\sin\omega_0 t\; u(t)$	$\dfrac{\pi}{2j}[\delta(\omega - \omega_0) - \delta(\omega + \omega_0)] + \dfrac{\omega_0}{\omega_0^2 - \omega^2}$
9.	$\cos\omega_0 t$	$\pi[\delta(\omega - \omega_0) + \delta(\omega + \omega_0)]$
10.	$\sin\omega_0 t$	$j\pi[\delta(\omega + \omega_0) - \delta(\omega - \omega_0)]$
11.	$e^{-at}\sin\omega_0 t\; u(t)$	$\dfrac{\omega_0}{(a + j\omega)^2 + \omega_0^2}$

12. $\frac{W}{2\pi} \text{Sa} \frac{(Wt)}{2}$	$G_W(\omega)$
13. $G_\tau(t)$	$\tau \text{Sa}\left(\frac{\omega\tau}{2}\right)$
14. $\left.\begin{matrix} 1 - \frac{\lvert t \rvert}{\tau} \text{ .... } \lvert t \rvert < r \\ 0 \text{ .... } \lvert t \rvert < r \end{matrix}\right\}$	$\tau\left[\text{Sa}\left(\frac{\omega\tau}{2}\right)\right]^2$
15. $e^{-a\lvert t \rvert}$	$\frac{2a}{a^2 + \omega^2}$
16. $e^{-t^2/2\sigma^2}$	$\sigma\sqrt{2\pi}e^{-\sigma^2\omega^2/2}$
17. $\delta_T(t)$	$\omega_0\delta_{\omega_0}(\omega) \quad \left(\omega_0 = \frac{2\pi}{T}\right)$

IMPORTANT PROPERTIES OF FOURIER TRANSFORMS

1. Linearity Property

Suppose we have two functions $f_1(t)$ and $f_2(t)$. Let their individual transforms be $F_1(\omega)$ and $F_2(\omega)$. Then for the combined signal $f(t) = f_1(t) \pm f_2(t)$, the Fourier transform is $F(\omega) = F_1(\omega) \pm F_2(\omega)$.

Like this we can generalise for more functions.

2. Scaling Property

If we have $F(\omega)$ for $f(t)$.

To find $f(at)$'s fourier transform, we use this

$$f(at) \leftrightarrow \frac{f}{|a|} F\left(\frac{\omega}{a}\right). \qquad ...(3.26)$$

This means time compression causes frequency expansion and vice versa.

3. Symmetry Property

Suppose we have a signal and its Fourier transform.

Then, if the waveform given by the transform is itself taken a Fourier transform, what would one get?

$f(t)$ given $F(\omega)$.

Now replace ω variable by x. Then take its F.T.

$$F\{F(x)\} = \int_{-\infty}^{\infty} F(x)e^{-j\omega x}\, dx \qquad (3.27i)$$

But

$$f(t) = \frac{1}{2\pi}\int_{-\infty}^{\infty} F(\omega)e^{j\omega t}\, d\omega \qquad (3.27ii)$$

Put $(-t)$ in this. Then

$$f(-t) = \frac{1}{2\pi}\int F(\omega)e^{-j\omega t}\,d\omega \qquad \text{...(3.27iii)}$$

In the above, the integrating variable can be replaced as x, without changing the value of the integral, *i.e.*, put $\omega = x$!

$$f(-t) = \frac{1}{2\pi}\int F(x)e^{-jxt}\,dx \qquad \text{...(3.27iv)}$$

Compare (*i*) and (*iv*). The variable is ω in (3.25*i*) and *t* in (3.25*iv*). Therefore

$$f(-t) = \frac{1}{2\pi}\int F(x)e^{-jxt}\,dx$$

$$= 2\pi\,\mathscr{F}\{F(x)\} \quad \text{(but as a time function } t\text{, not } \omega)$$

$$= 2\pi \times \text{Fourier transform of } \mathbf{F(\omega)}.$$

Thus when we take another Fourier transform of the F.T. of a signal, we get the time function back as $\frac{1}{2\pi}f(-t)$.

Thus,
$$f(t) \xrightarrow[\text{F.T.}]{} F(\omega) \xrightarrow[\text{F.T.}]{} \frac{1}{2\pi}f(-t) \qquad \text{...(3.27v)}$$

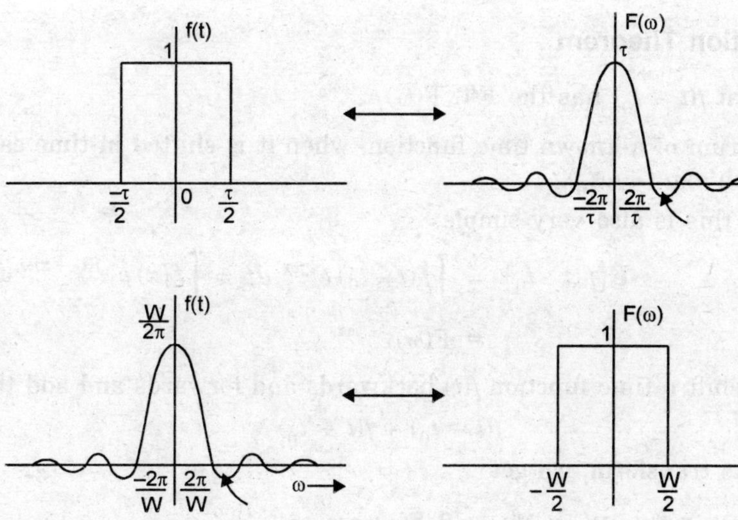

Fig. 18. Rectangular time function and its symmetrical transform pair.

Thus a rectangular time function and its F.T. one shown.

For the same function we can take the transform and get the time function as a rectangular time function.

4. Frequency Shift Theorem

If $f(t)$ has $F(\omega)$ as its transform, then $f(t)\,e^{j\omega_c t}$ has $F(\omega - \omega_c)$ as its transform.

This theorem can be easily proved and is left as an exercise. The meaning of this theorem is applicable in our modulation methods. $e^{j\omega_c t}$ and $f(t)$ multiplied means modulating the signal with the sinusoidal (since $\exp(e^{j\omega_c t}) = \cos \omega_c t + j \sin \omega_c t$) carrier frequency ω_c.

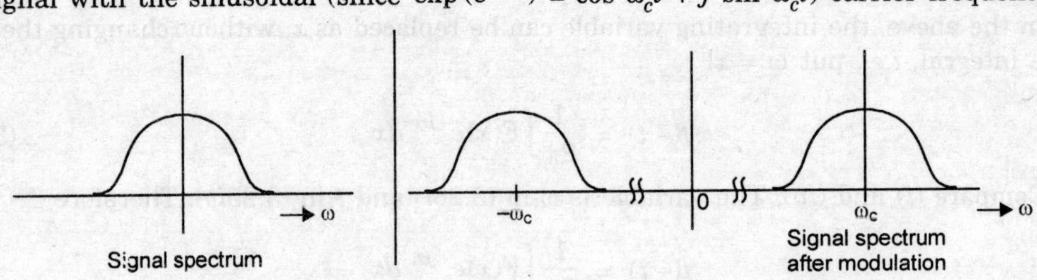

Fig. 19 (*a*) modulating signal and (*b*) its spectrum ω_c is carrier

The modulating signal's spectrum is centered around the $\pm\, \omega_c$ frequencies, as two *side-bands*. The modulated signal $f_m(t)$ is

$$f_m(t) = f(t) \cos \omega_c t = \frac{1}{2}\left\{f(t)\, e^{j\omega_c t} + f(t) e^{-j\omega_c t}\right\} \qquad \text{...(3.28}a\text{)}$$

Hence the spectrum of $f_m(t)$ is

$$\frac{1}{2}[\mathrm{F}(\omega + \omega_c) + \mathrm{F}(\omega - \omega_c)]$$

which are the two side bands shown above.

5. Time Translation Theorem

This says that $f(t - t_0)$ has the F.T. $\mathrm{F}(\omega)\, e^{-j\omega t_0}$

So, the spectrum of a known time function, when it is shifted in time can be found by multiplication with $\exp(-j\omega t_0)$.

The proof of this is also very simple.

$$\mathbf{F}\{f(t - t_0\} = \int f(t - t_0)\, e^{-j\omega t}\, dt = \int f(x)\, e^{-j(t_0 + x)\omega}\, dx \quad (x = t - t_0)$$

$$= \mathrm{F}(\omega)\, e^{-j\omega t_0} \qquad \text{...(3.28}b\text{)}$$

Suppose we shift a time function $f(t)$ backwards and forwards and add the two.

$$f(t - t_0) + f(t + t_0) \qquad \text{...(3.29)}$$

If we find its transform, we get

$$\mathrm{F}(\omega)\, e^{-j\omega t_0} + \mathrm{F}(\omega)\, e^{j\omega t_0} = 2\ \mathrm{F}(\omega) \cos \omega t_0$$

That means, it is the cosine modulation product which is got as the frequency domain function for the added value of two equally shifted signals $f(t)$.

TIME DERIVATIVE AND INTEGRAL

Given $f(t)$ and its F.T. as $\mathrm{F}(\omega)$, what will be the transform of $df(t)/dt$?

The Fourier transform actually expresses a function $f(t)$ in terms of a continuous sum of exponential functions of the form $e^{j\omega t}$. The derivative of $f(t)$ is therefore equal to the continuous sum of the derivatives of the individual exponential components. But the derivative of an exponential function $e^{j\omega t}$ is equal to $j\omega e^{j\omega t}$. Therefore the process of differentiation of $f(t)$ is equivalent to multiplication by $j\omega$ of each exponential component. Hence

$$\frac{df}{dt} \leftrightarrow j\omega F(\omega)$$

A similar argument applies to the integration.

Similarly $\int_{-\infty}^{t} f(\tau)\,d\tau$ has its F.T. as $\frac{1}{j\omega} F(\omega)$.

Thus multiplication by $j\omega$ is equivalent to differentiation and division by $j\omega$ as equivalent to integration in frequency domain.

Convolution Property : We start with definition of convolution :

$$y(t) = h(t) * x(t) = \int_{-\infty}^{\infty} x(s)\,h(t-s)\,ds \quad ...(3.30)$$

By the definition of the Fourier transform,

$$\hat{y}(\omega) = \int_{-\infty}^{\infty}\left[\int_{-\infty}^{\infty} x(s)h(t-s)\,ds\right] e^{-j\omega t}\,dt \quad ...(3.31)$$

Switching the order of integration,

$$\hat{y}(\omega) = \int_{-\infty}^{\infty} x(s)\left[\int_{-\infty}^{\infty} h(t-s)\,e^{-j\omega t}\,dt\right] ds \quad ...(3.32)$$

Letting $\sigma = t - s$,

$$\hat{y}(\omega) = \int_{-\infty}^{\infty} x(s)\left[\int_{-\infty}^{\infty} h(\sigma)\,e^{-j\omega(\sigma+s)}\,d\sigma\right] ds \quad ...(3.33)$$

$$= \int_{-\infty}^{\infty} x(s)\,e^{-j\omega s}\left[\int_{-\infty}^{\infty} h(\sigma)\,e^{-j\omega\sigma}\,d\sigma\right] ds$$

Then by the definition of the Fourier transform,

$$\hat{y}(\omega) = \int_{-\infty}^{\infty} x(s)\,e^{-j\omega s}\hat{h}(\omega)\,ds \quad ...(3.34)$$

$$= \hat{h}(\omega)\int_{-\infty}^{\infty} x(s)\,e^{-j\omega s}\,ds$$

$$= \hat{h}(\omega)\hat{x}(\omega)$$

This means that if there are two transform (frequency domain) functions $h(\omega)$ and $X(\omega)$, when we multiply them, they belong to a time signal which is the 'convolution' of their original time signals $h(t)$ and $x(t)$.

That means a product *here* is not a product *there*, but a 'convolution'.

We will see 'convolution' later is another chapter.

SIGNUM FUNCTION AND IMPULSIVE FUNCTION

$$\text{Sgn}(t) = 1 \text{ for } t > 0$$
$$\qquad -1 \text{ for } t < 0$$

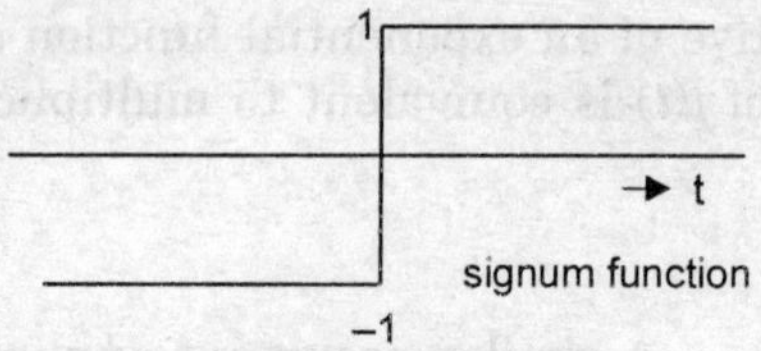

Fig. 20

It can be seen that

$$\text{Sgn}(t) = u(t) - u(-t)$$

where $u(t)$ is a step function, starting at $t = 0$.

Since there is a discontinuity at $t = 0$, we write

$$\text{Sgn}(t) = \underset{a \to 0}{\text{Lt}}\left[u(t)\, e^{-at} - e^{at}\, u(-t)\right] \qquad \text{...(3.35)}$$

Then finding the F.T. is easy, as

$$\underset{a \to 0}{\text{Lt}} \int_0^{\infty} e^{-at}\, e^{-j\omega t}\, dt + \int_0^{-\infty} e^{at}\, e^{-j\omega t}\, dt \qquad \text{...(3.36)}$$

$$= \lim_{a \to 0} \frac{-2j\omega}{a^2 + \omega^2}$$

$$= \frac{2}{j\omega}$$

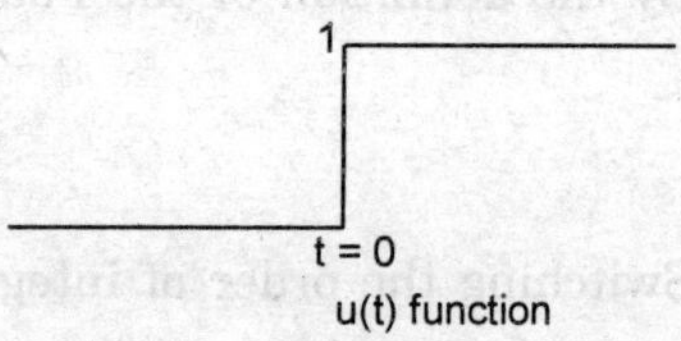

Fig. 21

The $u(t)$ function can be written as

$$u(t) = \frac{1}{2}[1 + \text{sgn}(t)]$$

Hence its F.T. is

$$\frac{1}{2}\text{F}(1) + \frac{1}{2} \cdot \frac{2}{j\omega}$$

Fourier transform of a constant (= 1) is $2\pi \times 1$, so we get

$$\text{F.T. } \{u(t)\} = \pi + \frac{1}{j\omega}.$$

Example: Find the F.T. of the Gaussian function

$$f(t) = e^{-\alpha t^2} \qquad \text{...}(i)$$

We start with the well-known definite integral

$$\int_{-\infty}^{\infty} e^{-\alpha t^2}\, dt = \sqrt{\frac{\pi}{\alpha}}. \qquad \text{...}(ii)$$

Differentiating n times with respect to α,

$$\int_{-\infty}^{\infty} t^{2n}\, e^{-\alpha t^2}\, dt = \frac{1.3 \ldots . (2n-1)}{2^n} \sqrt{\frac{\pi}{\alpha^{2n+1}}} \qquad \text{...}(iii)$$

Next we break the general Fourier integral into an exponential series :

$$F(\omega) = \int f(t) \sum \frac{(-j\omega t)^n}{n!}\, dt \qquad \left\{ e^{j\omega t} = \sum \frac{(-j\omega t)^n}{n!} \right.$$

$$= \sum_{n=0}^{\infty} (-j)^n . \frac{\omega^n}{n!} \int t^n f(t)\, dt \qquad \text{...}(iv)$$

We can replace n by $2n$ in the above without any loss of generality!

$$F(\omega) = \sum_{n=0}^{\infty} (-j)^{2n} . \frac{\omega^{2n}}{2n!} \int t^{2n} f(t)\, dt \qquad \text{...}(v)$$

If $f(t)$ is given by *(i)*, then, using *(iii)*,

$$= \sum (-j)^{2n} \frac{\omega^{2n}}{2n!} \frac{1.3 \ldots . (2n-1)}{(2\alpha)^n} \sqrt{\frac{\pi}{\alpha}} \qquad \text{...}(vi)$$

$$= \sqrt{\frac{\pi}{\alpha}}\, e^{-\omega^2/4\alpha} \qquad \text{...}(vii)$$

by a series sum of the exponential series.

So, we get the result in *(vii)*. Note this is also a Gaussian function.

THE HILBERT TRANSFORM

The Hilbert transform is defined as

$$H(f) = -j \ \mathrm{sgn}\ (f) \qquad \text{...}(3.37)$$

The $-$ sgn (f) is shown. The j term denotes a phase of $-$ 90° or $-$ $\pi/2$ for positive and $+\pi/2$ or 90° for negative frequencies.

Magnitude is constant for all ω.

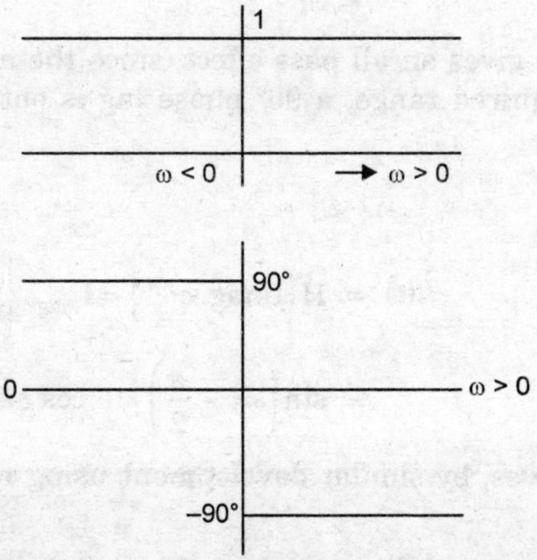

Fig. 22. Magnitude and phase of Hilbert transform.

In order to find the corresponding time domain function, for this Hilbert transform, let us define another function

$$G(f : \alpha) = \begin{cases} e^{-\alpha f} & \text{for } f < 0 \\ -e^{+\alpha f} & \text{for } f > 0 \end{cases} \quad \text{...(3.38)}$$

When we put $\alpha \to 0$ in the above, it becomes the signum function. Hence the time function for G is

$$g(t : \alpha) = F^{-1}[G(f : \alpha)]$$

$$= \int_0^{\infty} e^{-\alpha f} e^{j2\pi ft} df - \int_{-\infty}^{0} e^{\alpha f} e^{j2\pi ft} df$$

$$= \frac{4\pi j.t}{\alpha^2 + (2\pi t)^2} \quad \text{...(3.39)}$$

When $\alpha \to 0$ in the limit, we get $\frac{j}{\pi t}$.

This is the time function for sgn (f), but the Hilbert transform is $-j$ sgn (f), we get as (h denoting time domain Hilbert transform),

$$h(t) = \frac{-j^2}{n\pi t} = \frac{1}{\pi t}.$$

Note that the Hilbert transform is a function which operates on any function of t.

Given an input $m(t)$, the Hilbert transform is got as $\frac{1}{\pi}\int_{-\infty}^{\infty} \frac{m(\tau)}{t-\tau} d\tau$ by convolution of $m(t)$ with $h(t)$.

Thus the signal $s(t)$ has the Hilbert transform given by

$$\int_{-\infty}^{\infty} \frac{s(\tau)}{t-\tau} d\tau \quad \text{...(3.40)}$$

The Hilbert transform gives an all pass effect, since the amplitude of all frequencies is the same; but over the required range, a 90° phase lag is obtained.

Properties

1. **A sine wave**

$$h(t) = H\{\text{Imag.}\, e^{j\omega t}\} = I_{mag} \cdot \left\{\left(-\frac{j\pi}{2}\right) e^{j\omega t}\right\}$$

$$= \sin\left(\omega t - \frac{\pi}{2}\right) = -\cos \omega t.$$

A *cosine* wave gives, by similar development using real part of e^{jwt}, as its Hilbert transform

$$h(t) = \sin \omega t$$

Thus the *cos* and *sine* functions, which were noted as orthogonal in Chap. 1, form a Hilbert transform pair.

2. A number of physical signals based on properties of the medium are known to be Hilbert transforms of each other. The absorption (of light) by a specimen (a liquid, say) gives an absorption signal (over various frequencies of light). The dielectric property of the chemical also varies with frequency. It is known that absorption and dielectric constant are Hilbert transforms of each other.
3. The Hilbert transform of a square wave is shown.

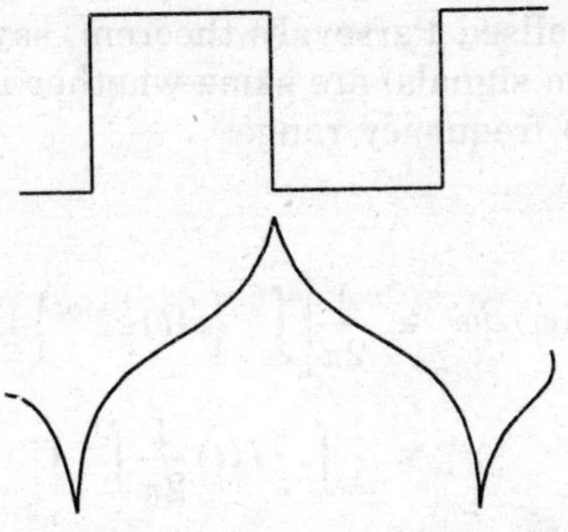

Fig. 23. The square wave and its Hilbert transform.

4. Hilbert transform can be realised by a digital filter. Using this, it is used to generate a single side-band amplitude modulated signal.

Example: Show that the Hilbert transform of the square wave

$$x(t) = A \quad (-\tau/2 < t < \tau/2)$$
$$= 0 \quad \text{for all other } t.$$

is a logarithmic function.

Let $\hat{X}(t)$ = Hilbert transform of $x(t)$.

$$\hat{X}(t) = x(t) * \frac{1}{\pi t} \quad \text{(Convolution)}$$

$$= \frac{1}{\pi}\int_{-\infty}^{\infty} \frac{x(\lambda)}{(t-\lambda)}\, d\lambda = \frac{1}{\pi}\int_{-\tau/2}^{\tau/2} \frac{A}{t-\lambda}\, d\lambda$$

$$= \frac{A}{\pi}\left[+\log_e (t-\lambda)\right]_{t-\tau/2}^{t+\tau/2} = -\frac{A}{\pi}\log_e\left(\frac{t-\tau/2}{t+\tau/2}\right).$$

Properties of H.T.

1. The Hilbert transform of a Hilbert transform is the negative of the original function.

Since the H.T. corresponds to a phase shift of $\pm \pi/2$, we note that the H.T. of $v(t)$ corresponds to the transfer function

$$[-j \operatorname{sgn}(f)]^2$$
$$-j^2 (\operatorname{sgn} f)^2 = -1 \text{ or } -\pi.$$

For example, Hence $H[H\{u(t)\}] = -u(t)$.

2. A function and its Hilbert transform are orthogonal over the infinite interval.

$$i.e. \quad \int_{-\infty}^{\infty} u(t)\, v(t)\, dt = 0 \qquad ...(i)$$

where $u(t)$ and $v(t)$ are H.T.

We note that

$$\int_{-\infty}^{\infty} u(t)\, v^*(t)\, dt = \int_{-\infty}^{\infty} U(f)\, V^*(f)\, df \qquad ...(ii)$$

This is by the generalised Parseval's theorem, saying that energy products (such as by multiplying two signals) are same whether it is evaluated over infinite time or over the (infinite) frequency range.

Parseval's Theorem

$$\frac{1}{2\pi}\int_{-\infty}^{\infty} V(\omega)\,.\,U^*(\omega)\, d\omega = \frac{1}{2\pi}\left[\int_{-\infty}^{\infty}\left\{v(t)\, e^{-j\omega t}\right\} U^*(\omega)\, d\omega\right]$$

$$= -\int_{-\infty}^{\infty} v(t)\frac{1}{2\pi}\int_{-\infty}^{\infty} U^*(\omega)\, e^{-j\omega t}\, d\omega$$

$$= \int_{t=-\infty}^{\infty} v(t)\, u^*(t)\, dt \qquad ...(iii)$$

Now take (i);

$$\int_{-\infty}^{\infty} v(t)\, u^*(t)\, dt = \int_{-\infty}^{\infty} (+\, j\, \text{sgn}\,(f))\, |U(f)|^2\, df$$

Note that the R.H.S. integral is an odd function; its $|U(f)|^2$ is always positive which is an *even* function, but j sgn (f) is odd function.

So, any integral which has an odd function and limits $-\infty$ and $+\infty$ would become zero. This proves that (i) is satisfied; hence orthogonal.

FOURIER TRANSFORM—ADDITIONAL PROBLEM

For each signal, find the Fourier transform, X(ω), and then plot | X(ω) | (note, you may want to use MATLAB for the plot in 3).

1.

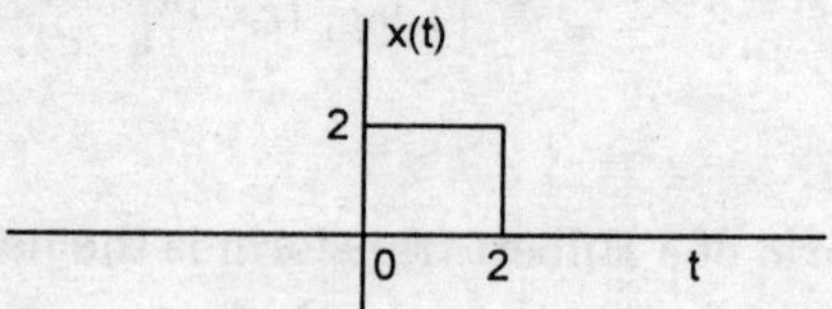

2.

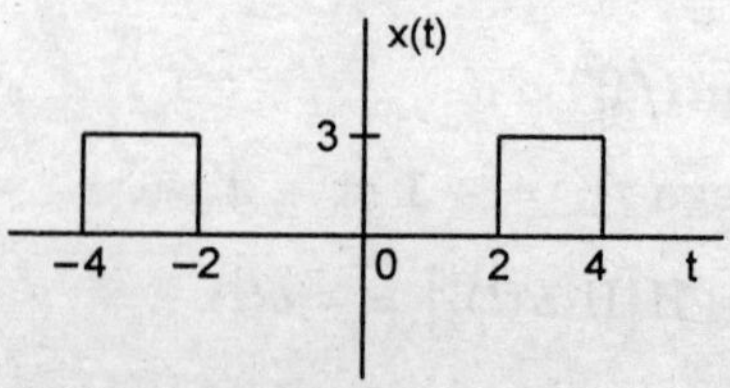

3. 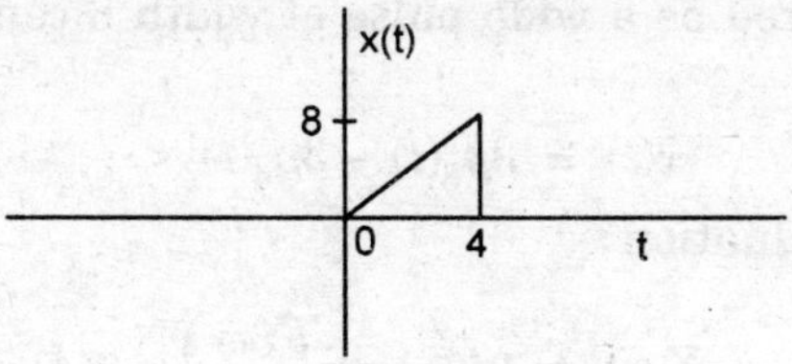

4. $$x(t) = \cos(200t)p_4(t)$$

5. $$x(t) = e^{-3t}\cos(10t)u(t)$$

Solution

1. The signal is a rectangular pulse of width 2, at $t = 1$ (centre of pulse). Recalling $p(t)$ as a pulse signal and that its F.T. is a 'sinc' function:

$$x(t) = 2\,p_2\,(t-1)$$

$$X(\omega) = 2 \cdot 2\,\text{sinc}\left(\frac{2\omega}{2\pi}\right)e^{-j\omega\cdot 1}$$

$$= 4\,\text{sinc}\left(\frac{\omega}{\pi}\right)e^{-j\omega}$$

$$|X(\omega)| = 4\,|\sin(\omega/\pi)|$$

The plot is a shown

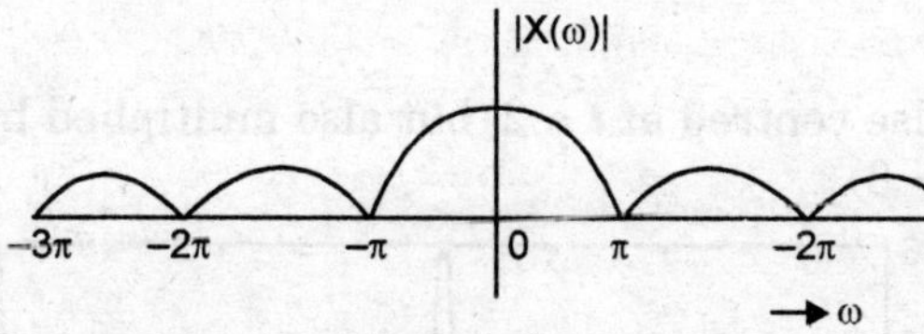

Fig. Problem 1.

2. Here there are two pulses.

Amplitude = 3 Centre = 3 Width = 2

Amplitude = 3 Centre = –3 Width = 2

$$x(t) = 3p_2(t-3) + 3p_2(t+3)$$

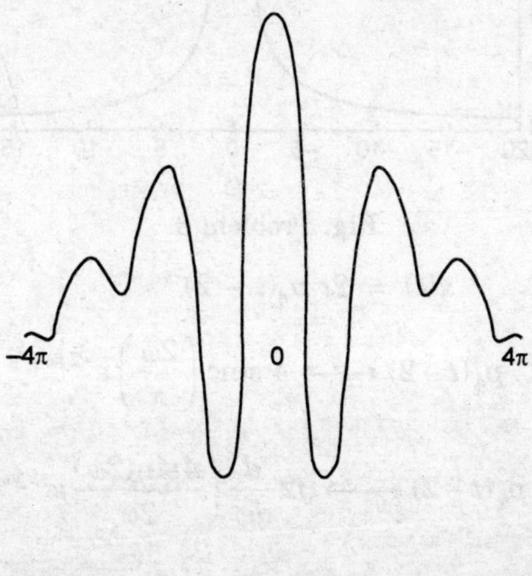

This can be also considered as a wide pulse of width 8 centred at $t = 0$, and subtracted by another of width 4.

$$x(t) = 3p_8(t) - 3p_4(t)$$

This makes for easy evaluation.

$$X(\omega) = 3(8)\,\text{sinc}\left(\frac{8\omega}{2\pi}\right) - 3(4)\,\text{sinc}\left(\frac{4\omega}{2\pi}\right)$$

$$= 24\,\text{sinc}\,(4\omega/\pi) - 12\,\text{sinc}\,(2\omega/\pi)$$

The first has its zero crossing at $\pi/4$.

The second has its zero crossing at $\pi/2$.

The plot is roughly sketched below.

It can be done by a computer program as with simple BASIC.

```
SCREEN 12: PI = 355/113   : I% = 1
FOR OMEGA = –4*PI   TO   4*PI   STEP 0–1
A = 4*OMEGA/PI : B = 2*OMEGA/PI
X = 24*SIN(A/A) : X = X – 12*SIN(B)/B
PSET (2*I%, 200 – 10*X)
I% = I% + 1
NEXT
```

3. In this we have a pulse centred at $t = 2$, but also multiplied by t, giving a ramp width is 4. Amplitude is $2(2 \times 4 = 8)$.

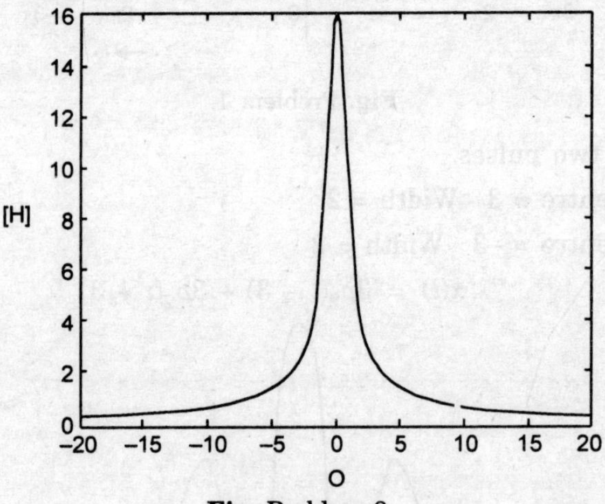

Fig. Problem 3.

$$x(t) = 2t\ p_4(t - 2)$$

$$p_4(t-2) \longleftrightarrow 4\ \text{sinc}\left(\frac{2\omega}{\pi}\right) e^{-2j\omega}$$

$$2t\ p_4(t-2) \longleftrightarrow j2\frac{d}{d\omega}\left(\frac{4\sin(2\omega)}{2\omega}e^{-2j\omega}\right)$$

$$= 4j\left[-\frac{\omega\,(2\cos(2\omega)\,e^{-2j\omega} - 2}{\omega^2} + \frac{\sin(2\omega)\,e^{-2j\omega} - \sin(2\omega)\,e^{-2j\omega}}{\omega^2}\right]$$

$$= \frac{4j}{\omega^2}e^{-2j\omega}\,(\omega 2\cos(2\omega) - 2j\omega\sin(2\omega) - \sin(2\omega))$$

$$= \frac{4j}{\omega^2}e^{-2j\omega}\left(\omega 2e^{-2j\omega} - \sin(2\omega)\right)$$

4. $$x(t) = \cos(200t)\,p_4(t)$$

$$p_4(t) \longleftrightarrow 4\,\text{sinc}\left(\frac{2\omega}{\pi}\right)$$

$$\cos(200t)p_4(t) \longleftrightarrow 2\,\text{sinc}\left(\frac{2}{\pi}(\omega - 200)\right) + 2\,\text{sinc}\left(\frac{2}{\pi}(\omega + 200)\right)$$

since $$\cos\theta = (e^{j\theta} + e^{-j\theta})/2.$$

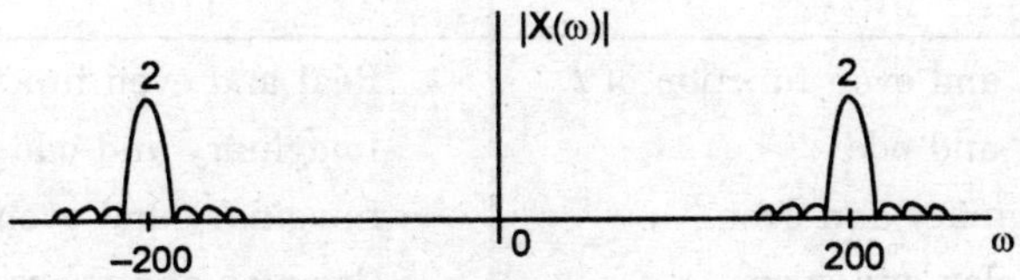

5. $$x(t) = e^{-3t}\cos(10t)\,u(t)$$

From page 83 F.T. table,

$$e^{-3t}\,u(t) \longleftrightarrow \frac{1}{j\omega + 3}$$

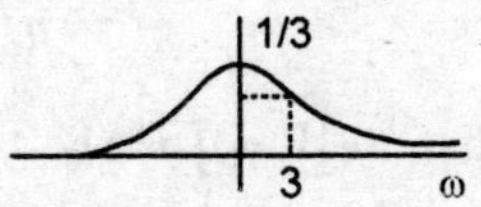

$$e^{-3t}u(t)\cos(10t) \longleftrightarrow \frac{1}{2}\left[\frac{1}{j(\omega - 10) + 3} + \frac{1}{j(\omega + 10) + 3}\right]$$

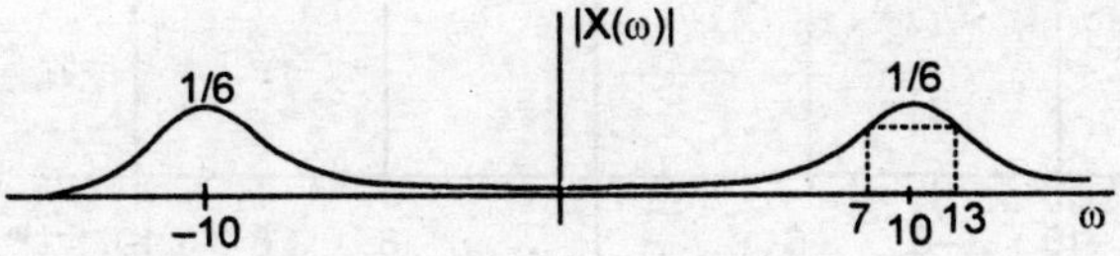

EXERCISES

1. A signal $x(t)$ lasts for 4.096 seconds which is digitised by an ADC at every 8 ms. Find the spacing between the frequency components of X(f), its Fourier transform. What is the highest feasible frequency in the spectrum?

   [**Hint** : $4.096/8 \times 10^{-3}$ = 512 points

   8 ms is spacing in time, corresponds to 1000/8 = 125 Hz. in frequency, and the highest frequency is 125/2 = **62.5 Hz.** The steps between frequencies = 125/512 = **0.25 Hz.**]

2. Show that when a signal has only a real component and no imaginary part, the Fourier transform has its real part as an *even* function and imaginary part an *odd* function.

   [**Hint :** Use the basic definition of Fourier transform and split into real, imaginary parts. Use the fact that sine function is odd, cosine even.]

3. Show that the Fourier transform of $f(t)$ may also be expressed as

$$F\{f(t)\} = F(\omega) = \int_{-\infty}^{\infty} f(t)\cos\omega t - j\int_{-\infty}^{\infty} f(t)\sin\omega t\, dt$$

Show also that if $f(t)$ is an even function of t, then

$$F(\omega) = 2\int_{0}^{\infty} f(t)\cos\omega t\, dt$$

and if $f(t)$ is odd, then

$$F(\omega) = -2j\int_{0}^{\infty} f(t)\sin\omega t\, dt$$

Thereby show that for the following $f(t)$, the $F(\omega)$ is :

$f(t)$	$F(\omega)$
Real and even function of t	Real and even function of ω
Real and odd	Imaginary and odd
Imaginary and even	Imaginary and even
Complex and even	Complex and even
Complex and odd	Complex and odd

4. A function $f(t)$ can be expressed as a sum of an odd part and even part :

$$f(t) = f_e(t) + f_0(t)$$

Show that $\text{Re}[F(\omega)]$ is the transform of $f_e(t)$

$j\text{Im}[F(\omega)]$ is the transform of $f_0(t)$

5. Determine $F(\omega)$ for the following time function.

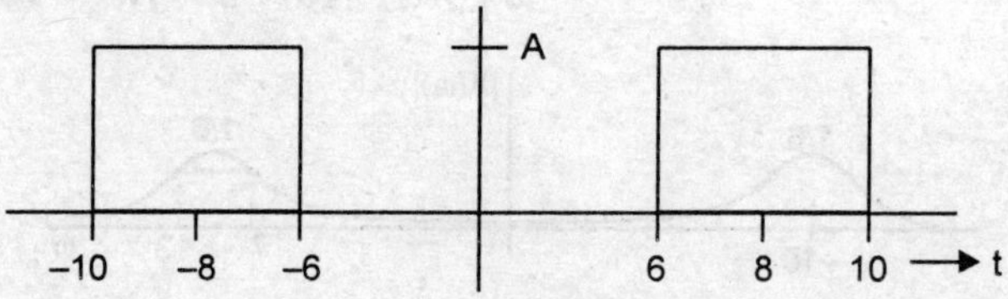

Fig. 24

6. Find the N-point Discrete Fourier Transform (DFT) of

   $h(n) = 1/3$ for $n = 0, 1, 2$, and zero otherwise. $\left[\textbf{Ans. } \frac{1}{3}e^{-j2\pi k/N}_{N=3}\left[1 + 2\cos\frac{2\pi k}{N}\right] = H(k)\right]$

7. Find the DFT of $x(n)$ if

$$x(n) = 1, \quad \text{for } n = 2 \text{ to } 6$$
$$x(n) = 0 \quad \text{for } n = 0, 1, 7, 8, 9.$$

Assume $x(n)$ is periodic beyond this interval 0–9.

$$\left[\textbf{Ans. } \exp\left[-j\frac{4\pi k}{5}\right]\sin\left(\frac{\pi k}{2}\right)\Big/\sin\left(\frac{\pi k}{10}\right)\right]$$

8. Find the response of the following system to the input:

$$x[n] = 2 + 2\cos\left(\frac{n\pi}{4}\right) + \cos\left(\frac{2\pi n}{3} + \frac{\pi}{2}\right)$$

System: $\qquad H(\omega) = e^{-j\omega}\cos(\omega/2).$

$$\left[\textbf{Ans. } y(n) = 2 + 1.84\cos\left(\frac{n\pi}{4} - \frac{\pi}{4}\right) + \frac{1}{2}\cos\left(\frac{2\pi n}{3} - \frac{\pi}{6}\right)\right]$$

9. The DTFT of a 6 sample signal $x[n]$ is shown below figure. Indicate the values of the DFT on this plot.

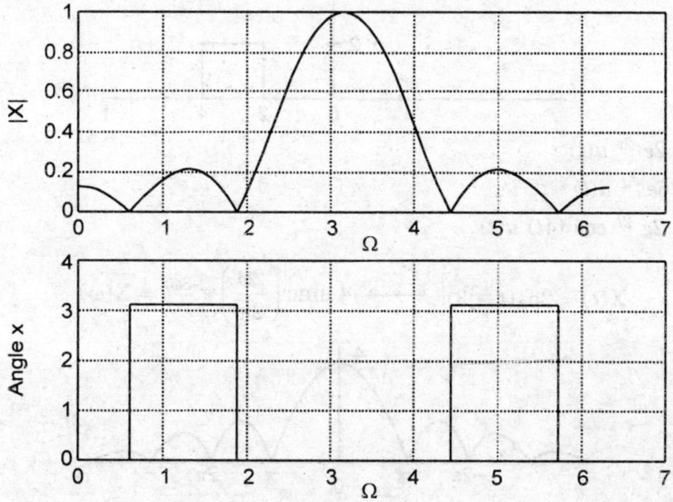

$\left[\right.$**Ans.** The DFT is a discretised version of the DTFT at $\omega = \dfrac{2\pi k}{6}$.

$\omega = 0$, $\omega = \pi/3$, $\omega = 2\pi/3$, $\omega = \pi$, $\omega = 4\pi/3$, $\omega = 5\pi/3$ are the points. $\left.\right]$

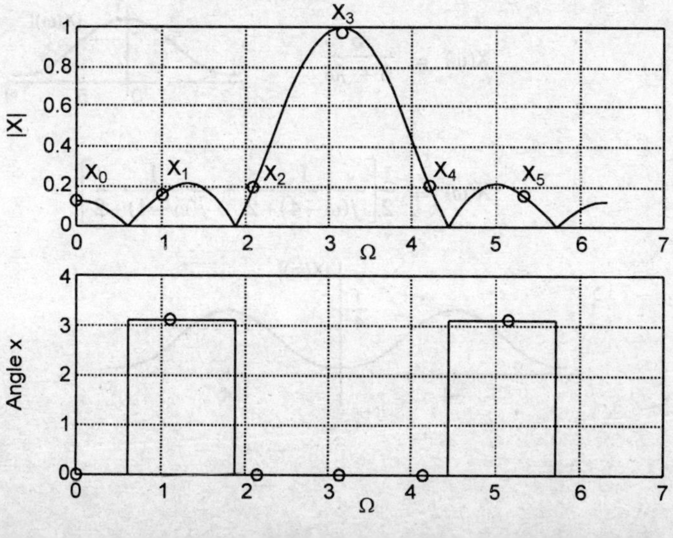

10. Show that the Hilbert Transform of $\exp(j\omega t)$ is $(-\operatorname{sgn} f)\exp(j\omega t)$

11. Using the formula

$$h(t) = \int_{-\infty}^{\infty} \frac{s(\tau)\, d\tau}{t - \tau}$$

Find the Hilbert transforms of

(*a*) $\sin \omega t$ $[-\cos \omega t]$

(*b*) $\cos \omega t$ $[\sin \omega t]$.

12. Find the Fourier transform of the following signals:

(*a*)

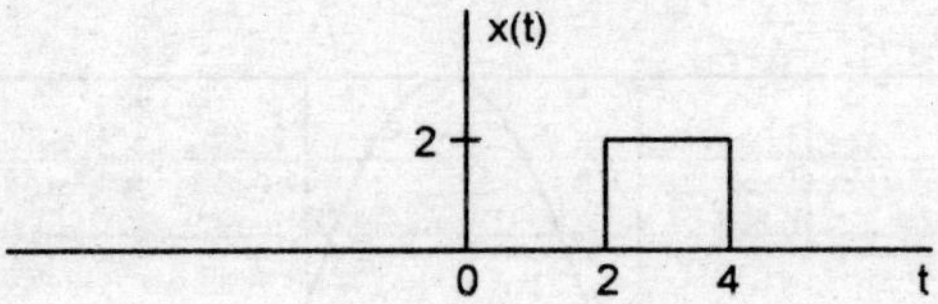

(*b*) $x(t) = 2e^{-2t}\, u(t)$

(*c*) $x(t) = 5e^{-5t}\, u(t)$

(*d*) $x(t) = 2e^{-2t} \cos(4t)\, u(t)$

Ans. (*a*) $X(t) = 2\rho_2(t-3) \longleftrightarrow 4 \operatorname{sinc}\left(\dfrac{2\omega}{2\pi}\right) e^{-3\omega j} = X(\omega)$

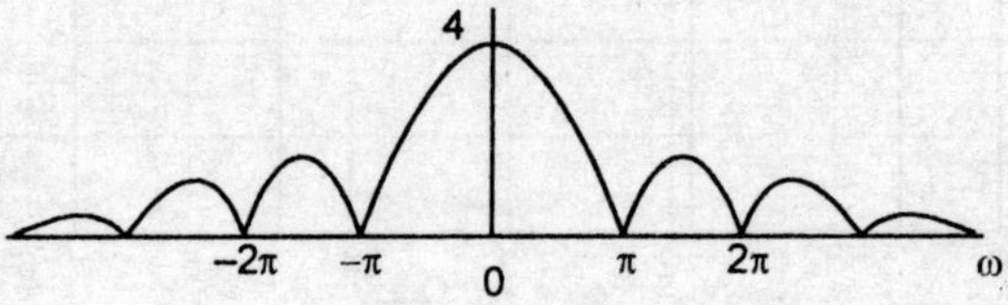

(*b*) $X(\omega) = \dfrac{2}{2 + j\omega}$

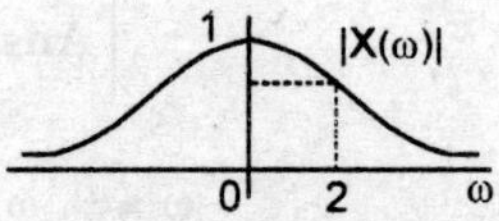

(*c*) $X(\omega) = \dfrac{5}{5 + j\omega}$

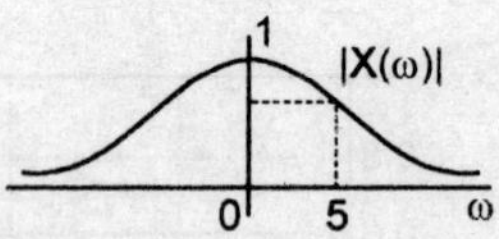

(*d*) $X(\omega) = \dfrac{1}{2}\left[\dfrac{1}{j(\omega-4)+2} + \dfrac{1}{j(\omega+4)+2}\right]$

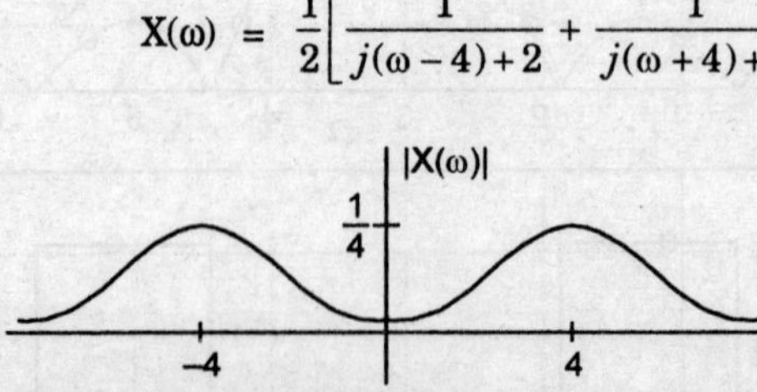

13. Match the time responses with the corresponding frequency responses.

1. .... 2. .... 3. .... 4. .... 5. ....

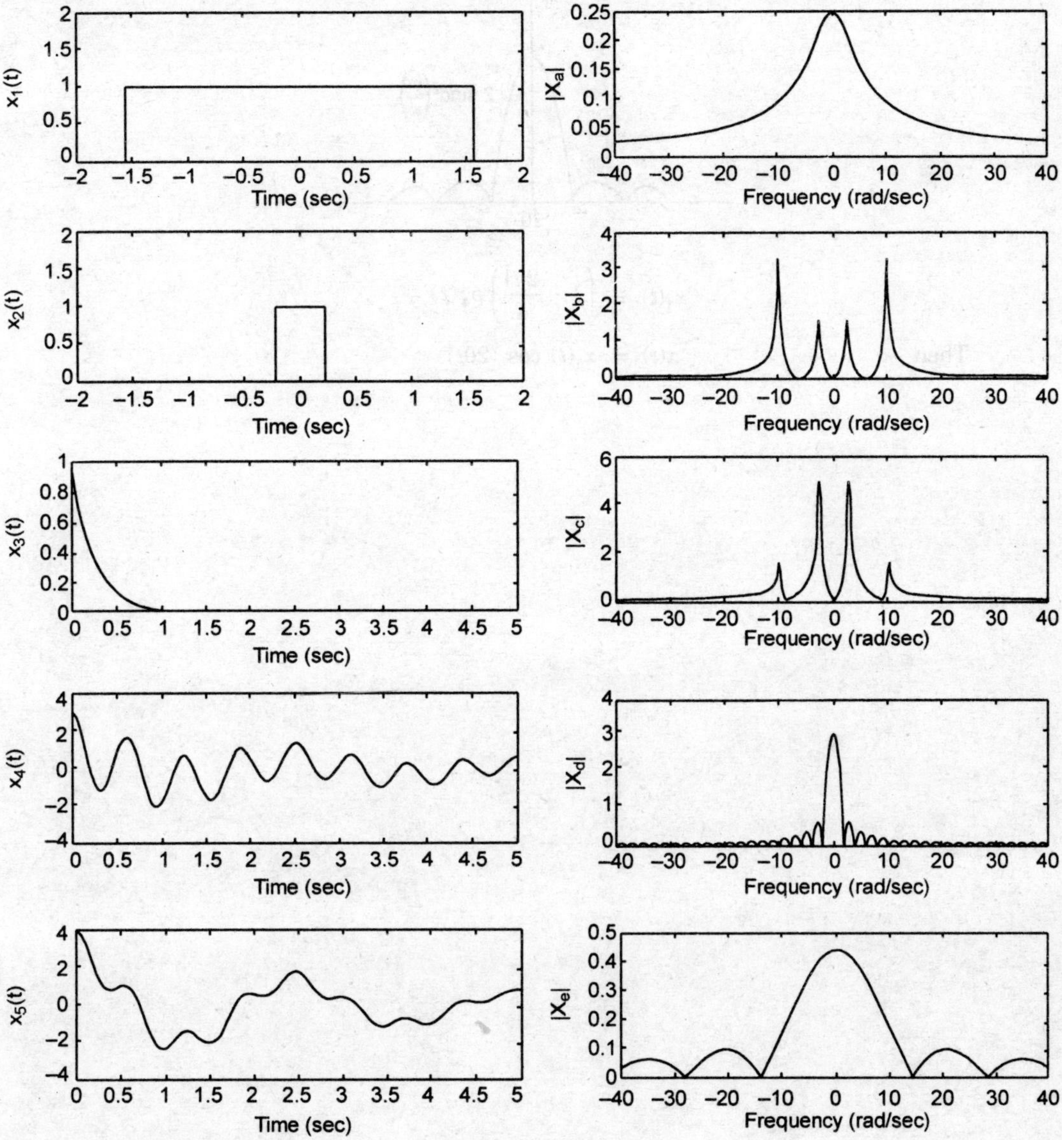

Ans. *d*, *e*, *a*, *b*, *c*.

14. Compute the F.T. of the following signal (figure).

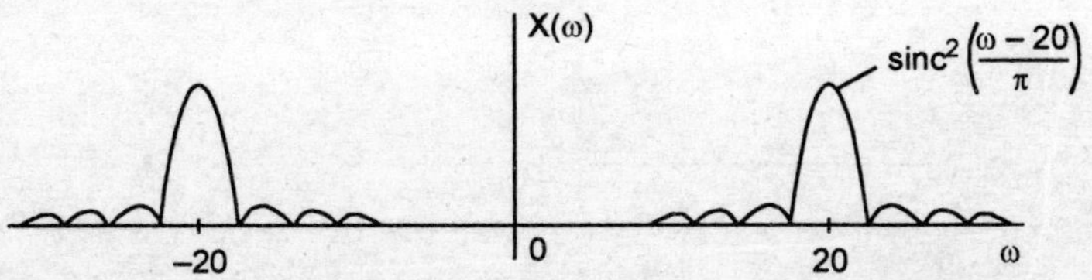

Ans. Consider

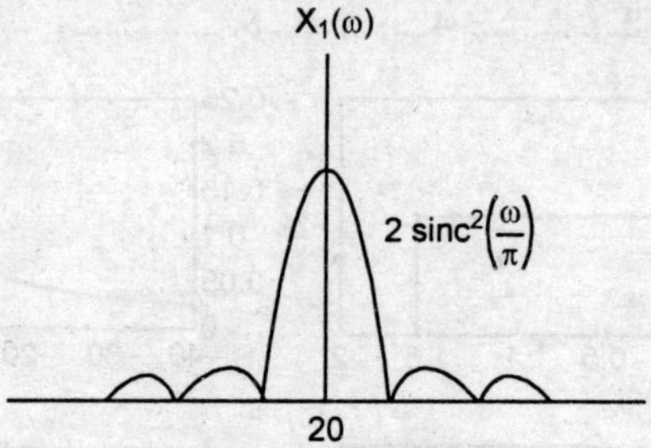

$$x_1(t) = \left(1 - \frac{2|t|}{4}\right) \rho_4(t)$$

Then
$$x(t) = x_1(t) \cos(20t)$$

4

SIGNAL TRANSMISSION THROUGH LINEAR SYSTEMS

LINEAR SYSTEM, IMPULSE RESPONSE

Characterizing the complete input-output properties of a system by exhaustive measurement is usually impossible. Instead, we must find some way of making a finite number of measurements that allow us to infer how the system will respond to other inputs that we have not yet ameasured. We can only do this for certain kinds of systems with certain properties. If we have that right kind of system, we can save a lot of time and energy by using the appropriate theory about the system's responsiveness. Linear systems theory is a good time-saving theory for *linear systems* which obey certain rules. Not all systems are linear, but many important ones are. When a system qualifies as a linear system, it is possible to use the responses to a small set of inputs to predict the response to any possible input. This can save the scientist enormous amounts of work, and makes it possible to characterize the system completely.

To get an idea of what linear systems theory is good for, consider some of the things in neuro-science that can be successfully modeled (at least, approximately) as shift-invariant, linear systems:

System	*Input*	*Output*
Passive neural membrane	Injected current	Membrane potential
Synapse	Pre-synaptic action potentials	Post-synaptic conductance
Cochlea	Sound	Cochlear microphonic
Optics of the eye	Visual stimulus	Retinal image
Retinal ganglion cell	Stimulus contrast	Firing rate
Human	Pairs of colour patches	Colour match settings

In addition, a number of neural systems can be approximated as linear systems coupled with simple nonlinearities (*e.g.*, a spike threshold).

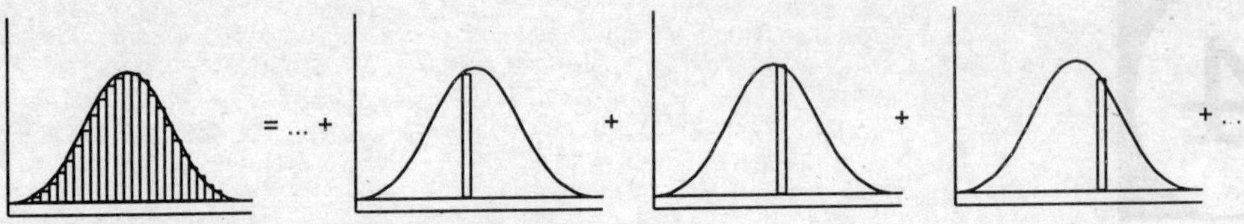

Fig. 1. Staircase approximation to a continuous time signal.

REPRESENTING SIGNALS WITH IMPULSES

Any signal can be expressed as a sum of scaled and shifted unit impulses. We begin with the pulse or "staircase" approximation $\tilde{x}(t)$ to a continuous signal $x(t)$, as illustrated in Fig. 1. Conceptually, this is trivial: for each discrete sample of the original signal, we make a pulse signal. Then we add up all these pulse signals to make up the approximate signal. Each of these pulse signals can in turn be represented as a standard pulse scaled by the appropriate value and shifted to the appropriate place. In mathematical notation:

$$\tilde{x}(t) = \sum_{k=-\infty}^{\infty} x(k\Delta)\, \delta_{\Delta}(t - k\Delta)\Delta.$$

As we let Δ approach zero, the approximation $\tilde{x}(t)$ becomes better and better, and the in the limit equals $x(t)$. Therefore,

$$x(t) = \lim_{\Delta \to 0} \sum_{k=-\infty}^{\infty} x(k\Delta)\, \delta_{\Delta}(t - k\Delta)\Delta.$$

Also, $\Delta \to 0$, the summation approaches an integral, and the pulse approaches the unit impulse:

$$x(t) = \int_{-\infty}^{\infty} x(s)\, \delta(t - s)\, ds. \qquad ...(4.1)$$

In other words, we can represent any signal as an infinite sum of shifted and scaled unit impulses. A digital compact disc, for example, stores whole complex pieces of music as lots of simple numbers representing very short impulses, and then the CD player adds all the impulses back together one after another to recreate the complex musical waveform.

Linear Systems

A *system* or *transform* maps an input signal $x(t)$ into an output signal $y(t)$:

$$y(t) = \mathrm{T}[x(t)]$$

where T denotes the transform, a function from input signals to output signals.

Systems come in a wide variety of types. One important class is known as *linear systems*. To see whether a system is linear, we need to test whether it obeys certain rules that all linear systems obey. The two basic tests of linearity are homogeneity and additivity.

Homogeneity

As we increase the strength of the input to a linear system, say we double it, then we predict that the output function will also be doubled. For example, if the current injected to a passive neural membrane is doubled, the resulting membrane potential fluctuations will double as well. This is called the *scalar rule* or sometimes the *homogeneity* of linear systems.

Additivity

Suppose we measure how the membrane potential fluctuates over time in response to a complicated time-series of injected current $x_1(t)$. Next, we present a second (different) complicated time-series $x_2(t)$. The second stimulus also generates fluctuations in the membrane potential which we measure and write down. Then, we present the sum of the two currents $x_1(t) + x_2(t)$ and see what happens. Since the system is linear, the measured membrane potential fluctuations will be just the sum of the fluctuations to each of the two currents presented separately.

Superposition

Systems that satisfy both homogeneity and additivity are considered to be linear systems. These two rules, taken together, are often referred to as the *principle of superposition*. Mathematically, the principle of superposition is expressed as:

$$T(\alpha x_1 + \beta x_2) = \alpha T(x_1) + \beta T(x_2) \quad ...(4.2)$$

Homogeneity is special case in which one of the signals is absent. Additivity is a special case in which $\alpha = \beta = 1$.

Shift-invariance

Suppose that we inject a pulse of current and measure the membrane potential fluctuations. Then we stimulate again with a similar pulse at a different point in time, and again we measure the membrane potential fluctuations. If we haven't damaged the membrane with the first impulse then we should expect that the response to the second pulse will be the same as the response to the first pulse. The only difference between them will be that the second pulse has occurred later in time, that is, it is shifted in time. When the responses to the identical stimulus presented shifted in time are the same, except for the corresponding shift in time, then we have a special kind of linear system called a *shift-invariant* linear system. Just as not all systems are linear, nor all linear systems are shift-invariant.

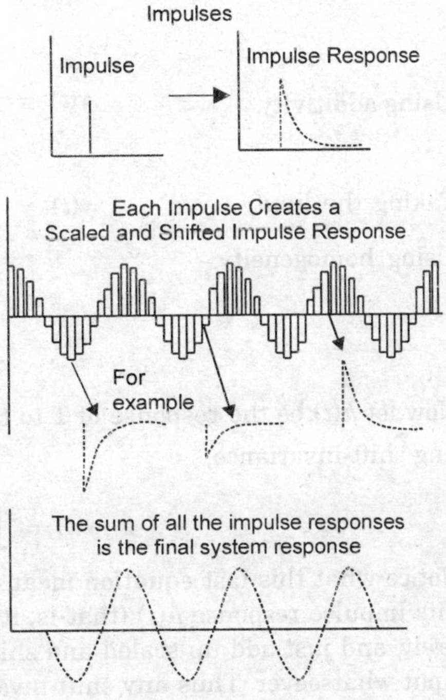

Fig. 2. Characterizing a linear system using its impulse response.

In mathematical language, a system T is shift-invariant if and only if:

$$y(t) = T[x(t)] \text{ implies } y(t - \alpha) = T[x(t - \alpha)] \quad ...(4.3)$$

In the above, α means the time shift.

In Laplace transform (frequency domain) notation, this becomes

$$Y(s) = T(s)\, X(s).$$

The way we use the impulse response function is illustrated Fig. 2. We conceive of the input stimulus, in this case a sinusoid, as if it were the sum of a set of impulse (Eq. 4.1). We know the response we would get if each impulse was presented separately (*i.e.*, scaled and shifted copies of the impulse response). We simply add together all of the (scaled and shifted) impulse responses to predict how the system will respond to the complete stimulus.

Now we will repeat all this in mathematical notation. Our goal is to show that the response (*e.g.*, membrane potential fluctuation) of a shift-invariant linear system (*e.g.*, passive neural membrane) can be written as a sum of scaled and shifted copies of the system's impulse response function.

The convolution integral. Begin by using (Eq. 4.1) to replace the input signal $x(t)$ by its representation in terms of impulses:

$$y(t) = \mathrm{T}[x(t)] = \mathrm{T}\left[\int_{-\infty}^{\infty} x(s)\, d(t-s)\, ds\right]$$

$$= \mathrm{T}\left[\lim_{\Delta \to 0} \sum_{k=-\infty}^{\infty} x(k\Delta)\, \delta_{\Delta}(t-k\Delta)\Delta\right].$$

Using additivity, $$y(t) = \lim_{\Delta \to 0} \sum_{k=-\infty}^{\infty} \mathrm{T}[x(k\Delta)\delta_{\Delta}(t-k\Delta)\Delta].$$

Taking the limit, $$y(t) = \int_{-\infty}^{\infty} \mathrm{T}[x(s)\,\delta(t-s)\, ds]$$

Using homogeneity,

$$y(t) = \int_{-\infty}^{\infty} x(s)\, \mathrm{T}[\delta(t-s)]\, ds$$

Now let $h(t)$ be the response of T to the unshifted unit impulse, *i.e.*, $h(t) = \mathrm{T}[\delta(t)]$. Then by using shift-invariance,

$$y(t) = \int_{-\infty}^{\infty} x(s)\, h(t-s)\, ds \qquad \text{...(4.4)}$$

Notice what this last equation means. For any shift-invariant linear system T, once we know its impulse response $h(t)$ (that is, its response to a unit impulse), we can forget about T entirely, and just add up scaled and shifted copies of $h(t)$ to calculate the response of T to any input whatsoever. Thus any shift-invariant linear system is completely characterized by its impulse response $h(t)$.

The way of combining two signals specified by (Eq. 4.4) is known as *convolution*. It is such a widespread and useful formula that it has its own shorthand notation, $*$. For any two signals x and y, there will be another signal z obtained by *convolving* x with y,

$$z(t) = x * y = \int_{-\infty}^{\infty} x(s)\, y(t-s)\, ds.$$

Convolution as a Series of Weighted Sums

While superposition and convolution may sound a little abstract, there is an equivalent statement that will make it concrete: a system is a shift-invariant, linear system if and only if the response are *a weighted sum of the inputs*. Fig. 3 shows an example: the output at each point in time is computed simply as a weighted sum of the inputs at recently past times. The choice of *weighting function* determines the behaviour of the system. Not surprisingly, the weighting function is very closely related to the impulse response of the system. In particular, the impulse response and the weighting function are time-reversed copies of one another, as demonstrated in the top part of the figure.

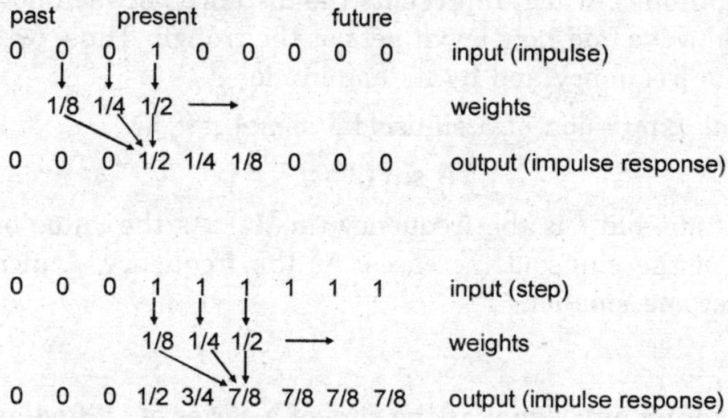

Fig. 3. Convolution as a series of weighted sums.

Properties of Convolution

The following things are true for convolution in general, as you should be able to verify for yourself with some algebraic manipulation:

$$\begin{aligned} x * y &= y * x && \text{commutative} \\ (x * y) * z &= x * (y * z) && \text{associative} \\ (x * z) + (y * z) &= (x + y) * z && \text{distributive} \end{aligned}$$

Example : Find the response to a step input if the impulse response is

$$h(t) = \frac{1}{\text{RC}} \exp\left(-\frac{1}{\text{RC}}\right).u(t)$$

$$\text{Response} = \int_0^t x(\lambda) \times h(t - \lambda)\, d\lambda \text{ for } t \geq 0. \text{ Here } x(t) = u(t)$$

$$y(t) = \int_0^t \frac{1}{\text{RC}} \exp \frac{(t-\lambda)}{\text{RC}}\, d\lambda = \frac{-\text{RC}}{\text{RC}} \left[e^{\lambda/\text{RC}}\right]_{\lambda=0}^{t}$$

$$= 1 - e^{-t/\text{RC}} \text{ for } i > 0.$$

FREQUENCY RESPONSE

Sinusoidal Signals

Sinusoidal signals have a special relationship to shift-invariant linear systems. A sinusoid is a regular, repeating curve, that oscillates above and the below zero. The sinusoid has a zero-value at time zero. The cosinusoid is a shifted version of the sinusoid; it has a value of one at time zero.

The sine wave repeats itself regularly, and the distance from one peak of the wave to the next peak is called the *wavelength* or *period* of the sinusoid and generally indicated by the Greek letter λ. The inverse of wavelength is frequency: the number of peaks in the signal that arrive per second. The units for the frequency of a sine-wave are Hertz, named after a famous 19th century physicist, who was a student of Helmholtz. The longer the wavelength, the shorter the frequency; knowing one we can infer the other. Apart from frequency, sinusoids also have various amplitudes, which represents the distance between how high their energy gets at the peak of the wave and how low it gets at the trough. Thus, we can describe a sine wave completely by its frequency and by its amplitude.

The mathematical expression of a sinusoidal signal is :

$$A \sin (2\pi ft)$$

where A is the amplitude and f is the frequency (in Hz). As the value of the amplitude, A, increases the height of the sinusoid increases. As the frequency, f, increases, the spacing between the peaks become smaller.

Fourier Transform

Just as we can express any signal as the sum of a series of shifted and scaled impulses, so too we can express any signal as the sum of a series of (shifted and scaled) sinusoids at different frequencies. This is called the *Fourier* expansion of the signal. An example is shown in Fig. 4.

The equation describing the Fourier expansion works as follows :

$$x(t) = \int_0^\infty A_\omega \sin (\omega t + \phi_\omega)\, d\omega \qquad ...(5)$$

where $\omega = 2\pi f$ is the frequency of each sinusoid, and $A\omega$ and $\phi\omega$ are the amplitude and phase, respectively, of each sinusoid. If we know the coefficients, $A\omega$ and $\phi\omega$, we can use this formula to reconstruct the original signal $x(t)$. If we know the signal, we can compute the coefficients by the method called the Fourier transform, a way of decomposing a complex stimulus into its component sinusoids.

FOURIER SERIES APPROXIMATIONS

Example: Stereos as shift-invariant systems

Many people find the characterization in terms of frequency response to be intuitive. And most of us have seen graphs that described performance this way. Stereo systems, for example, are pretty good shift-invariant linear systems. They can be evaluated by measuring the signal at different frequencies. And the stereo controls are designed around the frequency representation. Adjusting the bass alters the level of the low frequency components, while adjusting the treble alters the level of the high frequency components. Equalizers divide up the signal band into many frequencies and give you finer control.

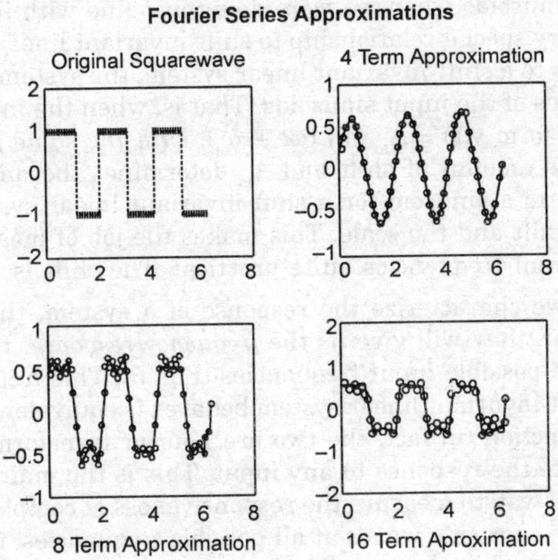

Fig. 4. Fourier series approximation of a square wave as the sum of sinusoids.

Shift-invariant linear systems and sinusoids

The Fourier decomposition is important because if we know the response of the system to sinusoids at many different frequencies, then we can use the same kind of trick we used with impulses to predict the response via the impulse response function. First, we measure the system's response to sinusoids of all different frequencies. Next, we take our input (*e.g.*, time-varying current) and use the Fourier transform to compute the values of the Fourier coefficients. At this point the input has been broken down as the sum of its component sinusoids. Finally, we can predict the system's response (*e.g.*, membrane potential fluctuations) simply by adding the response for all the component sinusoids.

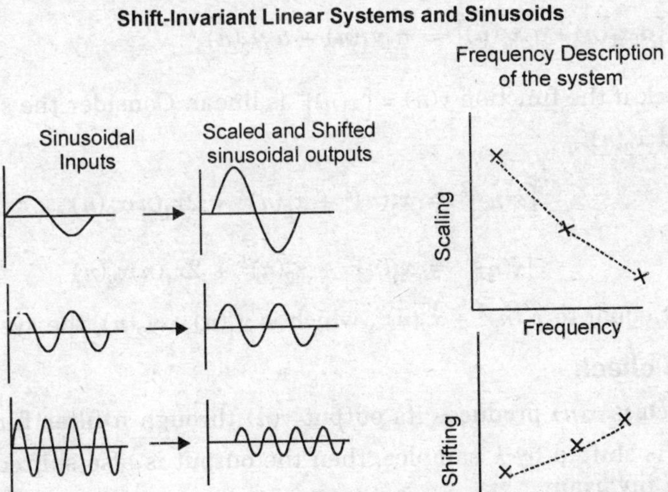

Fig. 5. Characterizing a system using its frequency response.

Why bother with sinusoids when we were doing just fine with impulses? The reason is that sinusoids have a very special relationship to shift-invariant linear systems. When we use a sinusoids as the inputs to a shift-invariant linear system, the system's responses are always (shifted and scaled) copies of the input sinusoids. That is, when the input is $x(t) = \sin(\omega t)$ the output is always of the form $y(t) = A_\omega \sin(\omega t + \phi_\omega)$, *with the same frequency as the input.* Here, $\phi\omega$ determines the amount of shift and A_ω determines the amount of scaling. Thus, measuring the response to a sinusoid for a shift-invariant linear system entails measuring only two numbers: the shift and the scale. This makes the job of measuring the response to sinusoids at many different frequencies quite practical. The shift is also the 'phase'.

Often then, when we characterize the response of a system, they will not tell us the impulse response. Rather, they will give us the *frequency response*, the values of the phase and scale for each of the possible input frequencies (Fig. 5). This frequency response representation of how the shift-invariant linear system behaves is equivalent to providing you with the impulse response function (in fact, the two are Fourier transforms of one another). We can use either to compute the response to any input. This is the main point of all this stuff: a simple, fast, economical way to measure the responsiveness of complex systems. If we know the coefficients of response for sine waves at all possible frequencies, then we can determine how the systems will respond to any possible input.

SUMMARY OF LTI PROPERTIES

1. Linearity implies superposition property

Hence a sequence $x(n)$ is linear if $y(n)$ is known for a given $x(n)$, then :
for a sequence which is $a\,x(n)$ (a = constant)
the y sequence is also a times $x(n)$.

In the same way, two sequences $x_1(n)$ and $x_2(n)$ are giving outputs $y_1(n)$ and $y_2(n)$, by a function that operates on the input the give the output $y(n)$ will also operate on the outputs accordingly.

$$F\left[a_1x_1(n) + a_2x_2(n)\right] = a_1y_1(n) + a_2y_2(n)$$

Example : Check if the function $y(n) = \left[x(n)\right]^2$ is linear. Consider the sequence x as split into two = $x_1(n)$ and $x_2(n)$.

Then $$\left[x(n)\right]^2 = x_1(n)^2 + x_2(n)^2 + 2x_1(n)x_2(n)$$

Then $$\left|x(n)\right|^2 = x_1(n)^2 + x_2(n)^2 + 2x_1(n)x_2(n)$$

Since this is not equal to $x_1(n)^2 + x_2(n)^2$, which is $y_1(n) + y_2(n)$, the system is not linear.

2. Time invariance check

A sequence which is $x(n)$ produces its output $y(n)$ through a filter function F.

If the sequence is shifted by k samples, then the output is also shifted by k samples of $y(n)$. Then it is time invariant.

Consider the example :

$$y(n) = -x(n)$$

Write the values in time sequence :

say $x(n) \rightarrow y(n)$, $x(n-k) \rightarrow y(n-k)$ — Time Invariance

Then,
$$x(n) = \{a_1, a_2, a_3, a_4 \ldots\}$$

$$x(-n) = \{\ldots a_4, a_3, a_2, a_1\}$$

If we introduce a delay of one sample in x, that will give

$$\{\ldots 0, a_4, a_3, a_2, a_1\}$$

A delay of the output will give

$$y_{\text{del}}(n) = \{0, \ldots a_4, a_3, a_2, a_1\}$$

Both are not same. So this system $y(n) = -x(n)$ is not a shift-invariant system.

Linear Time-Variant LTV Systems

For $x(n) \xrightarrow{F} (y(n))$.... A signal $x(n)$ through system F gives $y(n)$ at output

If $x(n-k) \xrightarrow{K} \neq y(n-K)$.... A shifted signal input by K does not similarly shift $y(n)$

Such a system is time-variant, though it can be linear

$$[a(x(n)) = a(y(n))]$$

In communication, there are some such variant systems also. Filters, differentiator etc. are linear, but modulator, folder, 'time' multiplier are such LTV systems.

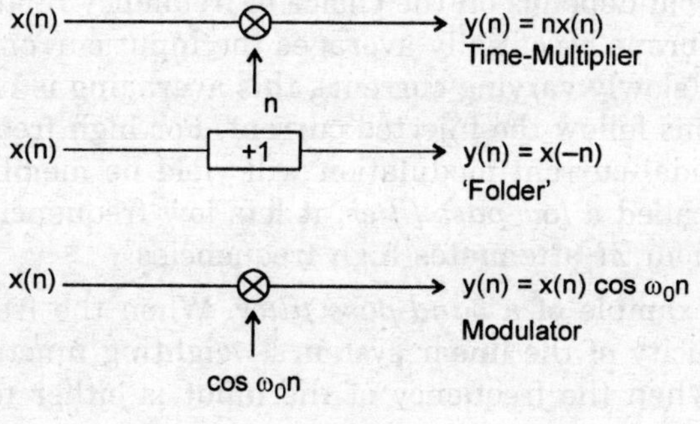

Fig. 6

For example, in the modulator, we have

$$y(n) = x(n) \cos \omega_0 n$$

Its response for a shifted input is

$$y(n, k) = x(n-k) \cos \omega_0 n$$

which is not $x(n-k) \cos \omega_0 (n-k)$ and hence it is an LTV system.

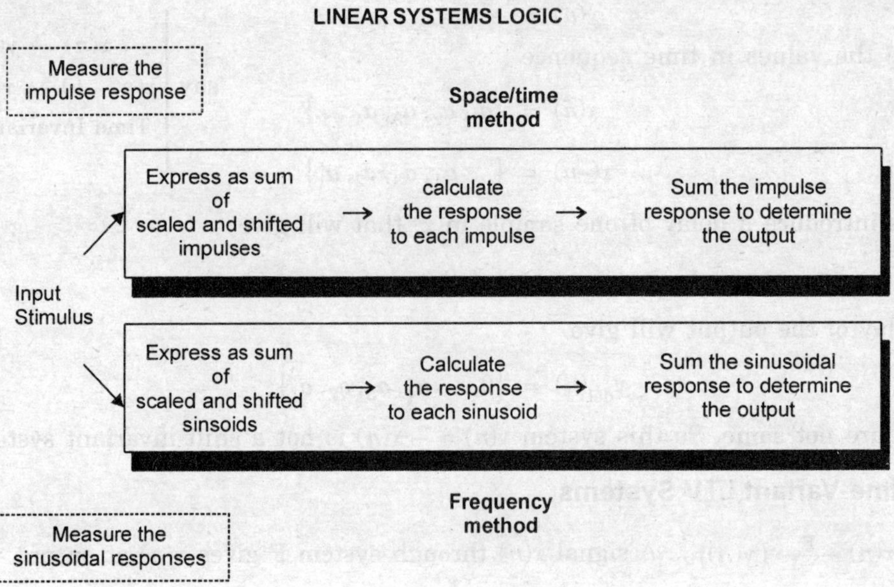

Fig. 7. Alternative methods of characterizing a linear system.

Linear Filters

Shift-invariant linear systems are often referred to as *linear filters* because they typically attenuate (filter out) some frequency components while keeping others intact.

For example, since a passive neural membrane is a shift-invariant linear system, we know that injecting sinusoidally modulating current yields membrane potential fluctuations that are sinusoidal with the same frequency (sinusoid in, sinusoid out). The amplitude and phase of the output sinusoid depends on the choice of frequency relative to the properties of the membrane. The membrane essentially averages the input current over a period of time. For very low frequencies (slowly varying current), this averaging is irrelevant and the membrane potential fluctuations follow the injected current. For high frequencies, however, even a large amplitude sinusoidal current modulation will yield no membrane potential fluctuations. The membrane is called a *low-pass filter*: it lets low frequencies pass, but because of its time-averaging behaviour, it attenuates high frequencies.

Figure 8 shows an example of a *band-pass filter*. When the frequency of a sinusoidal input matches the periodicity of the linear system's weighting function the output sinusoid has a large amplitude. When the frequency of the input is either too high or too low, the output sinusoid is attenuated.

Shift-invariant linear systems are often depicted with block diagrams, like those shown in Fig. 9. Fig. 9A depicts a simple linear filter with frequency response $\hat{h}(\omega)$. The equations that go with the block diagram are:

$$y(t) = h(t) * x(t)$$

or

$$Y(\omega) = H(\omega)\,X(\omega) \qquad \text{...(4.6)}$$

The first of these equations is the now familiar convolution formula (see Chap. 5) in which $x(t)$ is the input signal, $y(t)$ is the output signal, and $h(t)$ is the impulse response of the linear filter. The second equation says the same thing, but expressed in the Fourier domain: $\hat{x}(\omega)$ is the Fourier transform of the input signal, $\hat{y}(\omega)$ is the Fourier transform of the output signal, and $\hat{h}(\omega)$ is the frequency response of the linear filter (that is, the Fourier transform of the impulse response). At the risk of confusing you, it is important to note that $\hat{x}(\omega)$, $\hat{y}(\omega)$ and $\hat{h}(\omega)$ are complex-valued functions of frequency. The complex number notation makes it easy to denote both the amplitude/scale and phase/shift of each frequency component. We can also write out the amplitude and phase parts separately :

$$\text{amplitude}\left[Y(\omega)\right] = \text{amplitude}\left[H(\omega)\right]\text{amplitude}\left[X(\omega)\right]$$

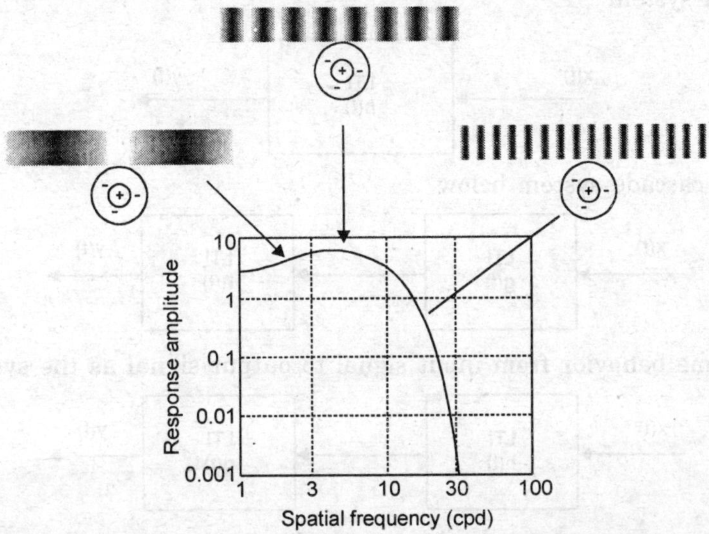

Fig. 8. Illustration of an idealized, retinal ganglion-cell receptive field that acts like a bandpass filter (redrawn from Wandell, 1995). This linear on-center neuron responds best to an intermediate spatial frequency whose bright bars fall on-center and whose dark bars fall over the opposing surround. When the spatial frequency is low, the center and surround oppose one another because both are stimulated by a bright bar, thus diminishing the response. When the spatial frequency is high, bright and dark bars fall within and are averaged by the center (likewise in the surround), again diminishing the response.

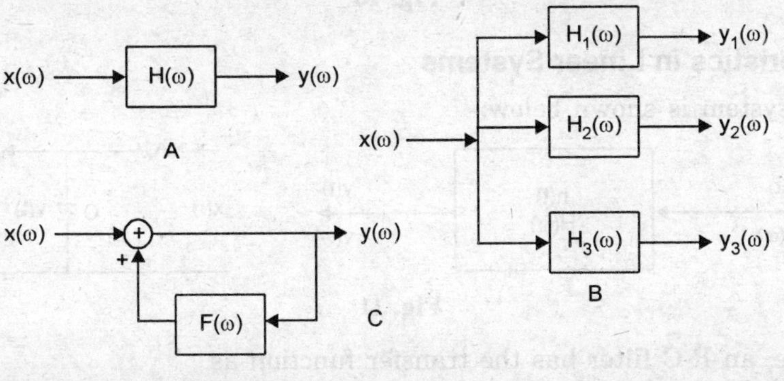

Fig. 9. Block diagrams of linear filters. A: Linear filter with frequency response H(ω). B: Bank of linear filters with different frequency responses. C: Feedback linear system.

$$\text{phase}\left[\hat{y}(\omega)\right] = \text{phase}\left[\hat{h}(\omega)\right] + \text{phase}\left[\hat{x}(\omega)\right]$$

For an input sinusoid of frequency ω, the output is a sinusoid of the same frequency, scaled by amplitude $\left[\hat{x}(\omega)\right]$ and it is shifted by phase$\left[\hat{x}(\omega)\right]$.

Fig. 9B depicts a bank of linear filters that all receive the same input signal. For example, they might be spatially-oriented linear neurons with different orientation preferences.

Cascading LTI Systems

Given an LTI system

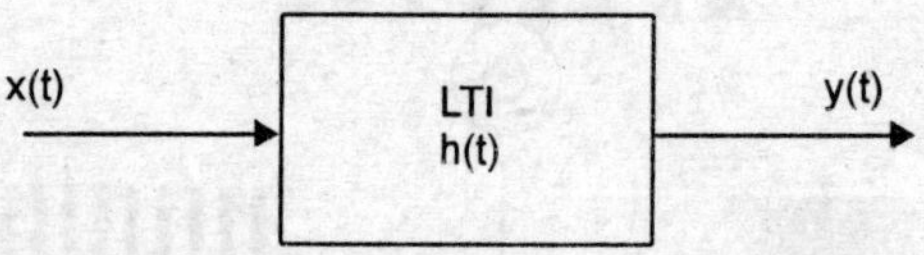

Consider the cascade system below

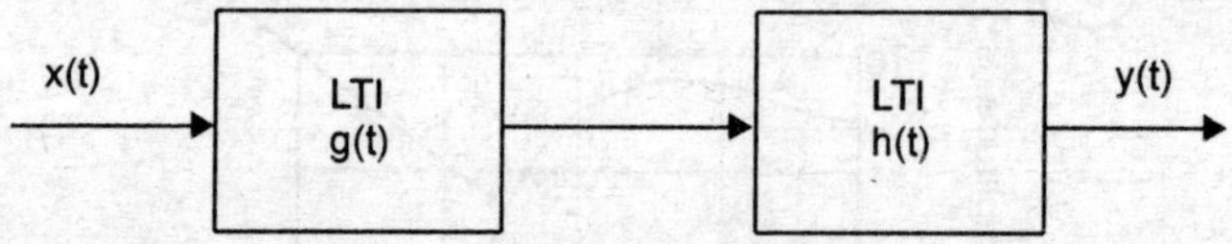

It has the same behavior from input signal to output signal as the system

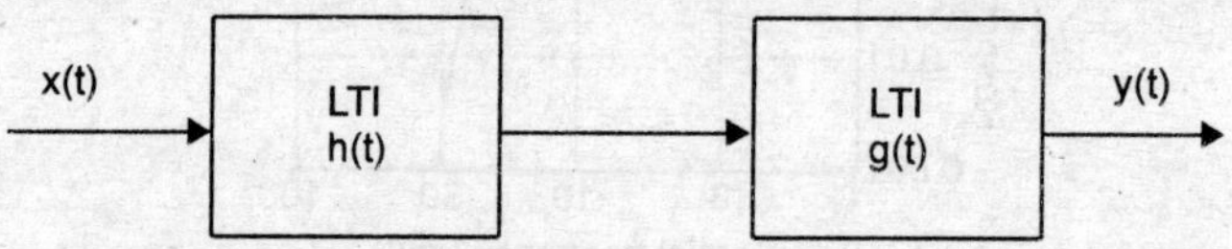

and, by associativity, either of these cascade systems can be represented as

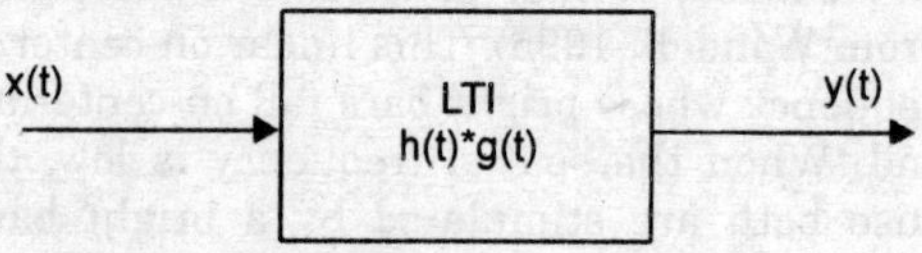

by the convolution of their impulse responses.

Fig. 10

Filter Characteristics in Linear Systems

A filter or system is shown below:

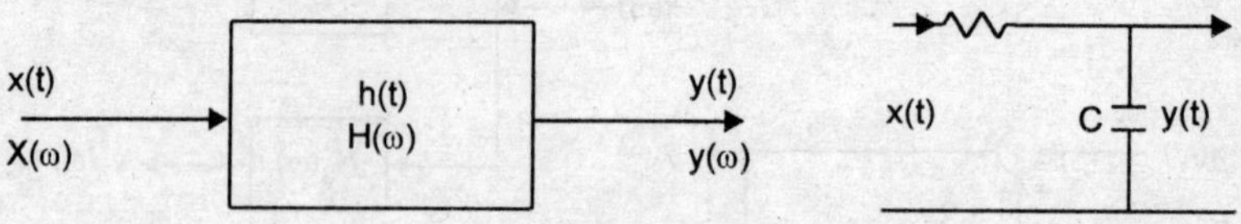

Fig. 11

For example, an R-C filter has the transfer function as

$$H(\omega) = \frac{1/j\omega C}{R + 1/j\omega C} = \frac{1}{1 + j\omega CR} \qquad ...(1)$$

The above function shows that the magnitude of the output to input is

$$H = \frac{|y(t)|}{|x(t)|} = \frac{1}{\sqrt{1+(\omega CR)^2}}$$

as given by the magnitude of the complex function (1).

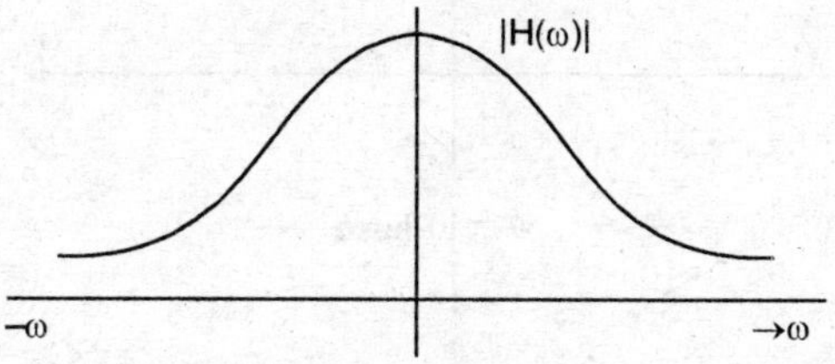

Fig. 12. Low pass filter.

This is plotted in figure. There is also a phase variation between output and input at any frequency given by $-\tan^{-1}(\omega CR)$.

This network is evidently a poor low-pass filter.

Suppose we apply a single square wave pulse to this network.

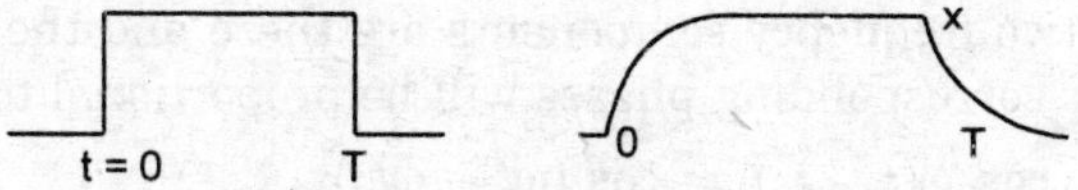

Fig. 13. Low pass response.

The response of the RC network to such a pulse shows delay in rising and falling edges. Since the input rise and fall are fast, that means high frequencies present; which are attenuated by RC network. That is why we get a rise and fall time delay.

Distortionless Transmission

The above network introduced a distortion to the input square wave. Since H(ω) attenuated higher frequencies, and did not have a constant magnitude for all frequencies, there was distortion. Apart from magnitude, phase response between output to input is also important. Even if amplitude did not vary, phase variation is enough to cause distortion.

If phase varies, the waveform of a given signal is totally different at the output.

Distortionless transmission through any (filter) network would give an output, for an input of $f(t)$, as

$$O(t) = k\, f(t - t_0) \qquad ...(a)$$

where k is the attenuation t_o is the delay.

Note however that output *function f* is the same, however.

The above is the condition for distortionless transmission. Taking transforms of (*a*),

$$O(\omega) = k\, F(\omega)\, e^{-j\omega t_0} \qquad ...(b)$$

by time-translation theorem. But general,

$$O(\omega) = F(\omega)\, H(\omega) \qquad ...(c)$$

So for a distortionless system,

$$H(\omega) = k\, e^{-j\omega t_0} \qquad ...(d)$$

If we plot the magnitude and phase of (d) for ω, we get it as below.

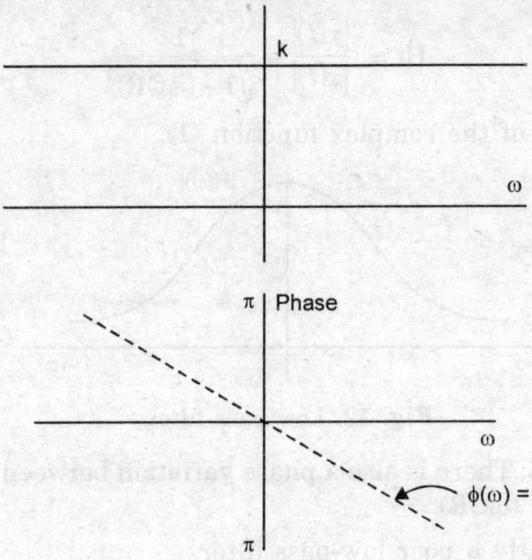

Fig. 14. Magnitude and phase of a delay system.

In this equation, if two frequency components are there and they are both shifted by the same time interval, their corresponding phases will be proportional to their frequencies. Since

$$\cos \omega(t - t_0) = \cos (\omega t - \omega t_0) \quad ...(4.7)$$

we note phase $(-\omega t_0)$ is proportional to ω.

That is shown in the phase graph shown. Such a graph is a "Linear phase shift".

Of course, for large frequencies, phase moves beyond π, beyond 2π, 3π, 4π and so on; hence phase is represented within $\pm\pi$, usually.

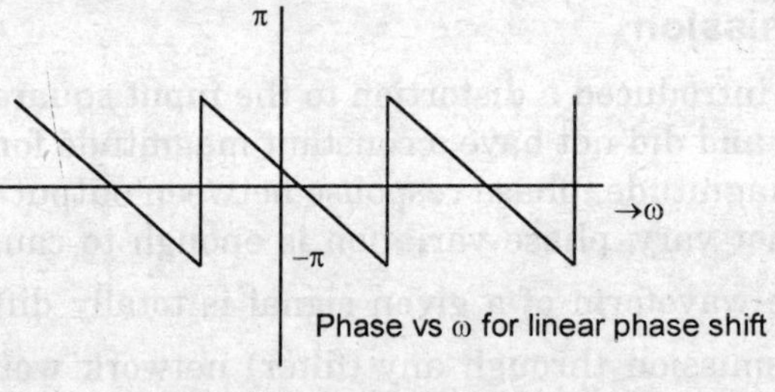

Fig. 15. Phase varies between –π and π.

This characteristic of filters is available in the so called F.I.R. (finite impulse response) filters.

Bandwidth and Half-power Bandwidth

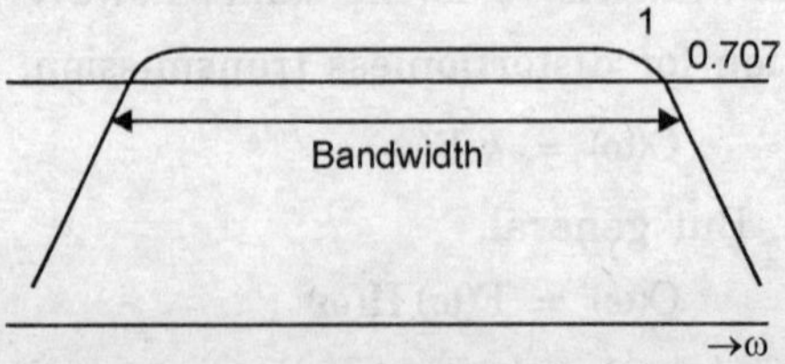

Fig. 16. Frequency response and bandwidth

The bandwidth of a system is calculated based on the 0.707 or $\left(\frac{1}{\sqrt{2}}\right)$ of the maximum (flat) frequency response amplitude, as shown in above figure.

Theoretically, for totally *no distortion*, we need an infinite bandwidth, which is practically impossible.

For a signal, where, in the frequency range useful information is present, there only the required bandwidth must be available. Hence the requirement varies with the signal's energy content.

An ideal low-pass signal filter would have a characteristic shown below. Its frequency spectrum is what we prescribe as a rectangle. The time-domain function $h(t)$ is also shown.

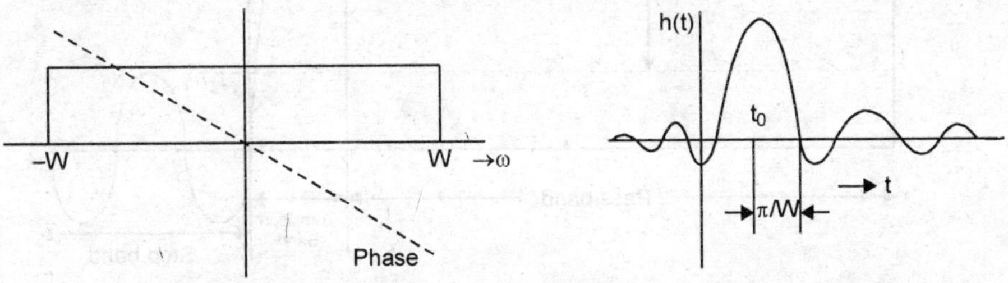

(*a*) (*b*) Impulse Response

Fig. 17. Low pass filter magnitude and phase.

Since the frequency function is a rectangle, the inverse Fourier transform giving $h(t)$ is

$$h(t) = \frac{W}{\pi} \frac{\sin W(t - t_0)}{W(t - t_0)} \quad ...(4.8)$$

Note that this is shown roughly in figure (*b*) above, and that it has an impulse response for even $t < 0$!

When we have applied the impulse itself at $t = 0$, how could this be? Does the system anticipates the impulse at $t = 0$?

The answer lies in the fact that a response like that is impossible. Such ideal filters are therefore only used as a starting point for designing practical filters.

Low, High and Band Filter

Fig. 18 shows the use of various filters, which could be implemented digitally, as ideal filters.

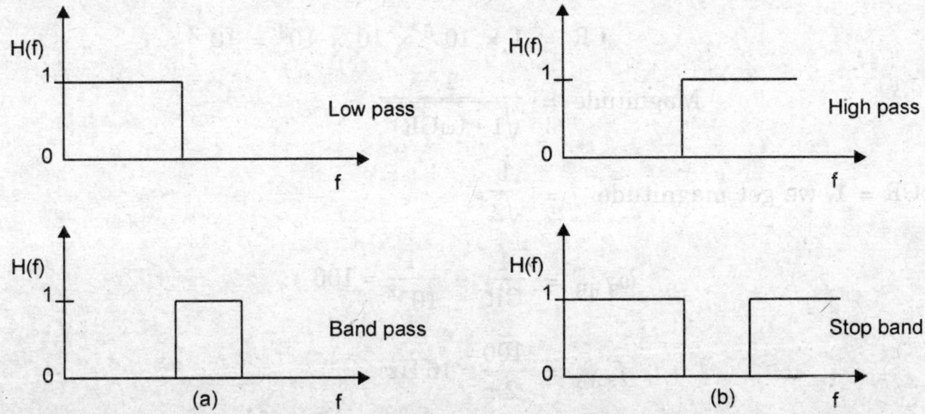

Fig. 18. (*a*) A digital filter implementation (*b*) Using some typical ideal filters.

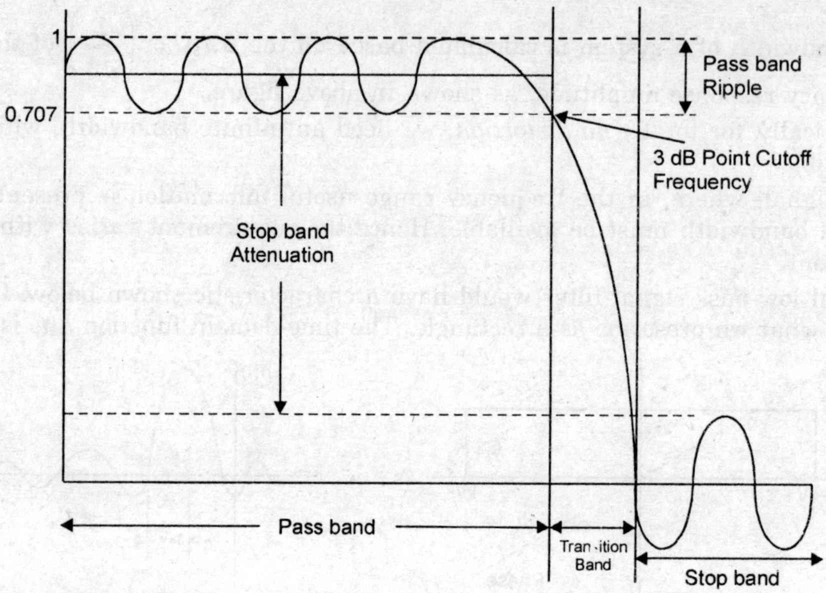

Fig. 18. (*c*) Key filter parameters—Defining the bands of a filter (low pass).

The key filter parameters as shown in Fig. 18(*c*) indicate the pass band, the stop band, the transition band for a low pass filter. The pass band of the filter characteristic may show ripples and the stop band attenuation is the difference between the gains at the pass and stop bands.

The 3 dB cut-off is same as the 0.707 amplitude ratio shown earlier. This is also called the *half power frequency* point. Thus

$$-3\text{dB} = 20 \log 0.707$$

Power at the 3dB point is $(0.707)^2$ of the peak power and is 1/2; hence the name.

Example : What is the 3 dB frequency point of a RC low pass filter if R = 10 K, C = 1 μF?

The transfer function of the R-C filter was shown as

$$\frac{1}{1 + j\omega CR}$$

$$CR = 1 \times 10^{-6} \times 10 \times 10^3 = 10^{-2}$$

$$\text{Magnitude} = \frac{1}{\sqrt{1 + (\omega CR)^2}}$$

If $\omega CR = 1$, we get magnitude $= \dfrac{1}{\sqrt{2}}$.

$$\therefore \quad \omega_{3\text{ dB}} = \frac{1}{CR} = \frac{1}{10^{-2}} = 100$$

$$f_{3\text{ dB}} = \frac{100}{2\pi} = 16\text{ Hz}$$

Causality

A system is *causal* if the output at $t = t_0$, namely $g(t_0)$, only depends on values of the input $f(t)$ for $t \leq t_0$. As an example, a capacitor C stores change and builds up voltage v when a current $i(t)$ is passing.

$$v(t) = \frac{1}{C}\int_{-\infty}^{t} i(\tau)\,d\tau$$

A system which does not satisfy this is said to be non-causal and is also unrealisable physically.

In terms of impulse response, an LTI system is causal if its impulse response is zero for all negative t values.

$$h(t) = 0 \text{ if } t < 0$$

Let us assume that the excitation of the system with the input signal starts at $t = t_0$.

$$f(t) = 0 \text{ for } t < t_0$$

Then the output is given by

$$g(t) = \int_{-\infty}^{\infty} f(\tau)\,h(t-\tau)\,d\tau = \int_{-\infty}^{t_0} f(\tau)\,h(t-\tau)\,d\tau$$

$$= 0 \text{ for } t < 0$$

and $$= \int_{t_0}^{t} f(\tau)\,h(t-\tau)\,d\tau \text{ for } t \geq 0. \qquad ...(4.9)$$

The output does not occur before the input and is composed only of present and past components of the input.

So, a LTI system is called *causal* if the output signal value at any time t depends only on input signal values for times less than t. It is easy to see from the convolution integral that if $h(t) = 0$ for $t < 0$, then the system is causal.

An LTI system is called *memoryless* if the output signal value at any time t depends only on the input signal value at that same time. Again from the convolution integral, if $h(t) = 0$ for all values of t (except $t = 0$), the system is memoryless. Of course in this situation the integrand is zero for all nonzero values of τ, and obviously the output signal will be identically zero unless $h(t)$ has an impulse at the origin. If $h(t) = a\,\delta(t)$, for some real constant a, then the system is memoryless. In other words, a constant-gain system is a memoryless system.

An LTI system is called *stable* if every bounded input signal produces a bounded output signal. A signal $x(t)$ is bounded if there is a finite constant b such that the absolute value of the signal satisfies $|x(t)| < b$, for all t. The following, short calculation shows that if the impulse response of an LTI system is *absolutely integrable*,

$$\int_{-\infty}^{\infty} |h(t)|\,dt = c < \infty \qquad ...(4.10)$$

then the system is stable. Assuming that there exists such b, c and using the fact that the absolute value of an integral is no greater than the integral of the absolute value, for any t we obtain the bound

$$|y(t)| = \left|\int_{-\infty}^{\infty} h(\tau)x(t-\tau)\,d\tau\right|$$

$$\leq \int_{-\infty}^{\infty} |h(\tau)x(t-\tau)|\,d\tau$$

$$= \int_{-\infty}^{\infty} |h(\tau)|\,|x(t-\tau)|\,d\tau$$

$$\leq \int_{-\infty}^{\infty} |h(\tau)|\,b\,d\tau$$

$$\leq bc$$

Example : Check if the digital filter given by the difference equation

$$y_n = 3x_{n-2} + 3x_{n+2}$$

is casual.

Causality requires that if two inputs $x_1(t)$ and $x_2(t)$ are given, such that

$$x_1(t) = x_2(t) \quad \text{for } t \leq t_0$$

$$x_1(t) \neq x_2(t) \quad \text{for } t > t_0$$

then their outputs $y_1(t)$ and $y_2(t)$ be equal for $t \leq t_0$.

The suffix n denotes (in the given equation), the instant nT, where T is the sampling time.

Assume that $x_1(n)$ and $x_2(n)$ satisfy the relation above, as

$$x_1(n) = x_2(n) \quad \text{for } n \leq k \quad (k \text{ corresponds to } t_0)$$

$$x_1(n) \neq x_2(n) \quad \text{for } n > k$$

For $n = k$, the outputs are :

$$y_1(k) = 3x_1(k-2) + 3x_1(k+2)$$

$$y_2(k) = 3x_2(k-2) + 3x_2(k+2)$$

But since $3x_1(k+2) \neq 3x_2(k+2)$ (n being greater than k)

$$x_1(n) \neq x_2(n) \text{ of } n = k.$$

Hence the system is non-causal.

Example : Check the following filter function for causality.

$$y_n = 3x_{n-1} - 3x_{n-2}$$

Let us give x_1 and x_2 to the filter.

$$y_1(n) = 3x_{1(n-1)} - 3x_{1(n-2)}$$

$$y_2(n) = 3x_{2(n-1)} - 3x_{2(n-2)}$$

If $n \leq k$, then $(n - 1)$ is also $< k$

$$(n-2) \text{ is also } < k$$

$$\therefore \quad x_{1(n-1)} = x_{2(n-1)}$$

$$x_{1(n-2)} = x_{2(n-2)}$$

Therefore, $$y_{1(n)} = y_{2(n)}$$

The filter is thus causal.

Paley-Weiner Criterion

In order that a filter function $h(t)$, with the transform $H(\omega)$ be physically realisable, one necessary condition is that

$$h(t) = 0 \text{ for } t < 0 \quad \text{(causality condition)}$$

Another is given by the frequency domain criterion. This was first stated by Raymond E.A.C. Paley and Norbert Wiener in 1934. We shall just give the relation here.

$$\int_{-\infty}^{\infty} \frac{\log_e |H(\omega)|}{1+\omega^2}\, d\omega < \infty \qquad \text{...(4.11)}$$

However, the function $H(\omega)$ must also be square integrable in the infinite range,

$$\int_{-\infty}^{\infty} |H(\omega)|^2\, d\omega < \infty$$

A system which has $|H(\omega)|$ not satisfying (*a*) above is said to be non-causal; response exists in the past prior to applying an input signal!

Because of the log function, $|H(\omega)|$ cannot have zero value even for a small range of ω.

Also $|H(\omega)|$ has to fall to zero only by an exponential order, like

$$|H(\omega)| \simeq \text{constant} \times \exp(-\alpha|\omega|)$$

and not any faster. For example, the slope of the low pass filter curve in the cut-off region cannot fall as fast as the square of frequency (*i.e.*, like a Gaussian function):

$$|H(\omega)| = k \exp(-\alpha\omega^2)$$

is not feasible. It will violate (*a*).

Because of the theoretical background given by (4.11), we always specify filter characteristics as shown in figure earlier, involving not too steep falls and not touching the baseline.

Bandwidth and Rise-time

When a signal changes sharply, high frequency components are present. So, a filter will filter-off these components. So, at the output we will not get such sharp changes. Any change is thus delayed. For a step input, the output rises exponentially for an RC filter. So, the rise-time will be more than at the input. It depends on the RC value. Since the cut-off (3 dB) frequency of this is 1/CR, the rise-time will depend on the cut-off frequency.

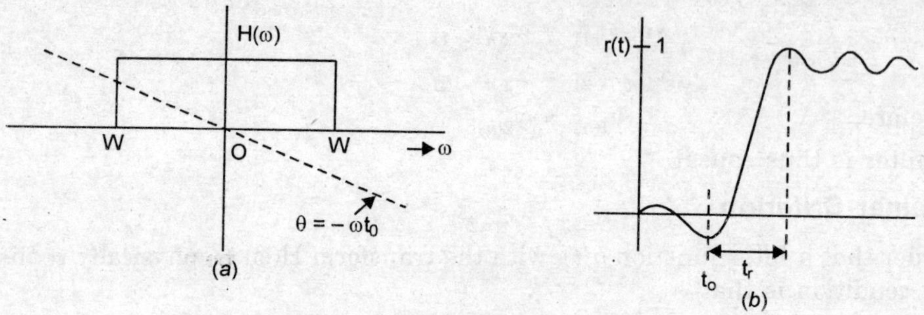

Fig. 19. (*a*) An ideal low pass filter and its phase function (phase change to per unit ω). (*b*) Its response to step input.

The low pass filter of gate width W is now considered. It has a linear phase of $\theta(\omega) = -\omega t_0$. So,

Transfer function of the low pass filter

$$H(\omega) = G_{2W}(\omega) \exp(-j\omega t_0) \text{ \{G means gate width 2W)} \quad ...(4.12)$$

The transform pair of the unit step function:

$$u(t) \leftrightarrow \pi\delta(\omega) + \frac{1}{j\omega} \quad ...(4.13)$$

The above was given in Ch. 3.

Let $r(t)$ be the time response of this filter to the **step input.** $R(\omega)$ is the corresponding Fourier transform.

$$R(\omega) = \left[\pi\,\delta(\omega) + \frac{1}{j\omega}\right] G_{2W}(\omega)\, e^{-j\omega t_0}$$

To calculate this, it is noted that

$$H(0) = 1 \text{ (from Fig. (a) above)}$$

Since $\delta(\omega)$ is an impulse at $t = 0$,

$$\delta(\omega)\, H(\omega) = \delta(\omega) H(0).$$

So, taking inverse transform,

$$r(t) = F^{-1}\left[\pi\,\delta(\omega) + \frac{1}{j\omega} G_{2W} \exp(-j\omega t_0)\right]$$

$$\left(F^{-1}(\pi\delta\omega)\right) = \frac{1}{2} \text{ from Fourier Transform Table P. 83 (No. 5).}$$

$$= \frac{1}{2} + F^{-1}\left[\left(\frac{1}{j\omega}\right) G_{2W} \exp(-j\omega t_0)\right] \quad ...(4.14)$$

The second term becomes : (using Inverse F.T. formula)

$$\frac{1}{2\pi}\int_{-\infty}^{\infty} \frac{G_{2W}(\omega)}{j\omega}\, e^{j\omega(t - t_0)}\, d\omega \quad ...(4.15)$$

Since G_{2W} has values upto W on either side, the integration limits become –W and +W.

$$\frac{1}{2\pi}\int_{-W}^{W}\frac{\exp(j\omega\,\overline{t-t_0})}{j\omega}\,d\omega \qquad ...(4.16)$$

Then we split the exp (*j*...) term into cos and sine.

$$\frac{1}{2\pi}\left\{\int_{-W}^{W}\frac{\cos\omega(t-t_0)}{(j\omega)}\,d\omega+\int_{-W}^{W}\frac{\sin(\omega\,\overline{t-t_0})}{\omega}\,d\omega\right\} \qquad ...(4.17)$$

Note that the *j* in the denominator is cancelled in the second integral (sine integral).

The first integral vanishes between its limits because $\frac{(\cos \omega x)}{\omega}$ is an odd function of ω.

The second term, the sine term becomes

$$\frac{1}{\pi}\int_{0}^{W}\frac{\sin(\omega\,\overline{t-t_0})}{\omega}\,d\omega = \frac{1}{\pi}\int_{0}^{W(t-t_0)}\frac{\sin x}{x}\,dx \qquad \left[\text{Put } x=(t-t_0)\omega\right]$$

Hence the total step response of the low pass filter is

$$r(t) = \frac{1}{2}+\frac{1}{\pi}\int_{0}^{W(t-t_0)}\text{Sa}(x)\,dx \qquad ...(4.18)$$

where Sa(*x*) is the $\frac{\sin x}{x}$.

This integral is tabulated as the *Sine Integral* by Jahnke and Emde, Table of Functions, Dover Publications, 1945, N.Y.

Such a function is roughly plotted in Fig. (b), dividing by π adding $\frac{1}{2}$, so that the response appears as in Fig. (*b*) above.

The mini. value occurs at $t_0-\frac{\pi}{W}$

The max. value occurs at $t_0+\frac{\pi}{W}$. ...(4.19)

Note that t_0, denote the linear phase change in Fig. (*a*), also called delay time.

Rise-time Calculation

Rise-time t_r is defined as :

The time required for the output (of any filter function) to reach from the minimum to the maximum value.

From Fig. (b) above, $t_0+\frac{\pi}{W}-\left(t_0-\frac{\pi}{W}\right)$, or as

$$t_r = \frac{2\pi}{W}=\frac{1}{B} \qquad ...(4.20)$$

where B is the bandwidth already noted as W from Fig. (*a*). The 2π is for ω to period relation:

$$\omega = \frac{2\pi}{T}$$

Thus the rise-time is inversely proportional to the bandwidth W.

Example : Find and sketch the response of an ideal low pass filter to a signal $f(t)$ given below. The cut-off frequency of the filter is 10 kHz. The delay time t_0 is 1 μs.

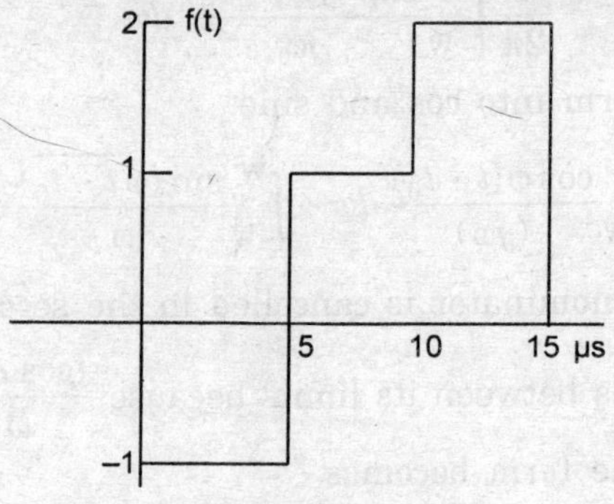

Fig. 20(*a*)

The $f(t)$ can be written, in terms of unit step functions.

$$f(t) = -u(t) + 2u(t-5) + 1u(t-10) - 2u(t-15)$$

The cut-off frequency is 10000 in Hertz, so

$$\omega_{\text{cut-off}} = W = 2\pi \times 10^3$$

$$r(t) = -\left[\frac{1}{2} + \frac{1}{\pi}\int_0^{W(t-t_0)} Sa(x)\,dx\right] + 2\left[\frac{1}{2} + \frac{1}{\pi}\int_0^{W(t-t_0-5)} Sa(x)\,dx\right]$$

$$+1\left[\frac{1}{2} + \frac{1}{\pi}\int_0^{W(t-t_0-10)} Sa(x)\,dx\right] - 2\left[\frac{1}{2} + \frac{1}{\pi}\int_0^{W(t-t_0-15)} Sa(x)\,dx\right]$$

$$= \left(-\frac{1}{2} + 2\times\frac{1}{2} + \frac{1}{2} - 2\times\frac{1}{2}\right) + \frac{1}{\pi}\left[\int + \int + \int + \int\right]$$

The value of $t_0 = 1$.

Using tables, the graph can be sketched.

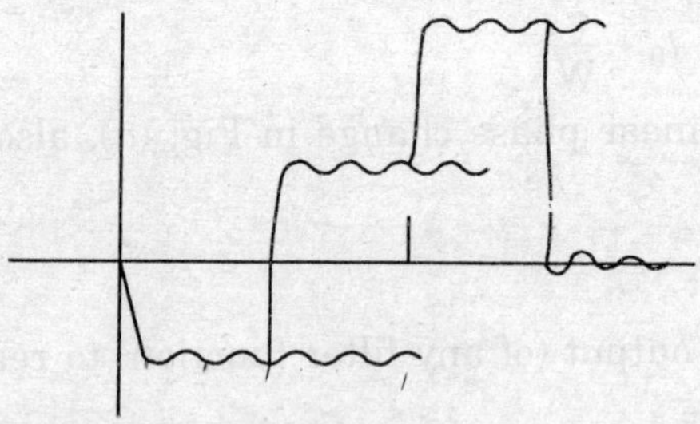

Fig. 20(*b*)

Linear Feedback and IIR Filters

Fig. 9(*c*) depicts a linear feedback system. The equation corresponding to this diagram is:

$$y(t) = x(t) + f(t) * y(t) \qquad ...(4.21)$$

Note that because of the feedback, the output $y(t)$ appears on both sides of the equation. The frequency response of the feedback filter is denoted by $\hat{f}(\omega)$, but the behavior of the entire linear system can be expressed as:

$$Y(\omega) = X(\omega) + F(\omega)Y(\omega) \quad ...(4.22)$$

Solving for $\hat{y}(\omega)$ in this expression gives :

$$Y(\omega) = H(\omega)X(\omega) = \frac{X(\omega)}{1 - F(\omega)}$$

where $H(\omega) = \frac{1}{[1 - F(\omega)]}$ is the effective frequency response of the entire linear feedback system. Using a linear feedback filter with frequency response F(ω) is equivalent to using a linear feed forward filter with frequency response H(ω).

There is one additional subtle, but important, point about this linear feedback system. A system is called *causal* or *nonanticipatory* if the output at any time depends only on values of the input at the present time and in past. For example, the system $y(t) = x(t - 1)$ and $y(t) = x^2(t)$ are causal, but the system $y(t) = x(t + 1)$ is not causal. Note that not all causal systems are linear and that not all linear systems are causal (look for examples of each in the previous sentence).

For Eq. (4.22) to make sense, the filter $f(t)$ must be causal so that the output at time t depends on the input at time t plus a convolution with *past* outputs. For example, if $f(t) = \frac{1}{2}\delta(t - 1)$ then :

$$y(t) = x(t) + \frac{1}{2} * y(t) = x(t) + \frac{1}{2} y(t - 1)$$

The output $y(t)$ accumulates the scaled input values at 1 sec. time intervals:

$$y(t) = x(t) + \frac{1}{2} x(t - 1) + \frac{1}{4} x(t - 2) + ...$$

Linear feedback systems like this are often referred to as *infinite impulse response* or *IIR* linear filters, because the output depends on the full past history of the input. In practice, of course, the response of this system attenuates rather quickly over time. A unit impulse input from ten seconds in the past contributes only 2^{-10} to the present response.

Worked Examples

System Properties

1. *Determine if the following systems are time-invariant, linear, causal, and/or memory-less?*

(a) $\frac{dy}{dt} + 6y(t) = 4x(t)$

(b) $\frac{dy}{dt} + 4ty(t) = 2x(t)$

(c) $y[n] + 2y[n-1] = x[n+1]$

(d) $yt = \sin(x(t))$

(e) $\frac{dy}{dt} + y^2(t) = x(t)$

(f) $y[n+1] + 4y[n] = 3x[n+1] - x[n]$

(g) $y(t) = \frac{dx}{dt} + x(t)$

(h) $y[n] = x[2n]$

(i) $y[n] = nx[2n]$

(j) $\frac{dy}{dt} + \sin(t)y(t) = 4x(t)$

(k) $\frac{d^2y}{dt^2} + 10\frac{dy}{dt} + 4y(t) = \frac{dx}{dt} + 4x(t)$

2. *The response of an LTI system to a step input, $x(t) = u(t)$ is $y(t) = 1 - e^{-2t}) u(t)$. What is the response to an input of $x(t) = 4u(t) - 4u(t-1)$?*

Solution. 1. (*a*) $\frac{dy}{dt} + 6y(t) = 4x(t)$

This is an ordinary differential equation with constant coefficients, therefore, it is linear and time-invariant. It contains memory and it is causal.

(*b*)
$$\frac{dy}{dt} + 4ty(t) = 2x(t)$$

This is an ordinary differential equation. The coefficients of $4t$ and 2 do not depend on y or x, so the system is linear. However, the coefficient $4t$ is not constant, so it is time-varying. The system is also causal and has memory.

(*c*)
$$y[n] + 2y[n-1] = x[n+1]$$

This is a difference equation with constant coefficients; therefore, it is linear and time-invariant. It is noncausal since the output depends on future values of x. Specifically, let $x[n] = u[n]$, then $y[-1] = 1$.

(*d*)
$$y(t) = \sin(x(t))$$

Check linearity:

$$y_1(t) = \sin(x_1(t))$$
$$y_2(t) = \sin(x_2(t))$$

Solution to an input of $a_1x_1(t) + a_2x_2(t)$ is $\sin(a_1x_1(t) + a_2x_2(t))$.

This is not equal to $a_1y_1(t) + a_2y_2(t)$.

As a counter example, consider $x_1(t) = \pi$ and $x_2(t) = \pi/2$, $a_1 = a_2 = 1$

The system is causal since the output does not depend on future values of time, and it is memoryless; the system is time-invariant.

(*e*)
$$\frac{dy}{dt} + y^2(t) = x(t)$$

The coefficient of y means that this is nonlinear; however, it does not depend explicitly on t, so it is time-invariant. It is causal and has memory.

(*f*) $$y[n+1] + 4y[n] = 3x[n+1] - x[n]$$

Rewrite the equation as $y[n] + 4y[n-1] = 3x[n] - x[n-1]$ by decreasing the index.

This is a difference equation with constant coefficients, so it is linear and time-invariant. The output does not depend on future values of the input, so it is causal. It has memory.

(*g*) $$y(t) = \frac{dx}{dt} + x(t)$$

Linear?

$$x_1(t) \longleftrightarrow \frac{dx_1}{dt} + x_1(t) = y_1(t)$$

$$x_2(t) \longleftrightarrow \frac{dx_2}{dt} + x_2(t) = y_2(t)$$

$$a_1x_1 + a_2x_2 \longleftrightarrow \frac{d(a_1x_1 + x_2a_2)}{dt} + a_1x_1 + a_2x_2$$

$$= \frac{a_1dx_1}{dt} + ax_1 + a_2\left(\frac{dx_2}{dt} + x_2\right)$$

$$= a_1y_1 + a_2y_2 \Rightarrow \text{linear}$$

Time-invariant?

$$x(t) \longleftrightarrow y(t) = \frac{dx(t)}{dt} + x(t)$$

$$x(t - t_1) \longleftrightarrow \frac{dx(t - t_1)}{dt} + x(t - t_1)$$

let $$\tau = t - t_1$$

$$d\tau = dt$$

$$\frac{dx(\tau)}{dt} = \frac{dx(\tau)}{d\tau}\frac{d\tau}{dt} = \frac{dx(\tau)}{d\tau}$$

So $$\frac{dx(t - t_1)}{dt} + x(t - t_1) = \frac{dx(\tau)}{d\tau} + x(\tau)$$

Now consider $$y(t - t_1) = y(\tau) = \frac{dx(\tau)}{d\tau} + x(\tau)$$

Gives same answer $\Rightarrow$ Time-invariant; by inspection, causal and has memory.

(*h*) $$y[n] = x[2n]$$

has memory since the output relies on values of the input at other than the current index n,

Causal? Let $x[n] = u[n-2]$, so $x[1] = 0$.

Then $y[1] = x[2] = 1$, so not causal.

Linear? Let $y_1[n] = x_1[2n]$ and $y_2[n] = x_2[2n]$.

The response to an input of

$$x[n] = ax_1[n] + bx_2[n] \text{ is } y[n] = ax_1[2n] + bx_2[2n],$$

which is $ay_1[2n] + by_2[2n]$, so this is linear.

Time-invariant: Let $y_1[n]$ represent the response to an input of $x[n - \text{N}]$, so $y_1[n] = x[2(n - \text{N})]$. This is also equal to $y[n - \text{N}]$, so the system is time-invariant.

(*i*) $$y[n] = nx[2n]$$

This is similar to part (*h*), except for the *n* coefficient. Similar to above, it is noncausal, has memory and is linear. Check time-invariance:

Let $y_1[n]$ represent the response to an input of $x[n - \text{N}]$,

so $$y_1[n] = nx[2(n - \text{N})].$$

This is not equal to $y[n - \text{N}] = (n - \text{N}) \times [2(n - \text{N})]$, so the system is time-varying.

(*j*) $$\frac{dy}{dt} + \sin(t)y(t) = 4x(t)$$

This is an ordinary differential equation with coefficients sin (*t*) and 4. Neither depends on *y* or *x*, so it is linear. However, the explicit dependence on *t* means that it is time-varying. It is causal and has memory.

(*k*) $$\frac{d^2y}{dt^2} + 10\frac{dy}{dt} + 4y(t) = \frac{dx}{dt} + 4x(t)$$

This is an ordinary differential equation with constant coefficients, so it is linear and time-invariant. It is also causal and has memory.

2. The response to $4u(t)$ is $4(1 - e^{-2t})u(t)$.

The response to $4u(t - 1)$ is $4(1 - e^{-2(t-1)})\, u(t - 1)$.

So the response to $x(t) = 4u(t) - 4u(t - 1)$ is $y(t) = 4(1 - e^{-2t})u(t) - 4(1 - e^{-2(t-1)})\, u(t-1)$.

EXERCISES

1. Given the impulse response of two systems as

$$h_1(n) = \left(\frac{1}{2}\right)^n u(n)$$

$$h_2(n) = \left(\frac{1}{4}\right)^n u(n)$$

Find their cascaded response. $\left[\textbf{Ans. } h_{12}(n) = \left(\frac{1}{2}\right)^n \left(2 - \frac{1}{2}^n\right) \text{ for } n \geq 0\right]$

2. Test if the following systems described by the difference equations are causal.

(1) $y(n) = \text{X}(2n)$ (non-causal)

(2) $y(n) = x(n) + 3x(n + 4)$ (non-causal)

(3) $y(n) = \sum_{u=-\infty}^{n} x(k)$ (causal)

(4) $y(n) = 5x_n$ (causal)

(5) $y(n) = nx_n$ (causal)

3. Test the following systems for time invariance.

(a) $y_n = x_n - x_{n-1}$ (time invariant)

(b) $y_n = nx_n$ (time variant)

(c) $y_n = x_n - bx_{n-1}$ (time invariant)

(d) $y_n = x_n + c$ (invariant)

(e) $y_n = nx^2_n$ (time variant)

(f) $y_n = \sum_{k=0}^{m} a_k x_{n-k} - \sum_{k=1}^{m} b_k y_{n-k}$ (invariant)

4. Find the response of the LTI system having the impulse response

$$h(n) = \{1, 2, 1, -1\}$$

to an input signal

$$x(n) = \{1, 2, 3, 1\}.$$

[**Ans.** {1, 4, 8, 8, 3, – 2, – 1}]

5. Test the causality of the following systems

(1) $y(n) = x(-n)$ (not causal)

(2) $y(n) = x(n) + 3x(n + 4)$ (not causal)

(3) $y(n) = x(2n)$ (not causal)

(4) $y_n = ax_n$ (causal)

(5) $y_n = x_n - x_{n-1}$ (causal)

6. Find, using Paley-Wiener criterion whether

$$H(f) = \exp(-\alpha f^2)$$

is a proper filter response for a causal LTI system (Not proper).

[**Hint:** $\int_{-\infty}^{\infty} H^2(f)\, df = \sqrt{\frac{\pi}{2\alpha}} < \infty$

$\int_{-\infty}^{\infty} \frac{\ln H(f)}{1+f^2}\, df = \alpha\left[f - \tan^{-1} f\right]_{-\infty}^{\infty} \to \infty$. Hence etc.]

7. Test for linearity, causality and memoryless condition:

(a) $\frac{dy}{dt} + 2ty(t) = 4x(t)$. [**Ans.** Linear, time variant, causal, has memory]

(b) $y(t) = 2\frac{dx}{dt} + 4x(t)$ [**Ans.** Linear, TV, causal, has memory]

8. The response of an LTI system to a step input is

$$y(t) = (1 - 2e^{-3t})u(t).$$

Find the response to the input

$$x(t) = 3u(t) - 3u(t-1).$$

[**Ans.** $3(1-2e^{-3t})u(t) - 3(1-e^{-3(t-1)})u(t-1)$]

5

CONVOLUTION AND CORRELATION OF SIGNALS

SIGNALS AND SYSTEMS

A continuous-time *signal* is a function of time, for example written $x(t)$, that we assume is real-valued and defined for all t,

$$-\infty < t < \infty$$

A continuous-time *system* accepts an input signal, $x(t)$, and produces an output signal, $y(t)$. A system is often represented as an operator "S" in the form

$$y(t) = \mathrm{S}[(x(t)] \qquad ...(5.1)$$

LTI Systems

A *linear* continuous-time system obeys the following property: For any two input signals $x_1(t)$, $x_2(t)$, and any real constant a, the system responses satisfy

$$\mathrm{S}[x_1(t) + x_2(t)] = \mathrm{S}[x_1(t)] + \mathrm{S}[x_2(t)] \qquad ...(5.2)$$

and

$$\mathrm{S}[ax_1(t)] = a\mathrm{S}[x_1(t)] \qquad ...(5.3)$$

A *time-invariant* system obeys the following time-shift invariance property: If the response to the input signals $x(t)$ is

$$y(t) = \mathrm{S}[x(t)]$$

then for any real constant T,

$$y(t - \mathrm{T}) = \mathrm{S}[x(t - \mathrm{T})] \qquad ...(5.4)$$

Examples of LTI Systems

Simple examples of linear, time-variant (LTI) systems include the constant-gain system,

$$y(t) = 3x(t)$$

and linear combinations of various time-shifts of the input signal, for example

$$y(t) = 3x(t) - 2x(t - 4) + 5x(t + 6)$$

The Laplace transform (Chap. 7) of say, the output of a typical filter can be considered as Y(s) corresponding to the time signal $y(t)$.

The input signal is $x(t)$ and the function of the filter is F(s).

Then $$Y(s) = F(s)\,X(s)$$

The theorem in Laplace transform says that

$$y(t) = \int_{-\infty}^{\infty} f(t-\tau)\, x(\tau)\, d\tau \qquad ...(5.5)$$

which is called the "Convolution Theorem". From this $y(t)$ is got for any $x(t)$.

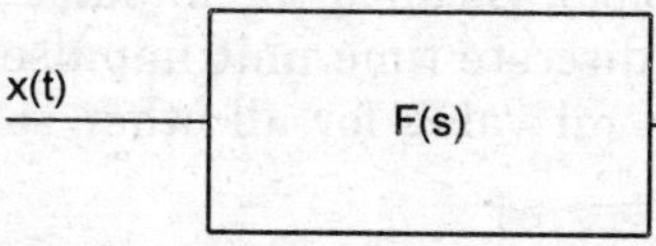

Fig. 1. A filter system, input and output.

Convolution Representation

A system that behaves according to the convolution integral

$$y(t) = \int_{-\infty}^{\infty} h(v)\, f(t-v)\, dv \text{ or } \int_{-\infty}^{\infty} h(t-v)\, f(v)\, dv \qquad ...(5.6)$$

where $h(t)$ is a specified signal, is a linear time-invariant system.

Essentially all LTI systems can be represented by such an expression for suitable choice of $h(t)$.

In the convolution expression, the integrand involves the product of two signals, both being functions of the integration variable, v. One of the signals, $f(t - v)$, involves a transformation of the integration variable and introduces t as a parameter.

Impulse Response

The signal $h(t)$ that describes the behavior of the LTI system is called the *impulse response* of the system, because it is the output of the system when the input signal is the unit-impulse, $f(t) = \delta(t)$. We also permit impulses in $h(t)$ in order to represent LTI systems that include constant-gain examples of the type shown above.

One can encounter well-behaved signals $x(t)$ and $h(t)$ for which the convolution integral

$$y(t) = \int_{-\infty}^{\infty} h(v)\, x(t-v) dv \qquad ...(5.7)$$

simply does not exist! For example take both signals to be identically 1. This is not surprising since the convolution integral is in general an improper integral, that is, integration over an infinite extent. We will not further mention this possibility, leaving it to the reader to perform sanity checks in particular examples. Of course there are mathematically possible signals that, for example, vary so rapidly that the Riemann integral is not well defined. We obviously do not consider these.

Beyond the technical issues, LTI system that cannot be represented by convolution are weird indeed. A typical example involves a system whose output is a constant that depends only on the infinite past:

$$y(t) = \lim t \to -\infty \{x(t)\} \qquad ...(5.8)$$

where the class of allowable input signals are those that have a limit as $t \to -\infty$. You are encouraged to check linearity and time invariance of such a system, but in any case we are happy to leave these out of engineering discussions.

The impulse response of a system F(*s*) is the response to a single impulse at time $t = 0$. Then, the output $y(n)$ (in samples), is called the impulse response of the filter system F. The input signal in this case is a discrete time unit impulse, having a value of 1 for the $t = 0$ sample, *i.e.*, for $n = 0$ and has nil value for all other sample instants.

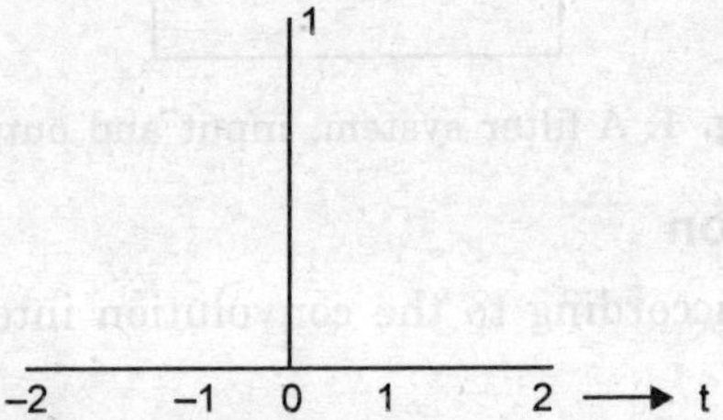

Fig. 2. A unit impulse (discrete time).

If the impulse response of the filter is $h(n)$, then for any other input signal, what will be the output ?

By superposition principle, we find this out. Suppose the input signal has $x(t)$ in continuous time and $x(n)$ as the sampled value in discrete system. (The nth time sample).

Then, $x(0)$ is the amplitude at time $t = 0$. The output for this $x(0).h(0$.

This $x(0)$ produces outputs at $t > 0$, given by the values of

$$x(0)h(1),\ x(0)\ h(2),\ ...\ x(0)h(n)...\ \text{etc.}$$

Now $x(1)$ is the sample of input at $t = t_1$ or for $n = 1$. This also produces outputs.

$$x(1)h(0),\ x(1)h(1),...\ x(1)h(n\text{–}1)\ ...\ \text{etc.}$$

The $n - 1$ in the above arises because the impulse response is that of the $(n - 1)$th coefficient of the impulse response function $h(t)$ in this case, as the sample came at a time slot $n = 1$.

In a similar way, there are outputs due to every one of the input samples, *i.e.*, for $x(2)$, $x(3)$... etc. The delayed impulse response for input which occurs not at $t = 0$ but k samples later is given by $h(n - k)$.

So, the output $y(n)$ comprises of such delayed responses for each of the input samples.

Hence, for any time t_n, or at sample instant n,

$$y(n) = \sum_{k=-\infty}^{\infty} x(k)\, h(n-k) \qquad ...(5.9)$$

This is the discrete form of the convolution of the two functions $x(t)$ and $h(t)$ given in continuous time by the integral relation given above.

The values of the outputs for each and every displaced impulse is summed up in this equation. This is done for all time from $-\infty$ but in practice, it will be summed up from the instant $t = 0$.

The above convolution equation can also be shown to be

$$y(n) = \sum_{k=-\infty}^{\infty} h(k)\, x(n-k) \qquad ...(5.10)$$

where the interchange of variables does not matter in the sum.

A graphical method of implementing such a convolution sum is given in Fig. 4 for an input and another function.

Here, a rectangular signal x is convolved with another of f, and the result of the integration in (5.10) above is shown as a net triangular signal.

$$\boldsymbol{x(t) * f(t) = \int x(\tau)\, f(t-\tau)}$$

The convolution is represented by a star in notation:

$$\boldsymbol{y(t) = x(t) * f(t)} \qquad ...(5.11)$$

where the * denotes the above integral or summation.

Convolution of a Function with a Unit Impulse Function

The convolution of a function $f(t)$ with a unit impulse function $\delta(t)$ yields the function $f(t)$ itself. This can be proved easily by using the sampling property

$$f(t) * \delta(t) = \int_{-\infty}^{\infty} f(\tau)\, \delta(t-\tau)\, d\tau$$

$$= f(t)$$

This result also follows from the time-convolution theorem and the fact that

$$f(t) \leftrightarrow \mathrm{F}(\omega) \text{ and } \delta(t) \leftrightarrow 1$$

Hence $\quad f(t) * \delta(t) \leftrightarrow \mathrm{F}(\omega)$

Consequently, $\quad f(t) * \delta(t) = f(t) \qquad ...(5.12)$

This result is also obvious graphically. Since the impulse is concentrated at one point and has an area of unity, the convolution integral in Eq. 5.12 yields the function $f(t)$. Thus the unit impulse function when convolved with a function $f(t)$ reproduces the function $f(t)$. A simple extension of Eq. (5.12) yields

$$f(t) * \delta(t-\mathrm{T}) = f(t-\mathrm{T}) \qquad (5.12a)$$

$$f(t-t_1) * \delta(t-t_2) = f(t-t_1-t_2) \qquad (5.12b)$$

$$\delta(t-t_1) * \delta(t-t_2) = \delta(t-t_1-t_2) \qquad (5.12c)$$

Example 1. Find graphically the convolution of $f_1(t)$ (Fig. 3a) with a pair of impulses of strength k each, as shown in Fig. 3b.

Following the procedure of graphical convolution to be described in what follows, we fold back $f_2(\tau)$ about the ordinate to obtain $f_2(-\tau)$. Since $f_2(\tau)$ is an even function of τ, $f_2(-\tau) = f_2(\tau)$. The convolution of $f_1(\tau)$ with $f_2(\tau)$ thus reduces to convolution of $f_1(\tau)$ with two impulses. From the property of an impulse function to reproduce the function by convolution (Eq. 5.12), it can be readily seen that each impulse produces a triangular pulse of height Ak at the origin (t = 0). Hence the net height of the triangular pulse is 2Ak at the origin. As the function $f_2(t - \tau)$ is moved farther in a positive direction, the impulse originally located at –T encounters the triangular pulse at τ = T and reproduces the triangular pulse of height Ak at t = 2T. Similarly, the impulse originally located at T reproduces a triangular pulse of height Ak at t = –2T. The final result of convolution is shown in Fig. 3(d).

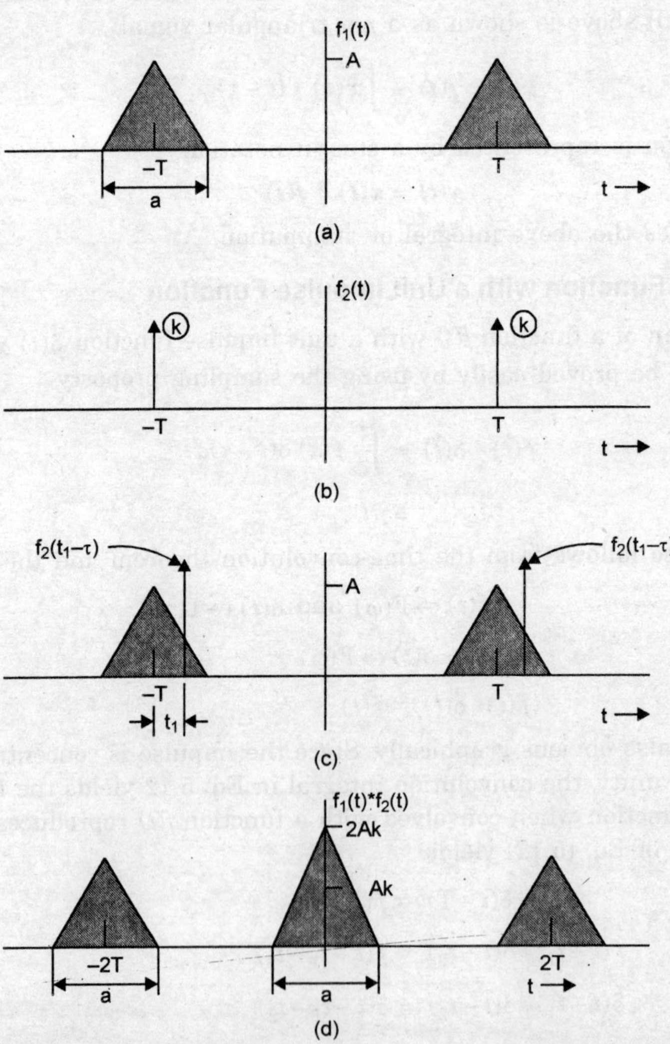

Fig. 3

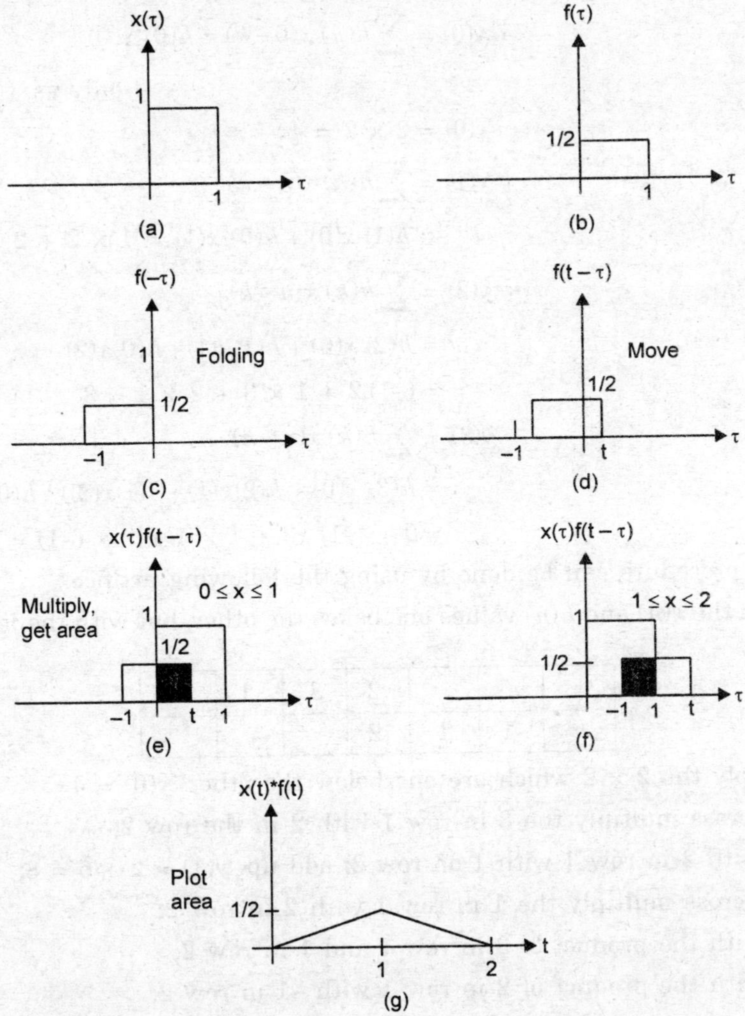

Fig. 4. Convolution-illustration using two sample functions.

Given $x(t)$ and $f(t)$ as per Fig. 6(a) and (b), find the output sequence.

Here the input signal $x(t)$ is known and also the system function $f(t)$. Hence, by convolution, we can find the output.

Using the equation

$$\boldsymbol{y(n) = \sum h(k)\, x(n-k)}$$

Example 2. Suppose an input signal has the samples at sequential sampling instants of :

2 3 1 –1.

If this sequence is applied to a system whose impulse response $h(\mathbf{n})$ is given by the sequence

$$h(0) = 2\ ;\ h(1) = 1;\ h(2) = -1.$$

The summation limits for k have to be first noted. Take for instance $y(0)$

$$y(0) = \sum h(k)\, x(0-k) = h(0)\, x(0)$$

only as $x(-k)$ are absent.

$$y(0) = 2 \times 2 = 4$$

$$y(1) = \sum h(k)\, x(n-k)$$

$$= h(1)\, x(0) + h(0).x(1) = 1 \times 2 + 2 \times 3 = 8$$

$$y(2) = \sum h(k)\, x(n-k)$$

$$= h(2)\, x(0) + h(1)x(1) + h(0)x(2)$$

$$= (-1).2 + 1 \times 3 + 2 \times 1 = 3$$

$$y(3) = \sum h(k)\, x(n-k)$$

$$= h(3)\, x(0) + h(2)x(1) + h(1)x(2)) + h(0)x(3)$$

$$= 0 + (-1)\; x3 + 1 \times 1 + 2 \times (-1) = -4.$$

The above procedure can be done by using the following artifice.

Write down the $x(n)$ and $h(n)$ values one below the other, but with the latter proceeding leftwards.

x	..	..	2	3	1	–1
h	–1	1	2			

Now multiply the 2 × 2 which are one below the other. $y(0) = 4$

1. Now cross multiply the 3 in row 1 with 2 in the row 2;
   add with 2 in row 1 with 1 on row 2; add up. $y(1) = 2 + 6 = 8$;
2. Then cross multiply the 1 in row 1 with 2 in row 2;
   add with the product of 3 in row 1 and 1 in row 2;
   add with the product of 2 in row 1 with –1 in row 2;
   $y(2) = 1 \times 2 + 3 \times 1 - 1 \times 2 = 3$
3. Then cross multiply the –1 in row 1 with 2 in row 2;
   add with the 1 in row 1 with the 1 in row 2;
   add with product of 3 in row 1 and –1 in row 2; $y(3) = -4$

Example 3. Two systems are cascaded. Their impulse responses can be described by $h_1(n) = 1/n$ for values of $n = 1, 2, 3, 4$; and $= 0$ elsewhere.

$$h_2(n) = n \text{ for same } n \text{ values.}$$

Find the overall impulse response.

If we give a unit impulse to the first, the output will $h_1(n)$. This becomes input to the second system and hence its output will be given by the convolution of $h_2(n)$ with it.

Convolution Table :

			1	1/2	1/3	1/4
4	3	2	1			

To convolve the two, use the above procedure

n	1	2	3	4
h_1	1	1/2	1/3	1/4
h_2	4	3	2	1

Now the values are found by the multiplications and additions.

$$1$$
$$1 \times 1/2 + 2 \times 1 = 2.5$$
$$1 \times 1/3 + 2 \times 1/2 + 3 \times 1 = 4.33$$
$$1 \times 1/4 + 2 \times 1/3 + 3 \times 1/2 + 4\ 1 = 6.147$$
$$2 \times 1/4 + 3 \times 1/3 + 4 \times 1/2 = 3.5$$
$$3 \times 1/4 + 4 \times 1/3 = 2.08$$
$$4 \times 1/4 = 1$$
$$= 0$$

1, 2.4, 4.333, 6.147, 3.5, 2.08, 1, 0, 0 is the required response h_{12}.

Example. 4 Calculate the convolution $y(n) = x(n) * h(n)$ of the pair of signals :

$$x(n) = \alpha^n u(n) \text{ for } 0 < d < 1$$

and
$$h(n) = u(n)$$

$u(n)$ is the step input sequence :

For $n \geq 0$

$x(k)\, h(n-k)$ is given by

$$x(k)\, h(n-k) = \alpha^k \quad (0 \leq k \leq n)$$
$$= 0 \quad \text{for all other } k \text{ such as } k > n.$$

So for $n > 0$,

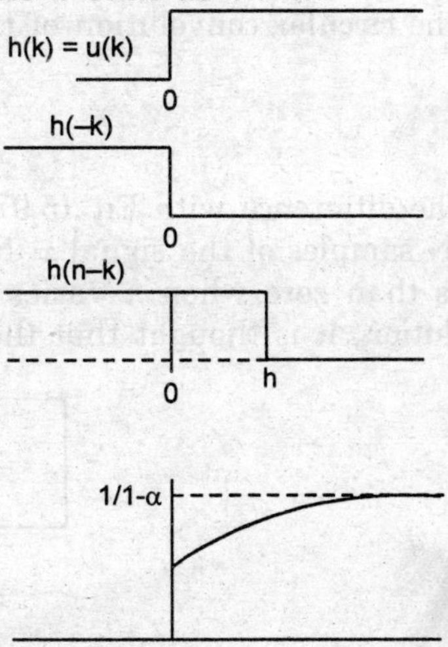

$$y(n) = \sum_{u=0}^{n} \alpha^k$$

If $\alpha = 1$, $y(n) = N$

If $\alpha \neq 1$, and < 1,

$$y(n) = \frac{1-\alpha^{n+1}}{1-\alpha}$$

Thus,
$$y(n) = \frac{1-\alpha^{n+1}}{1-\alpha}\, u(n)$$

Fast Convolution Using Inverse Fourier Transform

From (5.9), noting that

$$y(n) = x(n) * h(n) = h(n) * x(n)$$
$$Y(k) = X(k)Y(k)$$

where X(k) and Y(k) are the FFTs of $x(t)$ and $y(t)$.

If then we take inverse transform,

$$y(n) = \text{IFFT}\,(X(k)Y(k)) \qquad ...(5.13)$$

the method is called "fast Convolution".

Convolution in the Frequency Domain

In the previous case, the product of two transforms X(*k*)Y(*k*) is equivalent to the time domain convolution of the signals $x(t)$ and $h(t)$. (The input signal and the impulse response of a filter or system.)

$$X(k)\,H(k) <=> x(t)*y(t) \qquad ...(5.14)$$

If we consider two time signals just multiplied, we get the frequency domain convolution.

$$y(t) = x(t)\,s(t) <=> X(k)*S(k) \qquad ...(5.15)$$

For example, this method is used in FIR (finite impulse response) filters where $h(n)$ is the window function which truncates the otherwise infinite impulse response of the filter $h(n)$.

$$h'(n) = h(n)\,\omega(n) \qquad ...(5.16)$$

where

$h(n)$ = infinite impulse response of a typical filter

$\omega(n)$ = a truncating window function that makes the response finite (say, a rectangular window).

To find the transform of such an FIR filter, we must do a frequency convolution.

H(*k*) * W(*k*)

where H(*k*) = filter function in the freq. domain

W(*k*) = the transform of the window function.

Circular Convolution

The circular convolution of two signals, each with N samples can be written as

$$y(n) = \sum_{r=0}^{N-1} h(r)x(n-r)$$

The difference with Eq. (5.9) is in the limits. Here, it is from 0 to N – 1, because we have N samples of the signal x. Note that the parameter $(n - r)$ can become more than N or less than zero when n varies from 0 to any time t or any number n. So, in circular convolution, it is thought that the signal input $x(n)$ is periodic, *i.e.*, $x(k) = x(N + k)$... etc.

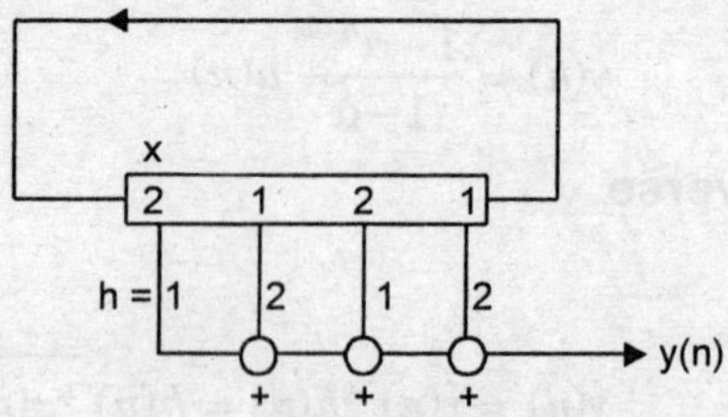

Fig. 5. Illustration of circular convolution.

This is seen from the sketch of Fig. 5, with N = 3, where y_0 is got by summing up the three products with the inputs x_0, x_1 and x_2. Rotating the input samples then gives the subsequent outputs.

Convolution of Discrete Time Systems

Example 1. *Perform the following convolution x[n]*v[n].*

(a) $x(n) = u(n) - u(n-4)$

$v(n) = 0.5^n\, u(n)$

Solution. $x(n) = u(n) - u(n-4)$

$v(n) = 0.5^{n\, u(n)}$

$$x(n) * v(n) = \sum_{u=-\infty}^{\infty} x(k)\, v(n-k)$$

$$= \sum_{u=-\infty}^{\infty} [u(k) - u(k-4)]\, 0.5^{n-k}\, u(n-k)$$

If $0 \le n \le 4$,

$$= \sum_{u=0}^{\infty} 0.5^{n-k} = 0.5^n \frac{1-2^{n+1}}{1-2}$$

$$= -(0.5^n - 2).$$

If $4 < n$

$$= \sum_{u=0}^{4} 0.5^{n-k} = 0.5^n \frac{1-2^5}{1-2}$$

$$= -0.5^n + 0.5^{n-5}.$$

(b) $x(n) = u(n),\ v(n) = 2(0.8)^n\, u(n)$

$y(n) = x(n) * v(n)$

$$= \sum_{k=-\infty}^{\infty} v(k)\, 2(0.8)^n\, u(n-k)$$

$$= \sum_{u=0}^{n} 2(0.8)^{n-k}$$

$$= 2(0.8)^n \sum_{u=0}^{n} 8^{-k}$$

$$= 2(0.8)^n \frac{1-1.25^{n+1}}{1-1.25}$$

$$= -8[0.8^n - 1.25] \quad \text{if } n \ge 0,$$

$$= -8(0.8)^n + 10 \quad \text{if } n \ge 0.$$

Circular Convolution Using Matrices

In this method, the h matrix is formed for its N samples as follows. Then it is multiplied by the x signal vector shown. The product gives the circular convoluted result y.

$$\begin{vmatrix} h(0) & h(N-1) & h(N-2) & \ldots h(2) & h(1) \\ h(1) & h(0) & h(N-1) & \ldots h(3) & h(2) \\ h(2) & h(1) & h(0) & h(4) & h(3) \\ \vdots & & & & \\ h(N-1) & h(N-2) & h(N-3) & h(1) & h(0) \end{vmatrix} \begin{vmatrix} x(0) \\ x(1) \\ x(2) \\ \vdots \\ x(N-1) \end{vmatrix} = \begin{vmatrix} y(0) \\ y(1) \\ \vdots \\ \vdots \\ y(N-1) \end{vmatrix}$$

Perform the circular convolution of the two sequences :

$$x_1(n) = \{2, 1, 2, 1\} \qquad y_1(n) = \{1, 2, 3, 4\}$$
$$\uparrow\uparrow \qquad\qquad \uparrow$$

We form the matrix for x_2 and vector for x_1.

$$\begin{bmatrix} 1 & 4 & 3 & 2 \\ 2 & 1 & 4 & 3 \\ 3 & 2 & 1 & 4 \\ 4 & 3 & 2 & 1 \end{bmatrix} \begin{bmatrix} 2 \\ 1 \\ 2 \\ 1 \end{bmatrix} = \begin{bmatrix} 14 \\ 16 \\ 14 \\ 16 \end{bmatrix}$$

$$x_1(n) * y_1(n) = \{14\ 16\ 14\ 16\}$$

Discrete Fourier Transform and Circular Convolution

The Fourier transform of the output signal is Y(k) and is got by

$$Y(k) = \sum_{r=0}^{N-1} \sum_{n=0}^{N-1} h(r)x(n-r) \exp(-j2\pi nk/N)$$

$$= \sum_{r=0}^{N-1} h(r) \sum_{n=0}^{N-1} x(n-r) \exp(-j2\pi nk/N)$$

Because the second sum is the Fourier transform of $x(n)$ circularly shifted by r samples, using the Shift relation theorem

$$Y(k) = \sum h(r)\, X(k) \exp(-j2rk/N)$$
$$= H(k)\, X(k) \qquad \ldots(5.17)$$

for all values of k from 0 to N – 1.

So, the circular convolution of two sequences, here h and $x(n)$, is got by multiplication of the frequency transform samples. Then, by taking an inverse discrete Fourier transform, the convolution of the two sequences is obtained.

$$y(k) = x(k) * h(k) \quad \ldots\text{convolution}$$
$$= F^{-1}\{X(k)\, H(k)\} \qquad \ldots(5.18)$$

Linear Convolution from Circular Convolution

Usually, we take signals of finite samples N and while doing a linear convolution with another function, say, the impulse response h function, we want to find the sum :

$$y(n) = \sum_{r=0}^{N-1} h(r)\, x(n-r)$$

For values of $n - r$ below zero and after N – 1, since the sequence $x(n)$ is not known, we take such values to be zero. For example, if N = 4, $n - r$ can have all values from –3 to +4 and hence $x(-3)$ to $x(-1)$ are taken as zero; this is equivalent to taking three zeroes before $x(0)$ to $x(3)$ for N = 4 and taking products as per the above equation.

The procedure to do fast linear convolution of two sequences will therefore require both sequences be padded with zeros to make them 2N samples. Then, after taking the Fourier transform of both, we multiply them term by term and then perform the inverse F.T. This gives the output $y(n)$ for n = 0 to 2N – 1.

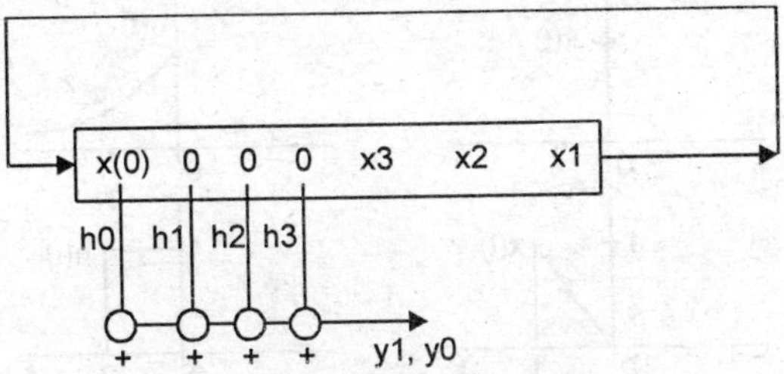

Fig. 6. Linear using circular convolution.

Example 5. Two sequences of values are given by :

$x(n)$: 0, cos 30°, cos 60°, cos 90°, cos 120°, cos 150°

$h(n)$: sin 30°, sin 60°, sin 90°, sin 120°, sin 150°, sin 0.

Find the convolution of the two sequences.

We find the values of the sequences as under :

$x(n)$: 1, 0.866, 0.5, 0, –0.5, –0.866

$h(n)$: 0.5, 0.866, 1, 0.866, 0.5, 0.

After padding, we get

$x(n)$: 1, 0.866, 0.5, 0, –0.5, –0.866, 0, 0, 0, 0, 0, 0

$h(n)$: 0.5, 0.866, 1, 0.866, 0.5, 0, 0, 0, 0, 0, 0, 0

Taking the Fourier transform of these 12 sequence signals can be done by further padding four zeros and performing a program on the computer to get X and H.

X(n): Real imaginary part:

1 + j0, 2.485 + j0.615, 2.72 – j1.725, 1.777 – j1.985, 0, 1.11 – j0.278, 0.2753 – j0.7247, 0.222 + j0.322, 1, 0.222 – j0.322, 0.2753 – j0.7247, 1.11 – j0.27, 0, 1.777 + j1.985, 2.72 + j1.725, 2.485 – j0.615

Y(n): 3.732 + j0, 2.3385 – j2.3386, 0 – 2.24j, –0.675 – j0.675, 0, 0.2615 – 0.2615, 0 – j0.2247, 0.0756 + j0.0756, 0.268, 0.075 – j0.075, 0 + 0.224j, 0.261 + j0.261, 0 + j0, –0.675 + j0.675, 0 + j0.2224, 2.334 + j2.334

After taking the products term by term,

3.73 + $j0$, 7.25 – $j4.373$, –3.837 – $j6.061$, –1.466 + $j1.22$, 0, 0.218 – j .364, –0.1628 – $j0.0618$, –0.0075 + 0.0411, –0.1628 + $j0.06186$, 0.218 + $j0.364$, 0, –1.462 – $j1.221$, –3.837 + 6.062j, 7.25 + $j4.373$

This is inverse transformed by the IDFT program in a computer : (there are no imag. Part terms, because they all become zero):

$y(n)$ = 0.5, 1.3, 2, 2.165, 1.5, 0, –1, –1.3, –1, –0.433, 0, 0, 0, 0, 0, 0 giving the desired sequence.

Continuous Time Convolution: Additional Examples

1. *Solve the following for $y(t) = x(t)*h(t)$*

$$x(t) = u(t) - u(t-4);\ h(t) = r(t)$$

2. *Convolve the following:*

(a)

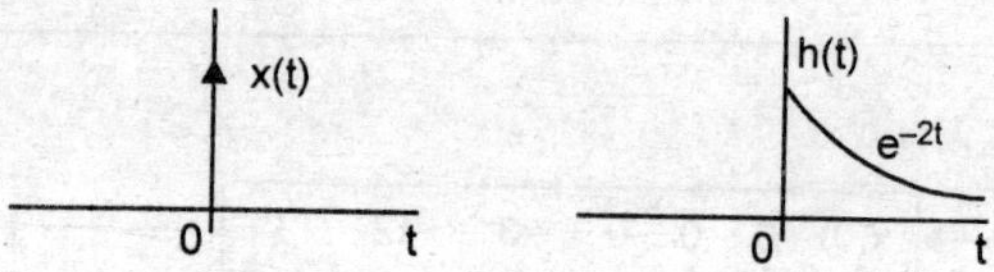

(b)

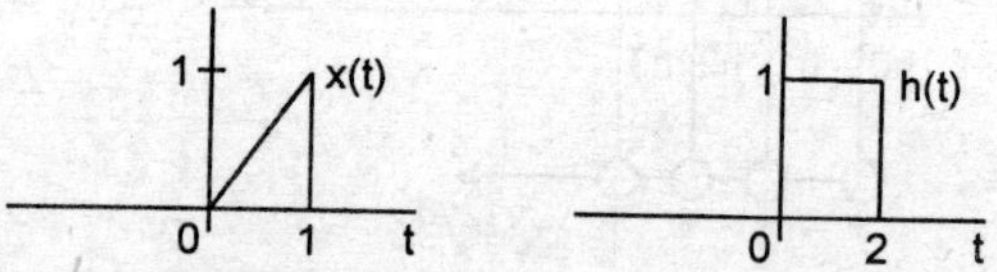

(c)

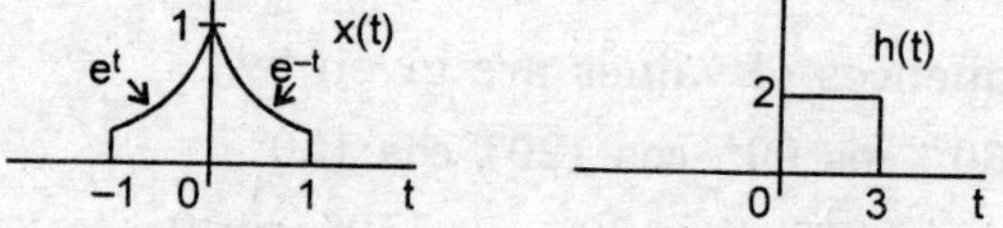

(d)

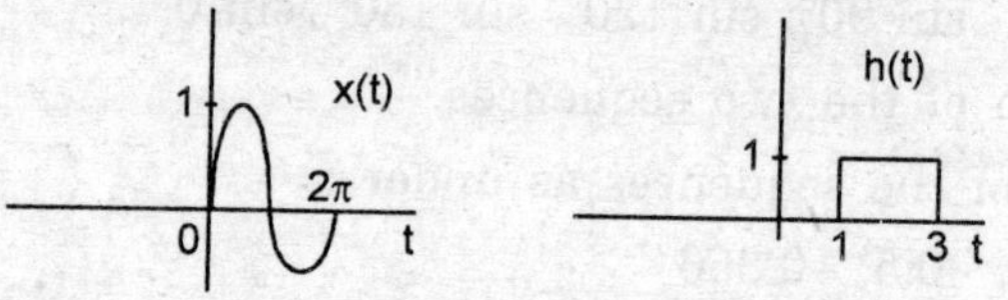

3. *Find the response of a system to an input of $x(t) = 2u(t-10)$ if $h(t) = \sin(2t)u(t)$.*

4. *A linear time invariant system has the following impulse response:*

$$h(t) = 2e^{-at}u(t)$$

Use convolution to find the response y(t) to the following input:

$$x(t) = u(t) - u(t-4)$$

Sketch y(t) for the case when $a = 1$.

5. *Determine $y(t) = x(t)*h(t)$ where $x(t) = u(t)$ and*

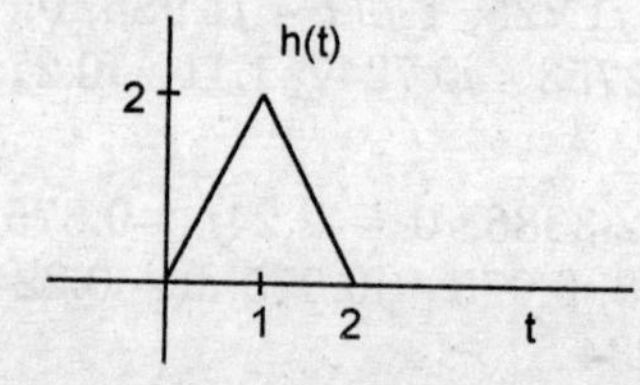

6. *Compute $x(t)*v(t)$*

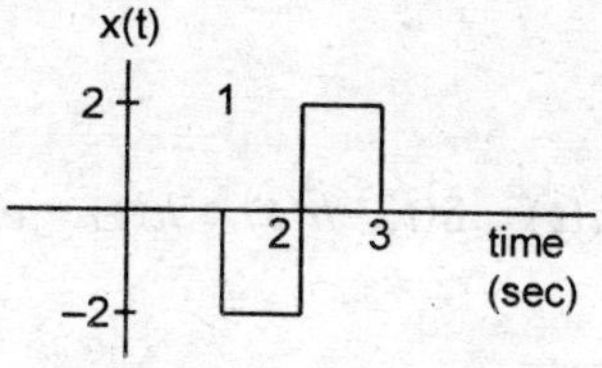

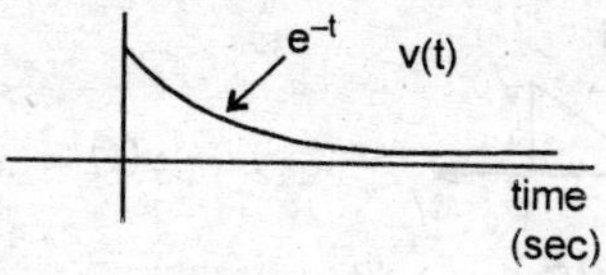

Solution.

1. $x(t) = u(t) - u(t-4)$ $h(t) = u(t)$ $h(\lambda)$

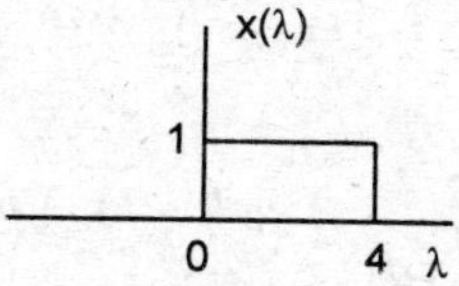

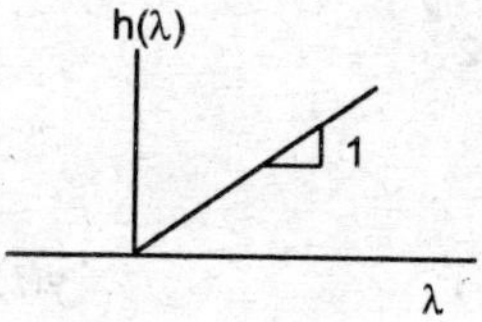

Since $h(t)$ is more complicated, flip and switch X instead of h, that is, compute

$$h(t)*x(t) = \int_{-\infty}^{\infty} h(\lambda)x(t-\lambda)d\lambda$$

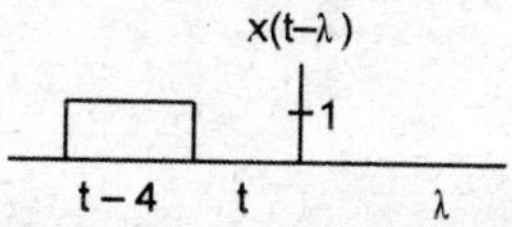

(*i*) $t < 0$, then $h(t)*x(t) = 0$

(*ii*)

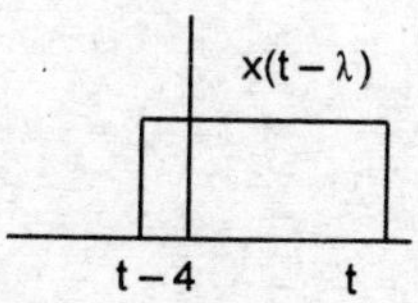

if $0 < t$ and $t - 4 < 0$

or $0 < t < 4$

$$h(t)*x(t) = \int_0^t \lambda d\lambda = \left.\frac{\lambda^2}{2}\right|_0^t = \frac{t^2}{2}$$

(*iii*)

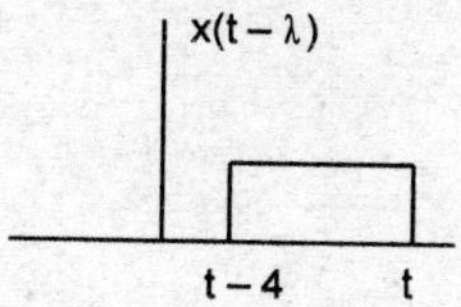

if $4 < t$

$$h(t)*x(t) = \int_{(t-4)}^{t} \lambda d\lambda = \left.\frac{\lambda^2}{2}\right|_{t-4}^{t} = \frac{t^2}{2} - \frac{(t-4)^2}{2}$$

$$h(t)*x(t) = \begin{cases} 0 & t < 0 \\ t^2/2 & 0 \le t < 4 \\ 4t - 8 & 4 \le t \end{cases}$$

2. (*a*) $x(t) = \delta(t),\ h(t) = e^{-2t}u(t)$

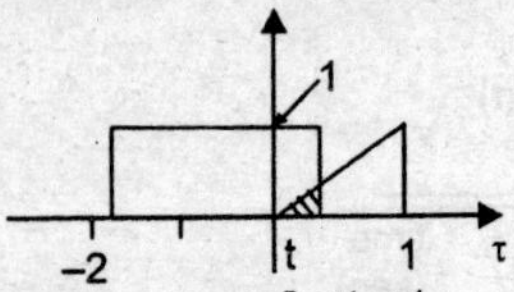

0 ≤ t < 1

$$y(t) = x(t) * h(t) = \delta(t) * h(t) = h(t) = e^{-2t}u(t)$$

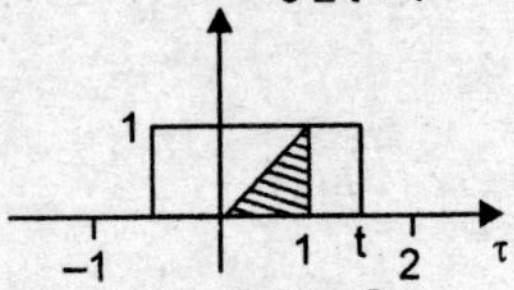

1 ≤ t < 2

(*b*) $0 \le t < 1,$

$$y(t) = \int_0^t x(\tau)d\tau = \int_0^t \tau\, d\tau = \frac{\tau^2}{2}\Big|_0^t$$

$$= \frac{t^2}{2}$$

$1 \le t < 2,$ $\qquad y(t) = \frac{1}{2}$

$2 \le t < 3,$

$$y(t) = \int_{t-2}^{1} x(\tau)d\tau = \frac{\tau^2}{2}\Big|_{t-2}^{t}$$

$$= \frac{1}{2} - \frac{(t-2)^2}{2}$$

elsewhere, $\qquad y(t) = 0$

(*c*) $-1 \le t < 0,$

$$y(t) = 2\int_{-1}^{t} e^{\tau}\, d\tau = 2e^{\tau}\Big|_{-1}^{t}$$

$$= 2e^t - 2e^{-1}$$

$$= 2(e^t - e^{-1})$$

$0 \le t < 1,$

$$y(t) = 2\int_{-1}^{0} e^{\tau} d\tau + e\int_0^t e^{-\tau} d\tau$$

$$= 2e^{\tau}\Big|_{-1}^{0} + (-2)e^{-\tau}\Big|_0^t$$

$$= 2 - 2e^{-1} - 2e^{-t} + 2$$

$$= 4 - 2e^{-1} - 2e^{-t}$$

1 £ t < 2, $\qquad y(t) =$

=

=

$2 \le t < 3,$

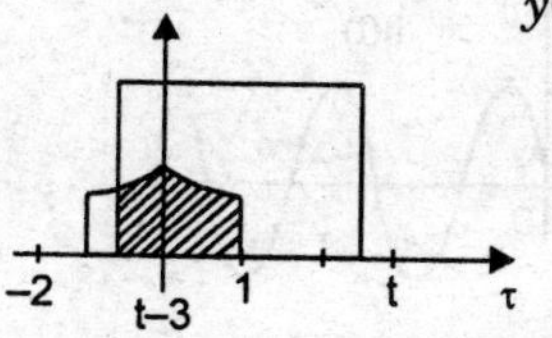

$$y(t) = 2\int_{t-3}^{0} e^{\tau} d\tau + 2\int_{0}^{1} e^{-\tau} d\tau$$

$$= 2(1 - e^{t-3}) + 2(1 - e^{-1})$$

$$= 4 - 2e^{-1} - 2e^{t-3}$$

$3 \le t < 4,$

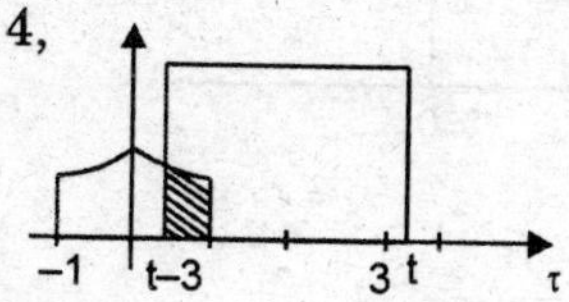

$$y(t) = 2\int_{t-3}^{1} e^{-\tau} d\tau = -2e^{-\tau}\Big|_{t-3}^{1}$$

$$= 2e^{-t+3} - 2e^{-1}$$

$$= 2(e^{3-t} - e^{-1})$$

elsewhere, $\quad y(t) = 0$

(*d*) $1 \le t < 3,$

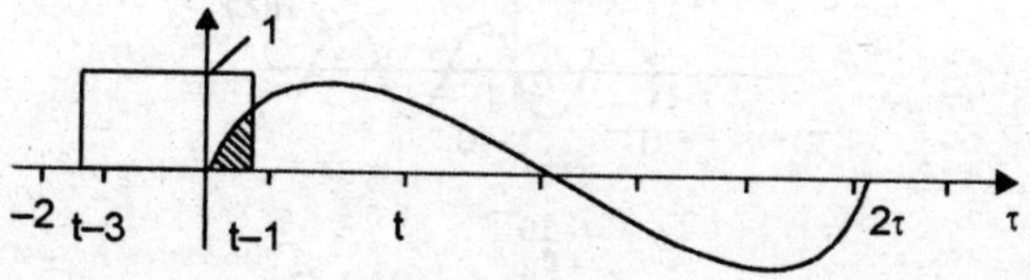

$$y(t) = \int_{0}^{t-1} \sin(\tau)\, d\tau = -\cos(\tau)\Big|_{0}^{t-1} = 1 - \cos(t-1)$$

$3 \le t < 2\pi + 1,$

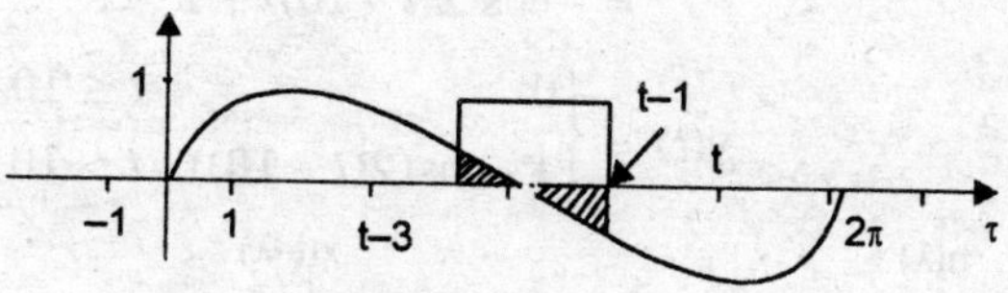

$$y(t) = \int_{t-3}^{t-1} \sin(\tau)\, d\tau = -\cos(\tau)\Big|_{t-3}^{t-1} = \cos(t-3) - \cos(t-1)$$

$2\pi + 1 \le t < 2\pi + 3,$

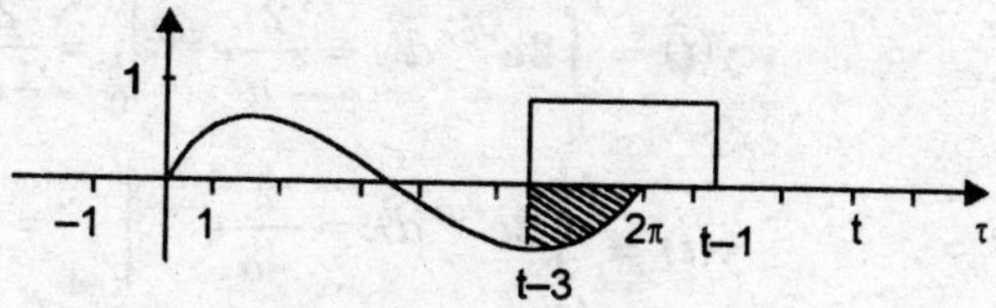

$$y(t) = \int_{t-3}^{2\pi} \sin(\tau)\, d\tau = -\cos(\tau)\Big|_{t-3}^{2\pi} = \cos(t-3) - 1$$

elsewhere, $\quad y(t) = 0$

3.

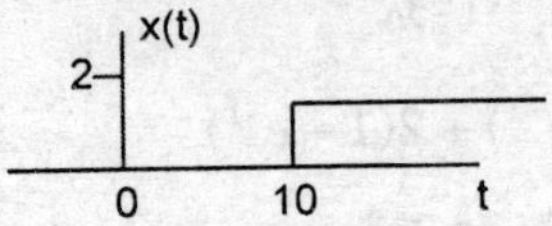

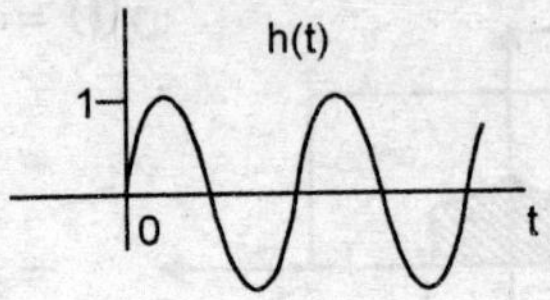

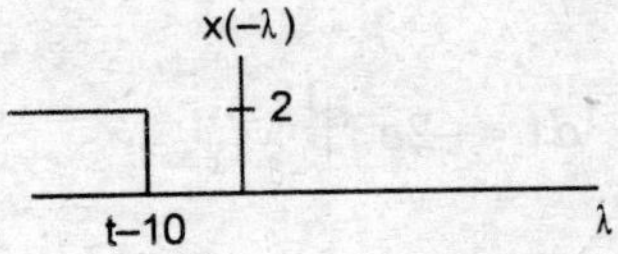

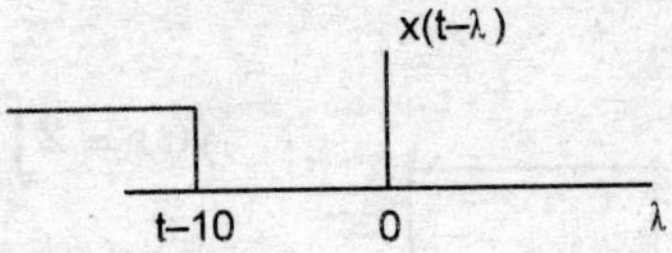

$$y(t) = x(t)*h(t)$$

$$y(t) = 0 \text{ if } t \leq 10$$

if $t \geq 10$

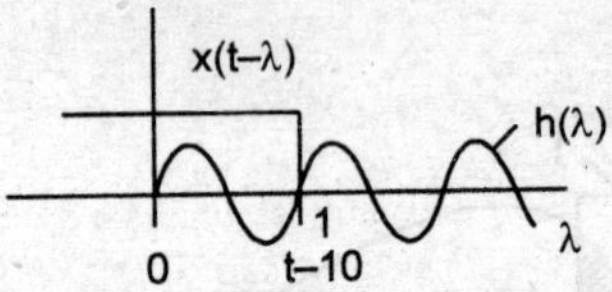

$$y(t) = \int_0^{t-10} 2\sin(2\lambda)d\lambda$$

$$= -\cos(2\lambda)\Big|_0^{t-10}$$

$$= -\cos(2(t-10)) + 1$$

$$y(t) = \begin{cases} 0 & t \leq 10 \\ 1 - \cos(2(t-10)) & t > 10 \end{cases}$$

4.

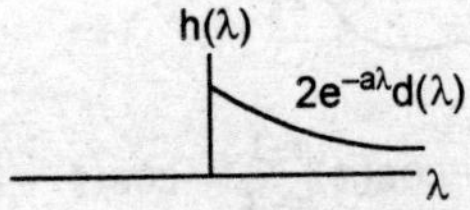

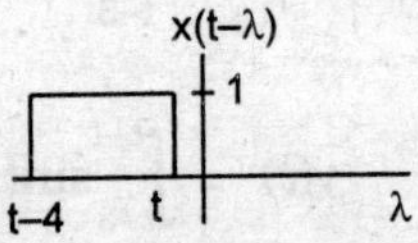

$t < 0$, $\quad y(t) = 0$

$0 \leq t < 4$, $\quad y(t) = \int_0^t 2e^{-a\lambda}d\lambda = \frac{2}{-a}e^{-a\lambda}\Big|_0^t = \frac{2}{-a}\left(e^{-at} - 1\right)$

$4 \leq t$, $\quad y(t) = \int_{t-4}^t 2e^{-a\lambda}d\lambda = \frac{2}{-a}e^{-a\lambda}\Big|_{t-4}^t = \frac{2}{a}\left(e^{-a(t-4)} - e^{-at}\right)$

$$y(t) = \begin{cases} 0 & t < 0 \\ \frac{2}{a}(1 - e^{-at}) & 0 \leq t < 4 \\ \frac{2}{a}(e^{-a(t-4)} - e^{-at}) & 4 \leq t \end{cases}$$

if $$a = 1$$

$$y(t) = \begin{cases} 0 & t < 0 \\ 2(1-e^{-t}) & 0 \le t < 4 \\ 2(e^{-a(t-4)} - e^{-t}) & 4 \le t \end{cases}$$

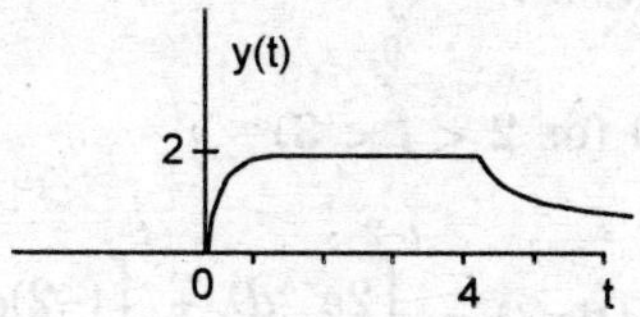

5.

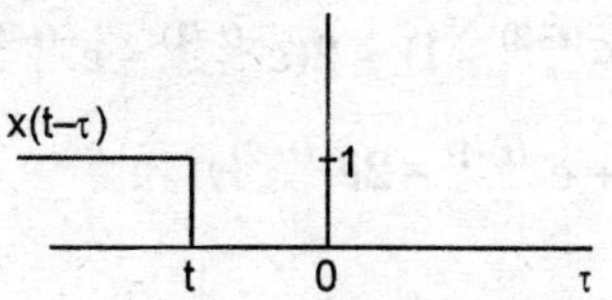

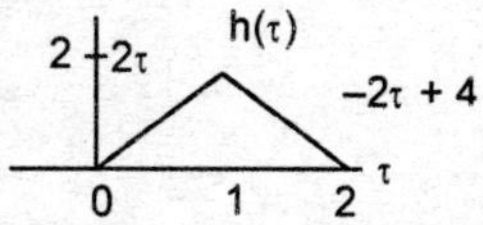

(*i*) $t < 0$, $$y(t) = 0$$

(*ii*) $0 \le t < 1$, $$y(t) = \int_0^t 2\tau\, d\tau = t^2$$

(*iii*) $1 \le t < 2$, $$y(t) = \int_0^1 2\tau\, d\tau + \int_1^t -2\tau + 4\, d\tau$$

$$= 1 + \left(-\tau^2 + 4\tau\right)\Big|_1^t = 1 - t^2 + 4t + 1 - 4$$

$$= -2 - t^2 + 4t$$

(*iv*) $2 \le t$, $$y(t) = \int_0^t 2\tau\, d\tau + \int_1^2 -2\tau + 4 d\tau$$

$$= 2$$

$$y(t) = \begin{cases} 0 & t < 0 \\ t^2 & 0 \le t < 1 \\ -t^2 + 4t - 2 & 1 \le t < 2 \\ 2 & 2 \le t \end{cases}$$

6. $v(t)*x(t)$ is easier to do

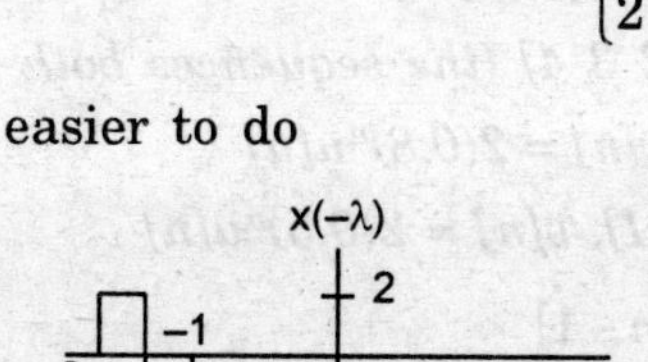

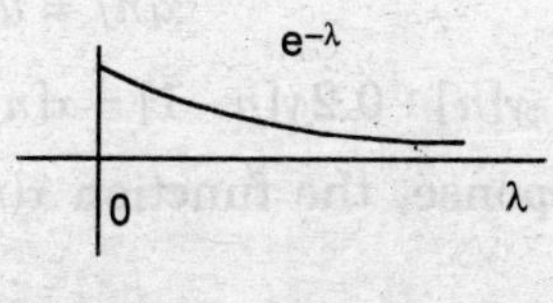

(1) If $t - 1 < 0$ (or $t < 1$)

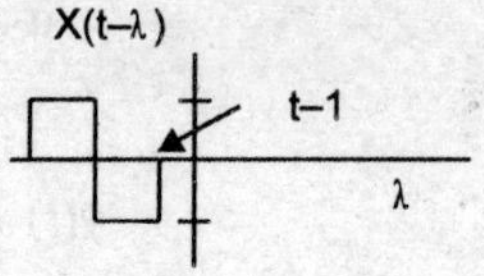

$$v(t)*x(t) = 0$$

(2) If $t - 1 > 0$ and $t - 2 < 0$ (or $1 < t < 2$)

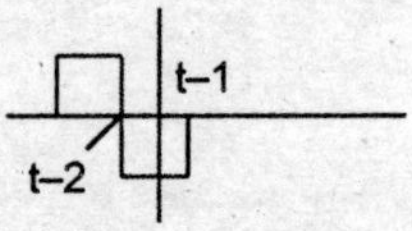

$$v(t)*x(t) = \int_0^{t-1} (-2)e^{-\lambda} d\lambda = 2e^{-\lambda}\Big|_0^{t-1} = 2(e^{-(t-1)} - 1)$$

(3) If $t - 2 > 0$ and $t - 3 < 0$ (or $2 < t < 3$)

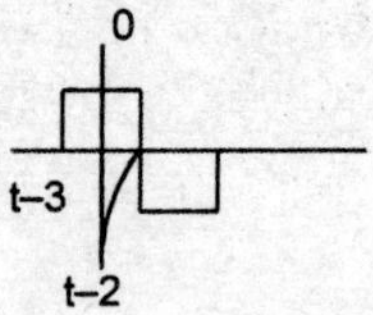

$$v(t)*x(t) = \int_0^{t-2} 2e^{-\lambda} d\lambda + \int_{t-2}^{t-1} (-2)e^{-\lambda} d\lambda$$

$$= -2(e^{-(t-2)} - 1) + 2(e^{-(t-1)} - e^{-(t-2)})$$

$$= 2(1 + e^{-(t-1)} - 2e^{-(t-2)})$$

(4) If $t > 3$

$$v(t)*x(t) =$$

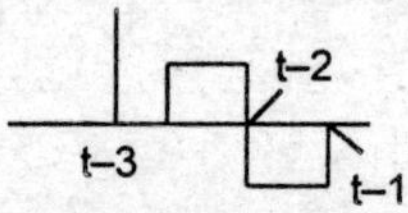

$$\int_{t-3}^{t-2} 2e^{-\lambda} d\lambda + \int_{t-3}^{t-1} (-2)e^{-\lambda} d\lambda = 2\left(e^{-(t-1)} - 2e^{-(t-2)} + e^{-(t-3)}\right)$$

$$x(t)*v(t) = \begin{cases} 0 & t < 1 \\ 2e^{-(t-1)} - 2 & 1 \le t < 2 \\ 2 + 2e^{-(t-1)} - 4e^{-(t-2)} & 2 \le t < 3 \\ 2e^{-(t-1)} - 4e^{-1(t-2)} + 2e^{-(t-3)} & t \ge 3 \end{cases}$$

Discrete-Time Convolution

1. *Find the impulse response for each of the following discrete-time systems :*

(a) $y[n] + 0.2y[n - 1] = x[n] - x[n - 1]$

(b) $y[n] + 1.2y[n - 1] = 2x[n - 1]$

(c) $y[n] = 0.24(x[n] + x[n - 1] + x[n - 2] + x[n - 3]$

(d) $y[n] = x[n] + 0.5x[n - 1] + x[n - 2]$

2. *Perform the following convolutions,* $x[n]*v[n]$

(a) $x[n] = u[n] - u[n - 4]$, $v[n] = 0.5^n u[n]$

(b) $x[n] = [1\ 4\ 8\ 2]$; $v[n]$

$= [0\ 1\ 2\ 3\ 4]$ *(the sequences both start at* $n = 0$*)*

(c) $x[n] = u[n]$, $v[n] = 2(0.8)^n u[n]$

(d) $x[n] = u[n - 1]$, $v[n] = 2(0.5)^n u[n]$

Solution. 1 (*a*) $y[n] + 0.2y[n-1] = x[n] - x[n-1]$

For impulse response, the function $x(n)$ is the inpulse delta function.

$\therefore$

$$h[n]+.2h[n-1] = \delta[n]-\delta[n-1]$$

$$h[0] = -.2h[-1]+\delta[0]-\delta[-1] = 1$$

$$h[1] = -.2h[0]+\delta[1]-\delta[0] = -.2,\ -1 = -1.2$$

$$h[2] = -.2h[1] = 0.24$$

$$h[3] = -.2h[2] = -0.048$$

$$\vdots$$

$$h[n] = (-.2)^{n-1}(-1.2) \text{ for } n \geq 1$$

(*b*)

$$y[n]+1.2y[n-1] = 2x[n-1]$$

$$h[n]+1.2h[n-1] = 2\delta[n-1]$$

$$h[0] = -1.2h[-1]+2\delta[-1] = 0$$

$$h[1] = -1.2h[0]+2\delta[0] = 2$$

$$h[2] = -1.2h[1]+2\delta[1] = -1.2(2)$$

$$h[3] = -1.2h[2] = (-1.2)^2\,(2)$$

$$\vdots$$

$$h[n] = (-1.2)^{n-1}\,(.2) \text{ for } n \geq 1$$

(*c*)

$$y[n] = 0.24(x[n]+x[n-1]+x[n-2]+x[n-3])$$

$$h[n] = 0.24(\delta[n]+\delta[n-1]+\delta[n-2]+\delta[n-3])$$

$$= \begin{cases} 0.24 & 0 \leq n \leq 3 \\ 0 & \text{otherwise} \end{cases}$$

(*d*)

$$y[n] = x[n] + .5x[n-1] + x[n-2]$$

$$h[n] = \delta[n]+.5\delta[n-1]+\delta[n-2]$$

$$h[n] = [1\quad .5\quad 1],\ h[n] = 0 \text{ all other } n$$

$$n = 0$$

2. (*a*)

$$x[n] = u[n] - u[n-4],\ v[n] = 0.5^n\, u[n]$$

$$x[n]*v[n] = \sum_{k=-\infty}^{\infty} x[k]\, v\,[n-k]$$

$$= \sum_{k=-\infty}^{\infty} (u[k]-u[k-u])0.5^{n-k}u[n-k]$$

if $0 \leq n \leq 4$

$$= \sum_{k=0}^{n} 0.5^{n-k} = 0.5^n \frac{1-2^{n+1}}{1-2} = -(0.5^n - 2)$$

if $4 < n$

$$= \sum_{k=0}^{4} 0.5^{n-k} = 0.5^n \frac{1-2^5}{1-2} = -0.5^n + 0.5^{n-5}.$$

(*b*) $x[n] = [1, 4, 8, 2], \; v[n] = [0, 1, 2, 3, 4]$

$x[n]$	1	4	8	2	0	0	0	0	0
$v[n]$	0	1	2	3	4	0	0	0	0
	0	1	2	3	4	16	32	8	
			0	4	8	12	24	6	
				0	8	16	4		
					0	2			
$y[n]$ =	[0	1	6	19	34	44	3Ɛ	8	0]
	$n = 0$								

(*c*) $x[n] = u[n], \; v[n] = 2(.8)^n \, u[n]$

$$x[n]*v[n] = \sum_{k=-\infty}^{\infty} u[k] 2\,.8^{n-k} \, u[n-k]$$

$$= \sum_{k=0}^{n} 2\,(.8)^{n-k}$$

$$= 2(.8)^n \sum_{k=0}^{n} (.8)^{-k} = 2(.8)^n \frac{1-1.25^{n+1}}{1-1.25}$$

$$= -8\,[0.8^n - 1.25], \; n \geq 0$$

$$= -8\,(.8)^n + 10, \; n \geq 0$$

(*d*)
$$y[n] = \sum_{k=-\infty}^{\infty} u[k-1]\, 2\,(.5)^{n-k} u[n-k]$$

$$= \sum_{k=1}^{n} 2\,(.5)^{n-k}$$

$$n \geq 1$$

$$= 2(.5)^n \sum_{k=1}^{n} 2^k,$$

$$= 2(.5)^n \left(\sum_{k=0}^{n} 2^k - 1 \right)$$

$$= 2(.5)^n \left[\frac{1-2^{n+1}}{1-2} - 1 \right]$$

$$= 2(.5)^n (-2 + 2^{n+1})$$

$$= -(.5)^{n-2} + 4, n \geq 1$$

CORRELATION OF SAMPLED DATA

The digital data acquired for signal processing will also include such things as :

(i) Correlation

(ii) Averaging

(iii) Power spectral density.

In statistical signal theory, we deal with signals which are deterministic or random. The latter is usually coming in due to noise or extraneous reasons. For evry time t, if $x(t)$ is known definitely, then we call it deterministic. We use random variables when the nature of the signal is not known. When we deal with a collection of time varying signals, we call it an Ensemble. Signals which have a correlation between the statistical properties are called "stochastic".

Average

One important statistical property is the well known average. Suppose we have a collection of signals

$$x_1(t),\ x_2(t),\ x_3(t) \ldots x_k(t). \qquad \ldots(5.19)$$

The average is given by

$$x'(t_1) = \sum_0^k x_i(t_1)/k \qquad \ldots(5.20)$$

This finds the average of the several signals at time t_1.

Similarly, the mean square value is

$$(x^2)' = (1/k)\sum_0^k x_i^2(t_1) \qquad \ldots(5.21)$$

Suppose we take another instant t_2 and find

$x'(t_2)$ and $x^2(t_2)$.

If now

$$x'(t_1) = x'(t_2) = x'(t_3)$$
$$(x^2)'(t_1) = (x^2)'(t_2) = (x^2)'(t_3) \qquad \ldots(5.22)$$

then it is called a "Stationary process".

We deal generally with these kinds of process.

Correlation Function

If x and y values are related, we define

$$\lim_{\tau=\infty}(1/2T)\int x(t)y(t+\tau)d\tau = \phi_{xy}(\tau) \qquad \ldots(5.23)$$

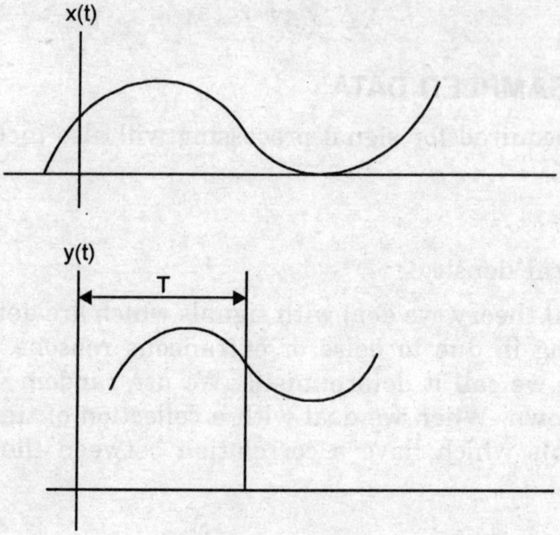

Fig. 7 (*a*) Correlating two functions with τ.

The value will change with τ. The further you move, lesser the correlation. (Fig. 7*a*).

The difference between the correlation and the convolution of x and y functions is in the value of the sign of τ in the second function within the integral sign and the intergration variable appers only in the y and not in the x.

As a comparative illustration to Fig. 4, Fig. 7(*b*) shows the case of two signals correlated.

For example, if = 0,

$$\phi_{xx}(0) = \lim \; (1/2\text{T})\int x^2(\tau)d\,\tau = x'^2 \;= \sigma^2$$

This is the auto correlation function, giving the standard deviation. We can show that

$$\phi_{xx}(0) > \;= \phi_{xx}(\tau) \text{ for any } \tau > 0.$$

For this first consider

$$\lim_{\tau\,\infty}\int_0^{\tau}\left(x(t)x-(t+\tau)\right)^2 dt \;=\; \int^{\tau}\left(x^2(t)+x^2(t+\tau)-2x(t)x(t+\tau)\right)dt \qquad \text{...(5.24)}$$

The first two square term integrals are equal and $= \phi_{xx}(0)$

$$= 2\,\phi_{xx}(0) - 2\,\phi_{xx}(\tau)$$

Since a square quantity is integrated, it must be positive.

Hence $\qquad \phi_{xx}(0) > \;= \phi_{xx}(\tau)$

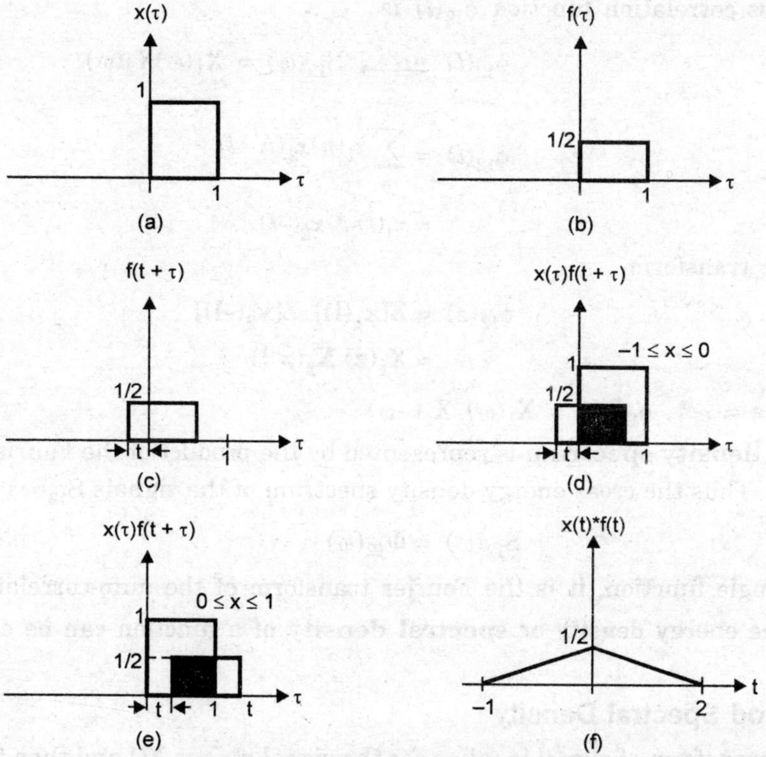

Fig. 7 (*b*) Correlation of the same two sample functions of Fig. 4.

Even Property

$$\phi_{xx}(\tau) = \phi_{xx}(-\tau)$$

This can be easily shown by replacing $t' = t + \tau$ in the integral (5.23).

Suppose we consider taking the Fourier transform of the above auto-correlation function.

For an even function, it was noted in chapter 3 that the Fourier transform is a real function and has no imaginary part therein.

Such a function is denoted by

$$\Phi(j\omega) = \text{Fourier Transform of } \phi_{xx}(\tau), \text{ as real function.}$$

This function is shown to be the power spectral density.

Correlation Theorem

If $x_1(n)$ and $x_2(n)$ are two sequences of time signals and if $X_1(\omega)$ and $X_2(\omega)$ their Fourier transforms;

$$x_1(n) \xrightarrow{F} X_1(\omega)$$

$$x_2(n) \longrightarrow X_2(\omega) \quad \text{...(5.25)}$$

The cross correlation function $\phi_{12}(l)$ is

$$\phi_{12}(l) \xrightarrow{F} S_{12}(\omega) = X_1(\omega)X_2(\omega)$$

Proof:
$$\phi_{12}(l) = \sum_{-\infty}^{\infty} x_1(n)x_2(n-l)$$
$$= x_1(l) * x_2(-l)$$

Taking z transform,

$$\phi_{12}(z) = \mathbf{Z[x_1(l)]\ Z[x_2(-l)]} \qquad \text{...(5.26)}$$
$$= \mathbf{X_1(z)\ X_2(z^{-1})}$$

Putting $z = e^{j\omega}$, $\phi_{12}(\omega) = X_1(\omega).\ X_2(-\omega)$

Energy density spectrum is represented by the product of the Fourier transform and its conjugate. Thus the cross energy density spectrum of the signals $S_{12}(\omega)$ is called $\Phi_{12}(\omega)$.

$$S_{12}(\omega) = \Phi_{12}(\omega)$$

For a single function, it is the Fourier transform of the auto-correlation function.

Thus, the energy density or **spectral density** of a function can be calculated in two ways :

Direct Method Spectral Density

Fourier transform of signal is taken for the signal $x(n)$ as X() and then S() is found from

$$S_{xx}(\omega) = |X(\omega)|^2$$

$$= \left| \sum_{n=-\infty}^{\infty} x(n)e^{-j\omega n} \right|^2 \qquad \text{...(5.27)}$$

Autocorrelation Method–Spectral Density

(*i*) The autocorrelation of the function is found first from the sequence.

$$\phi_{xx}(\omega) = \int x^*(t)\, x(t+\tau)\, dt \text{ between } -\infty \text{ to } \infty$$

(*ii*) The Fourier transform is taken for this function. That gives the spectral density.

$$S_{xx}(\omega) = \sum_{n=-\infty}^{\infty} \phi(n)e^{-j\omega n} \qquad \text{...(5.28)}$$

Discrete Auto Correlation

In discrete signal processing, the correlation integral becomes a sum, given as

$$\phi(m) = (1/N) \sum_{n=0}^{N-1-m} x(n)x(m+n) \text{ where } m \text{ is } 0 \le m \le N-1. \qquad \text{...(5.29)}$$

It is also true that $\phi(m) = \phi(-m)$

The power spectral density of a discrete sequence is defined to be (z-z-transform chapter 8)

$$S_N(z) = (1/N)\sum_{n=0}^{N-1}\sum_{m=0}^{N-1} x(n)x(m)z^{-(m-n)}$$

This is nothing but the sum of products of all discrete terms in sequence.

For example, if N = 3, we get

$$\begin{aligned} S_3(z) &= (1/3)[\{x(0)x(0)\,z^0 + x(0)x(1)z^{-1} + x(0)x(2)z^{-2}\} \\ &\quad + \{x(1)x(0)\,z^1 + x(1)x(1)z^0 + x(1)x(2)z^{-1}\} \\ &\quad + \{x(2)x(0)\,z^2 + x(2)x(1)z^1 + x(2)x(2)z^0\}] \\ &= (1/3)\,[z^0\{x(0)x(0) + x(1)x(1) + x(2)x(2)\} \\ &\quad + \{z^1 + z^{-1}\}\{x(0)x(1) + x(1)x(2)\} \\ &\quad + \{z^2 + z^{-2}\}\{x(0)x(2)\} \end{aligned} \qquad \text{...(5.30)}$$

From the above, the general formula can be inferred. It is

$$\begin{aligned} S_N(z) &= \sum_{m=0}^{N-1} \phi(m)(z^{-m} + z^m) - \phi(0) \\ &= \sum_{m=-(N-1)}^{N-1} \phi(m)z^{-m} \end{aligned} \qquad \text{...(5.31)}$$

where $\phi(m) = (1/N)\sum_{n=0}^{N-1-m} x(n)x(n+m)$ for $0 \le m \le N-1$ ...(5.32)

being the auto-correlation for the finite sequence.

Energy Density Spectrum

A useful parameter of $f(t)$ is its normalised energy.

$$E = \int_{-\infty}^{\infty} f^2(t)dt \qquad \text{...(5.32)}$$

The integral is finite if the signal is a proper energy signal.

Using Fourier transform,

$$\begin{aligned} E &= \int_{-\infty}^{\infty} f^2(t)dt = \int_{-\infty}^{\infty} f(t)\left[\frac{1}{2\pi}\int_{-\infty}^{\infty} F(\omega)e^{j\omega t}d\omega\right]dt \\ &= \frac{1}{2\pi}\int_{-\infty}^{\infty} F(\omega)\int_{-\infty}^{\infty} f(t)e^{j\omega t}dt\,d\omega \\ &= \frac{1}{2\pi}\int_{-\infty}^{\infty} F(\omega)F(-\omega)d\omega \end{aligned} \qquad \text{...(5.33)}$$

Since integral of $f(t)e^{j\omega t}dt$ is F(–ω)

$$= \frac{1}{2\pi}\int_{-\infty}^{\infty}|F(\omega)|^2 d\omega = \int_{-\infty}^{\infty}|F(\omega)|^2 df \quad ...(5.34)$$

$$= \text{area under } |F(\omega)|^2 \text{ curve.} \quad ...(5.35)$$

The plot of $|F(\omega)|^2$ is called the Energy density spectrum.

Since squaring is involved, this is an even function of ω. Hence the total energy can be got as twice the integral between 0 and ∞, as

$$E = 2\int_{0}^{\infty}|F(\omega)|^2 df \quad ...(5.36)$$

Power Density Spectrum for Periodic Signals

Since a periodic signal has existance for all time, it has infinite energy. Hence its energy is defined over its period, τ. Redefine $f(t)$ for its period τ,

$$f_\tau(t) = \begin{cases} f(t) & |t| < \tau/2 \\ 0 & \text{for all other } t. \end{cases} \quad ...(5.37)$$

Then its integral is a power integral, got by dividing by τ.

$$\frac{1}{\tau}\int_{-\tau/2}^{\tau/2} f^2(t)dt \quad ...(5.38)$$

This is also equal to Fourier transform based integral

$$\int_{-\infty}^{\infty}\lim_{\tau\to\infty}\frac{|F_\tau(\omega)|^2}{\tau}df \quad ...(5.39)$$

and the power density spectrum is defined as $\frac{|F_\tau(\omega)|^2}{\tau}$.

Power spectral density shows the strength of the variations (energy) as a function of frequency. In other words, it shows at which frequencies, variations are strong and at which frequencies weak.

The unit of PSD is energy for frequency (width) as a function of frequency. The total energy in a frequency band can be found by integrating the PSD within that range.

Calculation of PSD is done either by the fast Fourier transform or by the autocorrelation function and transforming it.

PSD is useful to identify oscillatory signals in time data. By looking at the PSD, we can find the major frequencies present in any vibration, for instance.

Correlation for Noise Corrupted Signals

Correlation is useful for identifying noise corrupted signals.

Let $y(n)$ be a signal vector which has $x(n)$, a periodic signal and $\omega(n)$ an additive random interference.

Let us take just M samples of the signal $y(n)$

Let N be the periodicity of $x(n)$. Here $\mu >> N$. That means, we observe many such periods of the total signal.

The auto-correlation sequence, after dividing by μ, is

$$\phi_{yy}(l) = \frac{1}{\mu}\sum_{n=0}^{\mu-1} y(n)y(n-l)$$

$$= \frac{1}{\mu}\sum [x(n)+\omega(n)][x(n-l)+\omega(n-l)]$$

$$= \frac{1}{\mu}\sum x_n x_{n-l} + \frac{1}{\mu}\sum x_n\omega_{n-l} + \omega_n x_{n-l} + \frac{1}{\mu}\sum \omega_n\omega_{n-l}$$

$$= \phi_{xx}(l) + \phi_{x\omega}(l) + \phi_{\omega x}(l) + \phi_{\omega w}(l)$$

The first term ϕ_{xx} is the autocorrelation of $x(n)$.

The same will be periodic, of period N.

So peaks will be noticeable in ϕ_{xx} at $l = 0$, N, 2N...

But it is not proper to compute ϕ_{yy} for $l > \mu/2$, since for these large values, $x(n)x(n-l)$ will be zero. (Since negative values of samples are zero.)

Consider the correlation products with noise : $\phi_{x\omega}$, $\phi_{\omega x}$. These are small only. That is because $x(n)$ has no relationship to $\omega(n)$. Only ϕ_{xx} is likely to have peaks for $l > 0$.

As an example, the sunspot activity over the past century was noted and plotted. Its auto-correlation function reveals that there is a period of 10-11 years in the activity (Fig. 8).

As another example, additive random white noise to a periodic signal is illustrated in Fig 9.

The assumption is that the signal sequence $x(n)$ has some unknown period that we wish to estimate from the observed noisy signal y_n.

Let x_n have a period of 10 samples. Then we should take the number of samples of the observed signal as 100. ($\mu >> N$).

If the signal to noise ratio is large, we cannot determine the periodicity of x_n by a look at it.

But the auto-correlation function shows the periodicity as being 10.

Two levels of noise are added and tested. Even for the lesser noisy signal, the periodicity is not observable on y_n. But it is seen in ϕ_{yy}.

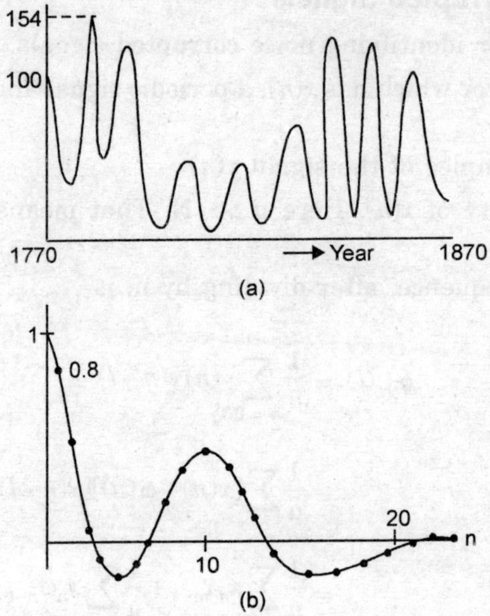

Fig. 8 (*a*) Showing a noisy signal example – Sunspot activity plotted over the years 1770 – 1870. Is there any periodicity in sunspot activity?
(*b*) Its autocorrelation function, shows a period of 10 – 11 years.

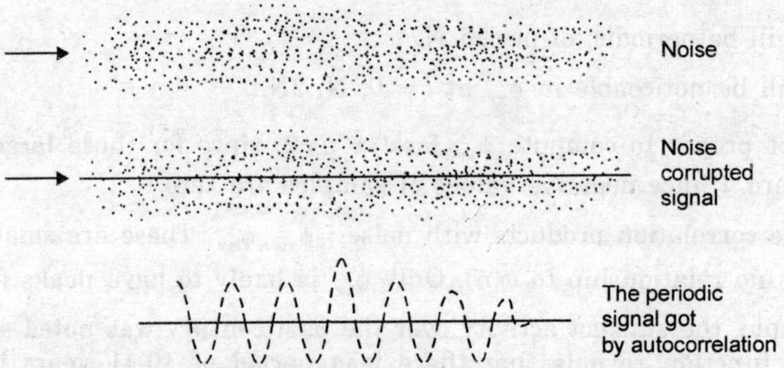

Fig. 9 Illustration for auto-correlation to eliminate noise.

Power Spectrum of Random Signals – The Welch Periodogram

An important class of signals classified as "Stationary random processes" do not possess a proper Fourier Spectrum, because they do not have a finite energy. However, they possess a finite average "Power" and are characterised by the Power Spectral Density rather than the Energy Spectral Density.

Since, the time processes though random possess an auto correlation function, given as a time function over the observation time $2T_0$, (a single realisation of the process),

$$R(\tau) = 1/(2T_0)\left\{\int_{-T_0}^{T_0} x^*(t)\, x(t+\tau)dt\right\} \quad ...(5.33)$$

If the process is "ergodic", which means that a single sample function of the variable is able to determine all the statistical properties, then the time correlation average is identical to the statistical average of the function.

Then $$\phi_{xx}(\tau) = \underset{T_0 \to \infty}{\text{Lt}} R_{xx}(\tau) \qquad ...(5.34)$$

Taking the Fourier transform of this function, we get

$$= \underset{T_0 \to \infty}{\text{Lt}} \int_{-T_0}^{T_0} R_{xx}(\tau) \exp(-j\omega\tau)\, d\tau$$

$$= \underset{T_0 \to \infty}{\text{Lt}} (1/2T_0) \int_{-T_0}^{T_0} \left\{ \int_{-T_0}^{T_0} x^*(t)x(t+\tau) \right\} \exp(-j\omega\tau)\, d\tau\, d\tau$$

$$= \underset{T_0 \to \infty}{\text{Lt}} (1/2T_0) \left| \int_{-T_0}^{T_0} x(t) \exp(-j\omega\tau)\, d\tau \right|^2 \qquad ...(5.35)$$

The actual power density spectrum (PDS) is the expected (mean) value of the above function, in the limit for T_0 tending to infinity.

Blackman-Tukey Spectral Estimate

The Blackman-Tukey estimation method consists of three steps. In the first step, the middle 2M + 1 samples of the autocorrelation sequence, $\phi_{xx}(m)$, $-M \le m \le M$ are estimated from the available N-point data record. The second step is to apply a data window to the estimated autocorrelation lags. Finally, the windowed autocorrelation estimate is Fourier-transformed to yield the Blackman-Tukey estimate. Mathematical formulation of this procedure can be given with the following equations :

$$r_{xx}(m) = 1/N \sum_{N} x^*(n)x(n+m), \quad -M \le m \le M$$

$$\bar{r}_{xx}(m) = w(m) r_{xx}(m)$$

$$\text{PBT}(F) = \sum_{m=-M}^{M} \bar{r}_{xx}(m) \exp(-j2\pi F m) \qquad ...(5.36)$$

For some data records and windows, the Power Spectral Density (PSD) estimate may contain negative values. When the window $w[n]$ is the rectangular window, the method of averaging periodograms is called Bartlett's procedure, and in this case it can be shown that

$$C_{ww}[m] = \begin{cases} L - |m|, & \text{for } |m| \le (L-1), \\ 0 & \text{otherwise,} \end{cases}$$

and, therefore, $$C_{ww}(e^{j\omega}) = \left(\frac{\sin(\omega L/2)}{\sin(\omega/2)} \right)^2 \qquad ...(5.37)$$

That is, the expected value of the average periodogram spectrum estimate is the convolution of the true power spectrum with the Fourier transform of the triangular sequence $C_{ww}[n]$ that results as the autocorrelation of the rectangular window. Thus, the average periodogram is also a biased estimate of the power spectrum.

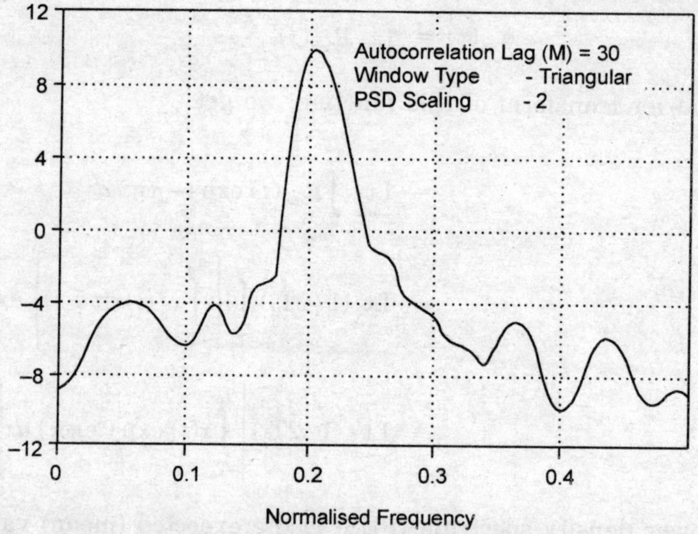

Fig. 10 Blackman-Tukey Power Spectral Density (PSD) Estimate Plot.

For each, after obtaining the periodogram, the average function is calculated.

Extraction of signal from Noise

Moving Average: A normal filter, say low pass filter, with proper cut-off, will be able to filter noise (Fig. 11). Methods based on averaging will work better if many sets of signals can be obtained.

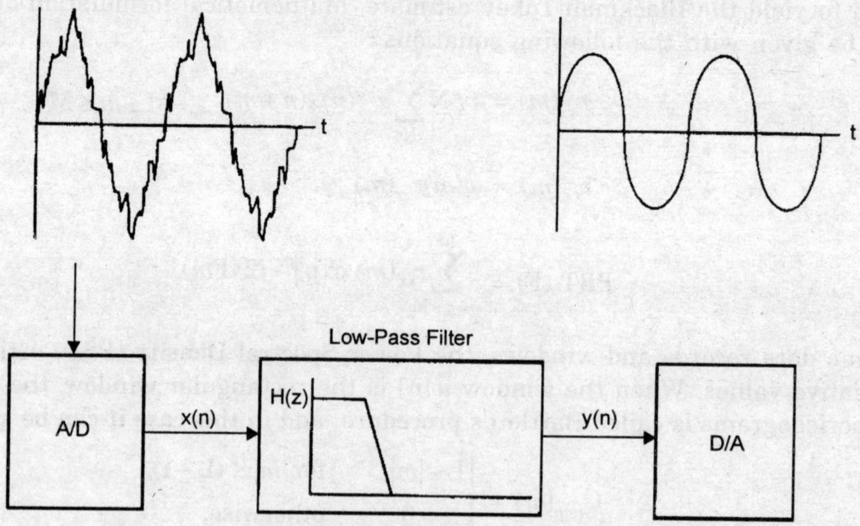

Fig. 11

In ensemble averaging, successive sets of data are collected and added for each point as an array in DSP. This process is called Coaddition. After the collection and summation is complete, the data are averaged by dividing the sum for each point by the number of scans performed. Fig. 12 shows the principle.

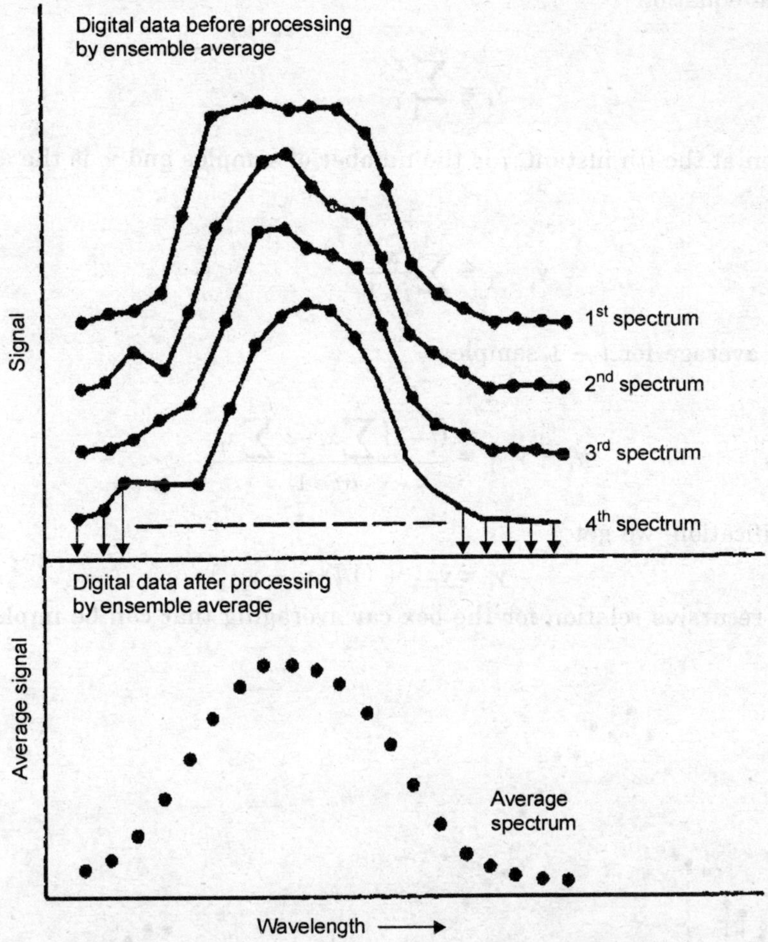

Fig. 12 Average of sets of (ensemble average) data.

Let the signal at a point x be S_x. After n sets of data have been summed up, the sum for x will be nS_x. Contained in S_x is the noise signal N_x, which also gets summed. Because the noise signals are random, however, they accumulate as the square root of n. Thus, the sum of the noise at x after n sets of data is equal to $\sqrt{n}N_x$. The signal to noise ratio (S/N) for the signal average is given by

$$S/N = nS_x/\sqrt{n}N_x = \sqrt{n}(S_x / N_x) \quad ...(5.38)$$

Thus, the signal to noise ratio increases by the square root of the number of times the data points are collected and averaged.

Moving Window Box Car Average

Fig. 13 shows the technique. The first point on the box car plot is the mean of the points 1, 2, 3; the second point is that of 2, 3, 4; and so forth. This averaging is often performed by DSP in real time (in contrast to ensemble averaging, which requires storage of data for later processing). The moving average is effective as a low pass filter, but removes noise.

Consider the equation

$$y_i = \sum_{i=1}^{i} \frac{x_i}{i} \qquad \text{...(5.39)}$$

x_i is sample taken at the ith instant, i is the number of samples and y_i is the average value of i samples.

$$y_{i-1} = \sum^{i-1} \frac{x_i}{i-1} \qquad \text{...(5.40)}$$

which gives the average for $i - 1$ samples.

$$y_i - y_{i-1} = \frac{(i-1)\sum^{i} x_i - i\sum^{i-1} x_i}{i(i-1)} \qquad \text{...(5.41)}$$

After simplification, we get

$$y_i = y_{i-1} + (1/i)(x_i - y_{i-1}) \qquad \text{...(5.42)}$$

which gives the recursive relation for the box car averaging that can be implemented in a DSP device.

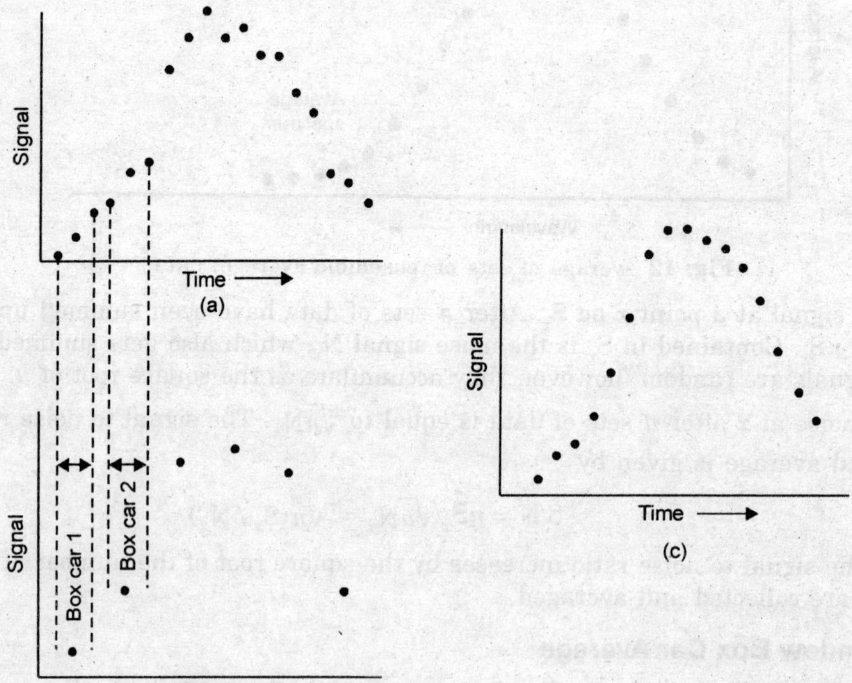

Fig. 13 Box car averaging technique.

EXERCISES

1. Find the cross correlation of the sequences

$$x_{1n} = [2, 0, 0, 1] \text{ and } x_{2n} = [4, 3, 2, 1]$$

Compare with their convolution.

0.5, 1, 1.5, 2.25, 0.5, 0.75, 1, 0;

8, 6, 4, 6, 3, 2, 1

2. Determine the response of the system which has an impulse response

$$\{1, \underset{\uparrow}{2}, 1, -1\}$$

if the input is

$$\{\underset{\uparrow}{1}, 2, 3, 1\} \qquad [1, \underset{\uparrow}{4}, 8, 8, 3, -2, -1]$$

3. Find the circular convolution of the two sequences :

{0.2, 0.4, 0.6, 0.8, 1.0, 1.2, 1.4, 1.6}

{0.1, 0.3, 0.5, 0.7, 0.9, 1.1, 1.3, 1.5}

[6, 6.48, 6.64, 6.48, 6, 5.2, 4.08]

4. Find the following convolutions:

(*a*) $x[n]$ = [1 4 8 2], $y[n]$ = [0 1 2 3 4] if both sequence start at $n = 0$.

[**Ans.** 0, 1, 6, 19, 34, 44, 38, 8, 0]

(*b*) $x[n] = u[n-1]$, $v[n] = 2(0.5)^n\, u[n]$ [**Ans.** $(-0.5^{n-2} + 4)$ for $n \geq 1$]

5. Find the mean square power levels and sketch the power density spectra of

(1) A cos 200 πt cos 2000 πt

(2) (A + sin 200πt)cos 2000 πt

(3) A $\sin^2$(200 πt)cos (2000 πt)

6. A signal $f(t) = 2e^{-t}\, u(t)$ is passed through an ideal low-pass filter with cut-off frequency 1 radian per second. Find the energy density spectrum of the output of the filter.

7. Find the linear convolution of the sequence

$$x_1(n) = \{1, 2\},\ x_2(n) = \{3, 4\} \qquad [3, 10, 8]$$

8. Explain how a noisy signal mixed with a periodic signal can be examined by correlation.

9. Solve the following for $y(t) = x(t) * h(t)$

$x(t) = u(t) - u(t - u)$; $h(t) = r(t)$.

[**Ans.** 0 for $t < 0$

$t^2/2$ for $0 < t < y$

$\dfrac{t^2}{2} - \dfrac{(t-y)^2}{2}$ if $y < t$]

10. Convolve the following functions

(*a*)

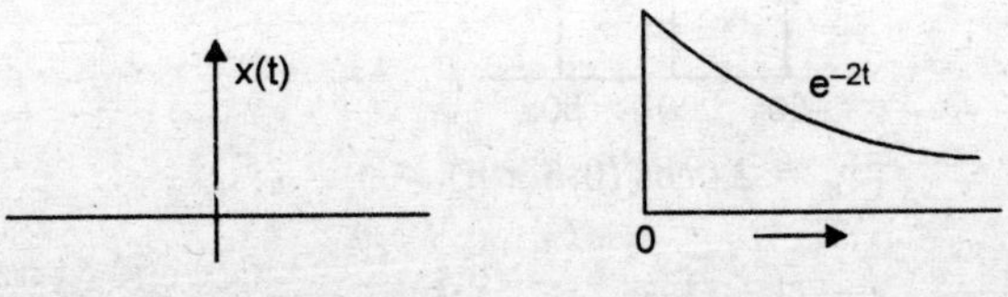

[**Ans.** $e^{-2t}\, u(t)$]

(b)

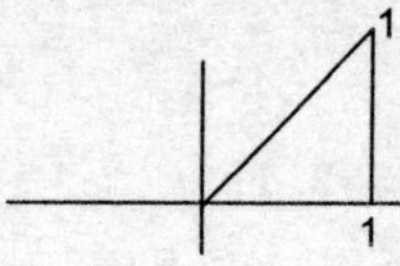
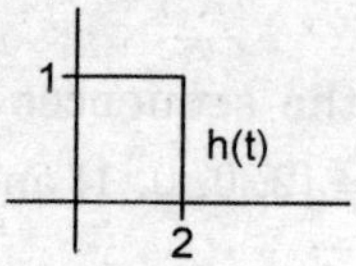

[**Ans.** $t^2/2, \dfrac{1}{2}, \dfrac{1}{2} - \dfrac{t-2^2}{2}$, for $t = 0 \to 1, 1 \to 2, 2 \to 3$]

(c) For 0 < t < 1

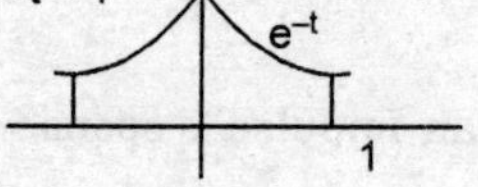
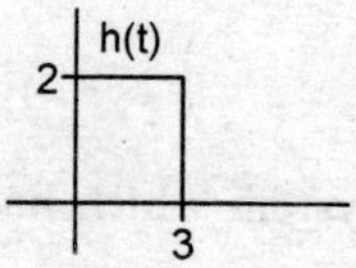

[**Ans.** $4 - 2e^{-1} - 2e^{-t}4(1-e^{-1})2(e^{3-t} - e^{1})$ $t = 0 \to 1, t = 1 \to 2, t = 3 \to 4$]

(d) For 1 < t < 3

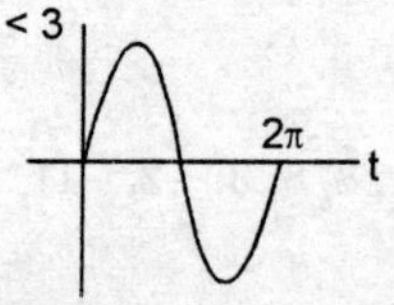
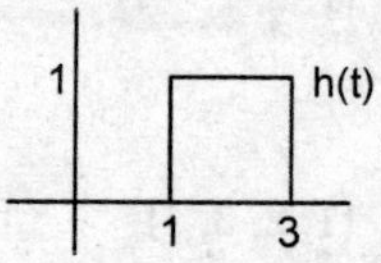

[**Ans.** $1 - \cos(t - 1)$ for $t : 1 \to 3$]

11. Find the response of system to an input to get $y(t)$.

$$x(t) = 2u\,(t - 10) \text{ if } h(t) = \sin\,(2t/u(t)$$

$[y(t) = 0$ for $t \le 10 = 1 - \cos 2(t - 10)\ t > 10]$.

12. A LT I system has the following impulse response.

$$h(t) = 2e^{-at}u(t)$$

Use convolution to find the response $y(t)$ to the following input : $x(t) = u(t) - u(t - 4)$

Sketch $y(t)$ for the case when $a = 1$

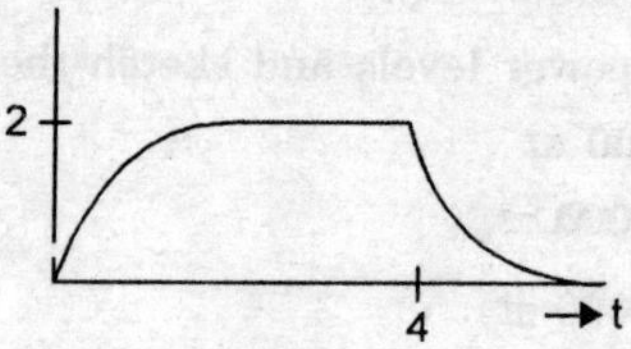

$y(t) = 0$ for $t < 0$

$y(t) = 2(1 - e^{t})$ for $0\ ^3\ t < y$

$2\{e^{-t-y} - e^{-t}\}$ for $y \le t$

13. Let $x(t) = 2 + \cos\,(50\pi t)$ and T = 0.01s

(a) Draw $|X_s(\omega)|$ where $x_s(t) = x(t)\,p(t)$. Determine if aliasing occurs.

(b) Determine the expression for $y(t)$

(c) Determine an expression for x_n.

Ans.

(a)

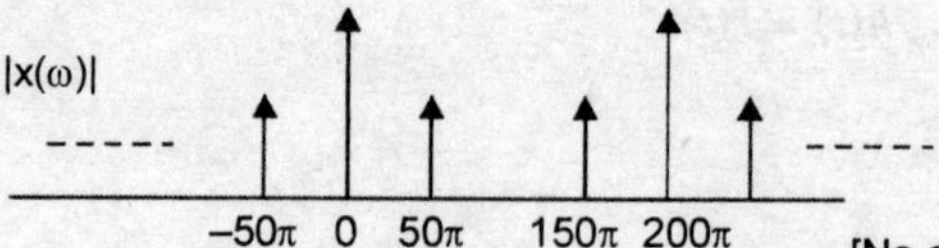

[No aliasing]

(b)

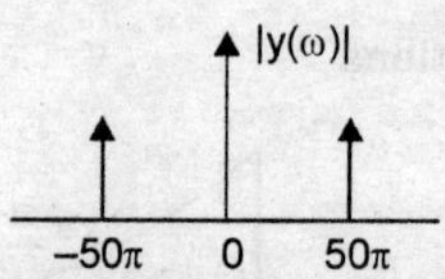

(c) $x_n = 2 + \cos\,(0.5\,\pi\,n)$.

6

SAMPLING, ALIASING IN DSP

SAMPLES, SAMPLING RATE, AND ALIASING

If there is a sine wave, it can be drawn fully if we knew its peaks, positive and negative. The rest of the wave can be drawn as a sinusoidal curve passing through these two peak points, one on the positive and another on the negative side. If its frequency is f_1, then we will need two such samples on the wave, so that we need $2f_1$ samples per second to represent it. Since any waveform can be broken down into a series of sinusoidal waves, it is possible to get the highest frequency component in it by sampling that waveform $2f_1$ times.

Talking in terms of the time-interval between samples, T,

$$T = \frac{1}{2f_h} \qquad ...(6.1)$$

Hence for going up to a frequency of 3 kHz, we need to take 6000 samples per second and T = 1/6000 = 0.16 ms. This is the sampling frequency and is called f_s, which is 600 Hz.

It is convenient to represent, say, an audio signal, in terms of its "spectrum". This is an X-Y plot of the components of the frequencies in it, where x-axis represents frequency and y-axis the amplitude of such a frequency component in the signal. A pure or single sine wave audio signal, such as from a fine tuning fork, will contain only one frequency component as a vertical line (Fig. 1(*a*)) while a general signal in the audio range will have its amplitudes varying in its range, as a curve Fig. 1(*b*). Treating a frequency as a sinusoidal function,

$$A \cos \omega t = A/2 \exp (j\omega t) + A/2 \exp (-j\omega t)$$

it can be considered as two components, one real and one imaginary. This speaks of its representation on both the negative and positive x-axis. This curve of the spectrum is symmetrical about the y-axis.

Suppose we draw the spectrum by evaluating the Fourier components of the signal, using its samples. We do not use the wave fully, but only use its samples.

Fig. 2(*a*) shows how if the samples are too slow, the wave cannot be inferred properly. The point A is at a negative maximum but the point B is not at the next maximum, but on the second wave. Actually these two points would appear to us to have been found from a single sine wave as in Fig. 2(*b*). It is often said that two points suffice to infer a single sine wave. But this is true only if the sampling time is half its period or the sampling frequency is twice the signal frequency.

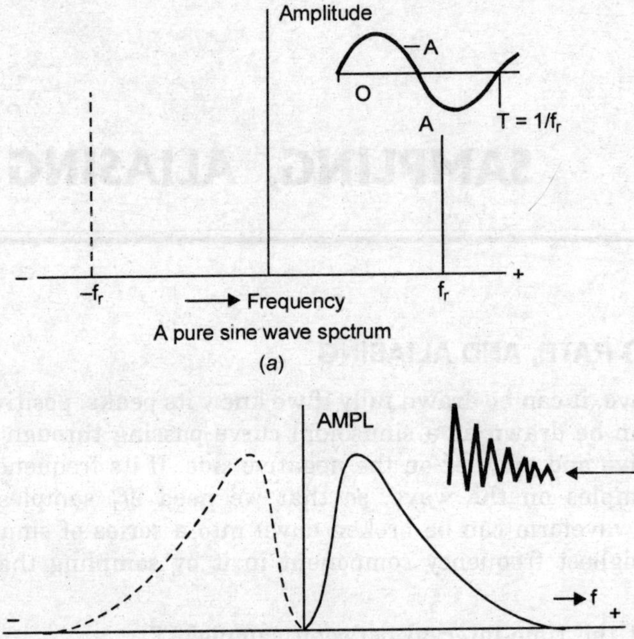

A pure sine wave spctrum

(*a*)

An audio signal spectrum

(*b*)

Fig. 1 Frequency spectrum of a single frequency signal and general audio signal (*a*) Single frequency (*b*) General signal.

Since f_h is the highest frequency of the spectrum, when we take samples at T=1/2f_h, we will therefore get several replica of the spectrum at frequencies f_h to $3f_h$, $3f_h$ to $5f_h$ etc. (Fig. 2(*c*)).

When one takes samples at intervals of T, the value of

$$f_h = 1/2T \qquad ...(6.3)$$

is called the "folding-frequency" because the total spectrum folds itself multiply as the Fig. 2(*c*) shows. (f_h is also denoted as f_0 in later chapter.)

In digital signal processing, therefore, we are confined to the frequency f_h=1/2T as the highest, and we cannot get any information of high frequency components of the signal than f_h. Suppose the signal actually does contain any such higher than f_h components, they will be considered to be only within the ± f_h range just as Fig. 2(*b*) shows that we consider the higher frequency sine wave of Fig. 2(*a*) as if it were a lower frequency one.

In other words, signals of $f>f_h$ will show up in effect as components of $f/3$, $f/5$ etc., causing a wrong spectrum to be shown for the signal (Aliasing).

The maximum frequency f_h contained in a signal is called the "Nyquist frequency" and the rate of sampling at $2f_h$ is called the "Nyquist rate". If samples are taken at any other rate, say f_s, then the folding frequency will be $f_s/2$. If samples are taken at the Nyquist rate, then, the folding frequency will also be equal to the highest frequency f_h in the signal.

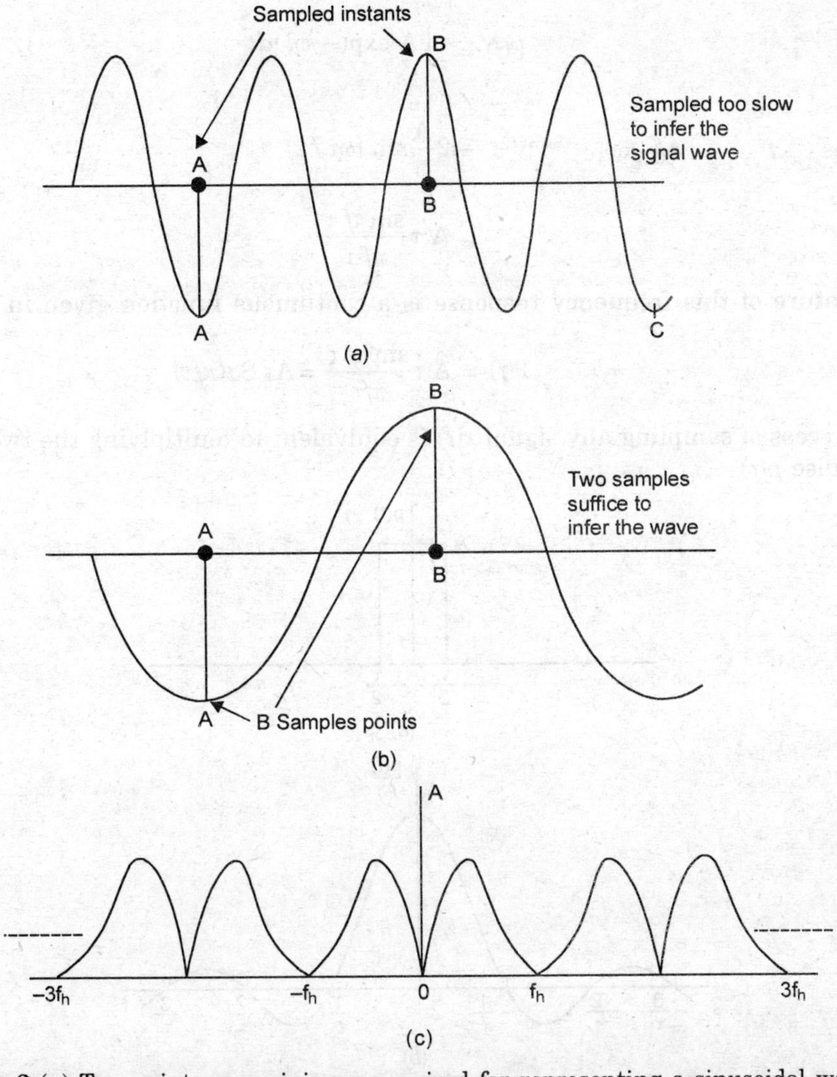

Fig. 2 (*a*) Two points are minimum required for representing a sinusoidal wave. Here the points are insufficient. (*b*) The same two points just show a wave. (*c*) Spectrum repeats after folding frequency either side till infinity.

If aliasing due to higher than folding frequency is to be eliminated, one has to remove all such frequencies above f_h even at the input. Such an input filter is called anti-aliasing filter.

FOURIER TRANSFORM OF A SAMPLING PULSE

Let us consider how a typical signal is sampled; samples are of time τ, spaced at intervals of T, the sampling time. When we study the transform of a sampled signal, we need to know the transform of the sampling pulse first. A typical sampling pulse in DSP will be a pulse, non-periodic by itself, and having a width of time τ. (Fig. 3)

It is noted that the sample pulse is of duration extending over the limits $-\tau/2$ to $+\tau/2$. Hence, it is convenient to write

$$P(f) = \int_{-\tau/2}^{\tau/2} A \exp(-j\omega t)dt \qquad ...(6.4)$$

which gives

$$A/(-j\omega)[e^{-j\omega t}]_{-\tau/2}^{\tau/2} = 2\frac{A}{\omega}\sin\{\omega\tau/2\} \qquad ...(6.5)$$

$$= A\,\tau\frac{\sin \pi f \tau}{\pi f \tau} \qquad ...(6.6)$$

The nature of this frequency response is a continuous function given in Fig. 3.

$$P(f) = A\,\tau\frac{\sin \pi f \tau}{\pi f \tau} = A\tau\, Sa(\pi f\tau) \qquad ...(6.7)$$

The process of sampling any signal $x(t)$ is equivalent to multiplying the two signals $x(t)$ with the pulse $p(t)$.

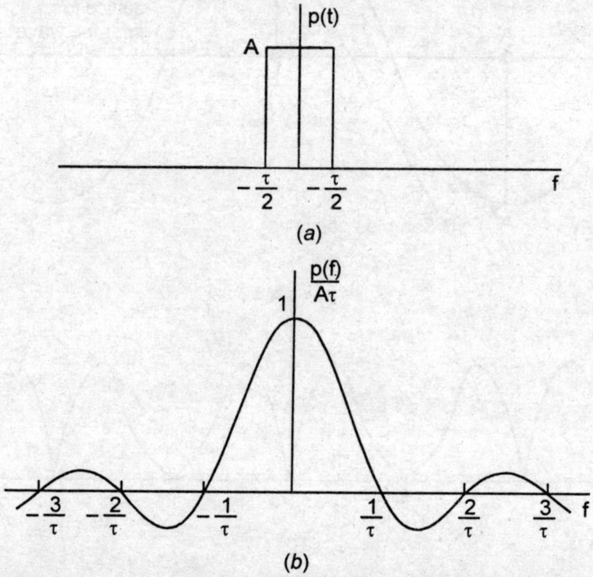

Fig. 3 Fourier spectrum of a pulse (non-periodic), width τ (*a*) pulse (*b*) spectrum.

Let $x^*(t)$ represent the sampled data signal of the continuous time signal $x(t)$.

The sampled data signal is shown in Fig. 4. When we find its frequency transform, we get

$$x^*(t) = x(t)\,p(t) \qquad ...(6.8)$$

where $x^*(t)$ represents the sampled data signal and $x(t)$ is the original continuous signal and $p(t)$ is the pulse that samples it.

$$p(t) = \sum_{-\infty}^{\infty} c_m \exp(jm\omega_s t) \qquad ...(6.9)$$

where $\omega_s = 2\pi f_s = 2\pi/T$

$$x^*(t) = \sum_{-\infty}^{\infty} c_m\, x(t)\exp(jm\omega_s.t) \qquad ...(6.10)$$

The spectrum can be found out by taking the Fourier transforms of both sides. Each term of the series on the right may be transformed by the formula:

If X(f) is the transform of $x(t)$, the transform of $x(t)e^{j\alpha t}$ is X(f – α).

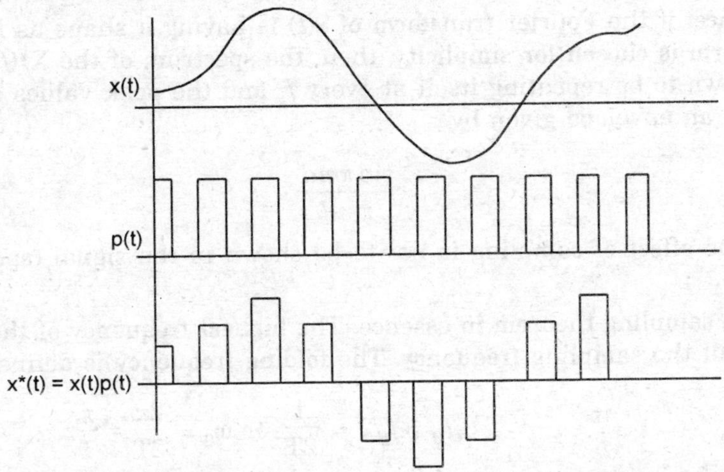

Fig. 4 The sampled data signal $x^*(t)$ is got by sampling $x(t)$.

The spectrum is determined by the Fourier transform.

$$X^*(f) = \sum_{-\infty}^{\infty} c_m X(f - mf_s) \qquad \text{...(6.11)}$$

Typical sketches of X(f) and X*(f) appear as in Fig. 5.

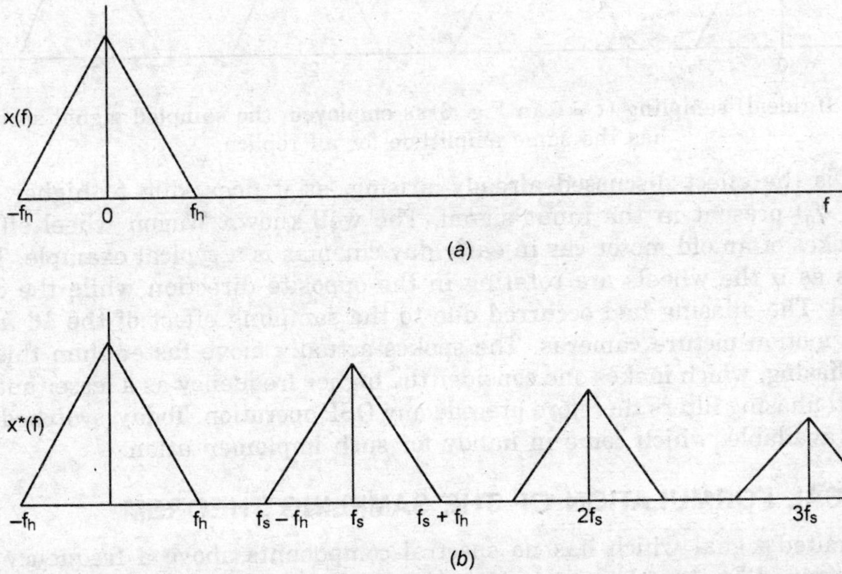

Fig. 5 (*a*) Frequency spectrum of $x(t)$ (*b*) of $x^*(t)$. Note the amplitude drops for the successive replica of spectra at f_s, $2f_s$...

This figure shows only a small section on the negative frequency side. The spectrum repeats itself at every f_s. The amplitudes of the succeeding spectral replicas due to sampling effect diminish progressively. When $f_s > 2f_h$, the signal and the replica due to the sampling effect overlap, giving rise to a misinterpretation.

For instance, if the Fourier transform of $x(t)$ is having a shape as in Fig. 6.5, where a triangular form is chosen for simplicity, then, the spectrum of the $X^*(f)$ of sampled data function is shown to be repeating itself at every f_s and the peak values of the subsequent spectra follows an envelope given by

$$\frac{\sin \pi md}{\pi md}$$

Thus, if the effect of sampling is not to be shown in the signal (spectra), f_s must be $> 2f_h$.

This is the sampling theorem in essence. The highest frequency of the signal f_h should be less than half the sampling frequency. The folding frequency is defined as

$$f_0 = f_s/2 = \frac{1}{2T} \text{ or } \omega_0 = \frac{2\pi}{2T} = \frac{\pi}{T}$$

If the pulse of sampling is considered to be an impulse, then the Fourier transform of the sampled data signal shows no diminution of the amplitudes of the successive replica of spectra. (Fig. 6)

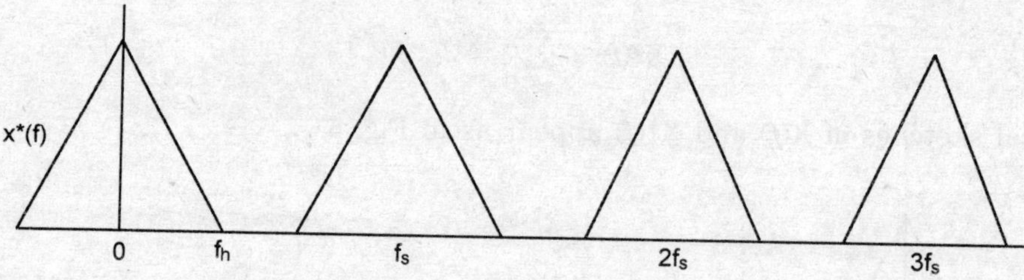

Fig. 6 If (ideal) sampling ($\tau = 0$ in Fig. 6) is employed, the sampled signal spectra has the same amplitude for all replica.

Aliasing is the effect discussed already, arising, as it does, due to higher frequency components ($>f_0$) present in the input signal. The well-known Wagon wheel effect which shows the spokes of an old motor car in early day cinemas is a typical example. There, the show appears as if the wheels are rotating in the opposite direction while the car moves really forward. The aliasing had occurred due to the sampling effect of the 16 frames per second of the motion picture cameras. The spokes actually move faster than this, thereby introducing aliasing, which makes one consider the higher frequency as a lesser and opposite frequency. Anti-aliasing filters therefore precede any DSP operation. Today, switched capacitor filter ICs are available, which come in handy for such implementation.

MATHEMATICAL FORMULATION OF THE SAMPLING THEOREM

A bandlimited signal which has no spectral components above a frequency f_m Hz is uniquely determined by its values at uniform intervals less than $1/2f_m$ second apart.

We use frequency convolution to prove it.

Let $f(t)$ be a bandlimited signal (Fig. 7(a)). It has no frequencies above f_m.

So, F(ω) is zero for all $\omega > \omega_m$. ($\omega_m = 2\pi f_m$). (Fig. 7(*b*))

Now multiply $f(t)$ by a *periodic* delta function or impulse function $\delta_T(t)$. (Fig. 7(*c*).)

The frequency response of $\delta_T(t)$ is also impulse, with separation $\omega_0 = 2\pi/t$. (Fig. 7(*d*))

The product $f(t)\ \delta_T(t)$ is a sequence of varying amplitude as in Fig. 6.7(*e*). This sampled function is denoted as $f_s(t)$.

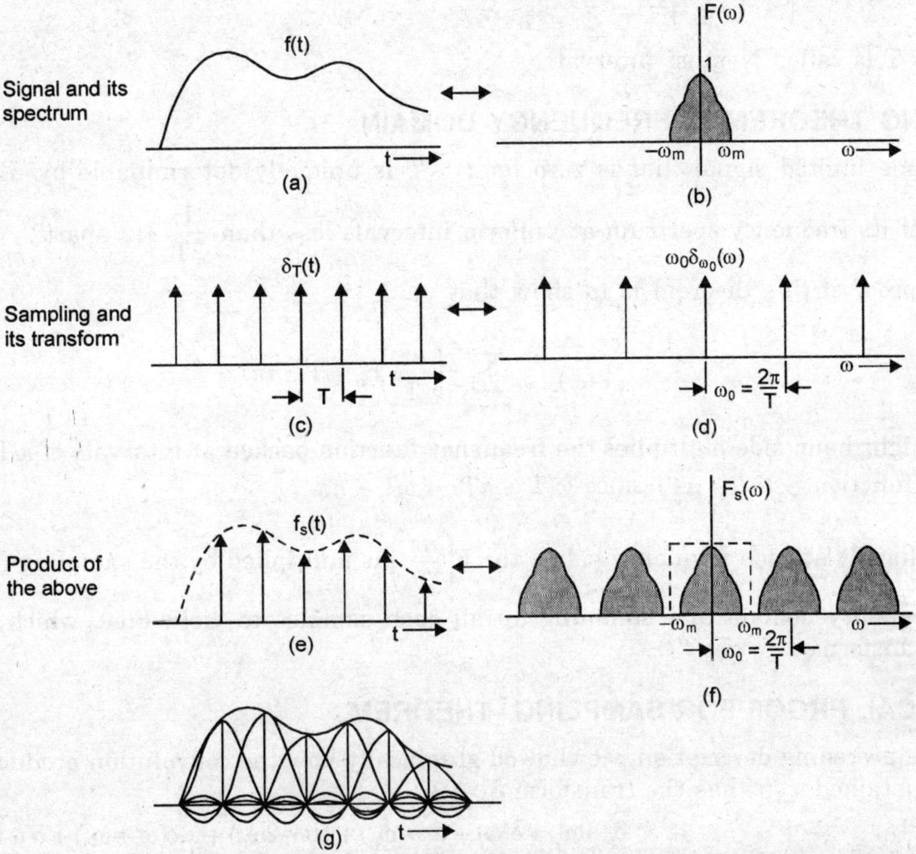

Fig. 7 See text. The shape of two F(ω) spectrum is only representative. (In Fig. 6 it was shown triangular)

The F.T. of $f_s(t)$ will be the convolution of the Fourier transforms of Fig. 7 (*b*) and 7 (*d*).

Since the impulses are spaced at a distance $\omega_0 = \dfrac{2\pi}{T}$, the convolution gives F(ω) repeating itself as in (*f*).

Call this function $F_s(\omega)$, so that

$$f_s(t) \leftrightarrow F_s(\omega)$$

If the $f(\omega)$ repetitions should not overlap (touch and cross),

$$\frac{2\pi}{T} \geq (2\pi f_m) \times 2$$

The above is by a look at Fig. 7 (*f*).

$$\therefore \qquad T \le \frac{1}{2f_m} \qquad ...(6.13)$$

This means the sampling time must be faster (less) than the half the maximum frequency's period $\left[\frac{1}{2}\times\left(\frac{1}{f_m}\right)\right]$.

Thus T is called Nyquist interval.

SAMPLING THEOREM IN FREQUENCY DOMAIN

"A time limited signal that is zero for $t > T$ is uniquely determinable by using the samples of its frequency spectrum at uniform intervals less than $\frac{1}{2T}$ Hz apart".

The proof of this theorem is to show that

$$F(\omega) \approx \sum_{n=-\infty}^{\infty} F\left(\frac{n\pi}{T}\right) S_a(\omega T - n\pi) \qquad ...(6.14)$$

The right hand side multiplies the frequency function packed at intervals of π/T by the sampling function $S_a(\omega\overline{T - nT})$ since $\omega(T - nT) = \omega T - n\pi$.

Graphically at each frequency value, the $F\left(\frac{n\pi}{T}\right)$ is multiplied by the sampling function in the frequency domain and summing up all such samples to yield $F(\omega)$, which can be inverse transformed to get $f(t)$.

ANALYTICAL PROOF FOR SAMPLING THEOREM

In the preceding description, we showed graphically how the convolution product of $f(t)$ $\delta_T(t)$ was obtained. $\delta_T(t)$ has the transform $\delta_{\omega_0}(\omega)$.

$$\delta_{\omega_0}(\omega) = \delta(\omega) + \delta(\omega-\omega_0) + \delta(\omega-2\omega_0) + ...\delta(\omega-n\omega_0) + \delta(\omega+\omega_0) + ... + \delta(\omega + n\omega_0)$$

$$= \sum_{n=-\infty}^{\infty} \delta(\omega - n\omega_0)$$

Since the sampled signal $f_s(t)$ is got by the inverse of:

$$f_s(t) \leftrightarrow \frac{1}{T}[F(\omega)\times\delta_{\omega 0}(\omega)]$$

The transform of $f_s(t)$:

$$F_s(\omega) = \frac{1}{T}[F(\omega) * \delta_\omega(\omega)] = \frac{1}{T}[F(\omega) * \sum_{n=-\infty}^{\infty} \delta(\omega - n\omega_0)]$$

$$= \frac{1}{T}\sum_{-\infty}^{\infty} F(\omega) * \delta(\omega - n\omega_0)$$

Since $\phi(t) * \delta(t - T) = \phi(t - T)$

$$F_s(\omega) = \frac{1}{T}\sum_{n=-\infty}^{+\infty} F(\omega - n\omega_0)$$

The right hand side means the function F(ω) repeatedly added at every ω_0 :– ω_0, $2\omega_0$..., $n\omega_0$ etc.

This is exactly the result shown in Fig. 7 (*g*).

It must be noted that F(ω) will repeat periodically without overlap as long as $\omega_0 > 2\omega_{max}$ or

$$\frac{2\pi}{T} \geq 2(2\pi f_{max})$$

or
$$T \leq \frac{1}{2f_{max}}$$

Therefore, as long as we sample *f*(*t*) at regular intervals less than $1/(2 f_{max})$ seconds apart, $F_s(\omega)$, the transform of *f*(*t*), will be a periodic replica of F(ω) and hence contain all that information in *f*(*t*), the signal.

RECOVERING THE SIGNAL FROM ITS SAMPLE

Consider sampling at intervals of T.

The maximum frequency f_{max} which can be retrieved is:

$$\frac{1}{T} = f_s$$

$$f_{max} = \frac{f_s}{2} = \frac{1}{2T};\ \omega_0 = \omega_{max} \times 2 = 2f_{max} \times 2\pi = \frac{2\pi}{T} \quad ...(6.15a)$$

Since the sampling process repeats the frequencies of the original signal every ω_0(look at Fig. (*d*)) ($\omega_0 = 2/\pi T$)

$$F_s(\omega) = \frac{1}{T}\sum_{n=-\infty}^{\infty} F(\omega - 2\omega_m n) \quad ...(6.15b)$$

where $F_s(\omega)$ is the spectrum of the sampled signal.

We use a low pass filter. It filters the sampling pulses.

Let us use a gate function G.

Its frequency response is G(ω).

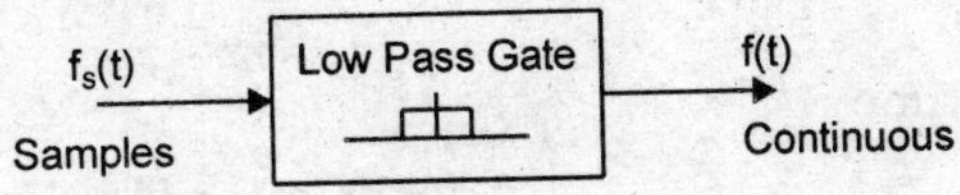

Fig. 8 Reconstruction using low pass filter.

then
$$F(\omega) = F_s(\omega)\, G(\omega).\, T \quad ...(6.15c)$$

The time function corresponding to G(ω) is $\frac{\omega_c}{\pi} S_a(\omega_c t)$

where ω_c is cut-off frequency of low pass filter.

Now do a convolution of eqn. 6.15 (c) above.

F(ω) becomes $f(t)$

$$F_s(\omega) \Rightarrow f(n)\,\delta(t - nT)$$

$$f(t) = \sum_n f(n)\delta(t - nT) * S_a(\omega_c t)$$

$$= \sum_n f(n) S_a(\omega_c\, \overline{t - nT})$$

$$= \sum_n f(n) S_a\{\omega_c t - n\pi\} \quad \text{...(6.16)}$$

Each of the sampling function $S_a(\omega_c t - n\pi)$ are shown in Fig. 7 (*g*) added together, whose envelope gives the reconstructed signal. The above is called an "interpolation formula".

Example. A Doppler echo signal is having a maximum blood blow frequency up to 10 kHz and one is required to take a spectrum of the same by sampling. Computer used to do the Fourier transform takes 1 ms for 1024 samples (using FFT → Fast Fourier Transform Algorithm).

Find the Analog-Digital Converter (ADC) speed.

Find the time over which the samples are to be taken.

Find the frequency resolution.

Solution. Maximum frequency present is signal = 10000 Hz

Minimum sampling rate will be 2 × 10000 or 20000 samples/sec.

This is the ADC speed.

$$\text{Time taken for 1024 samples} = \frac{1024}{20000} = 50 \text{ ms}$$

This is the time over which samples are to be taken per spectrum.

$$\text{Time between spectrum samples} \Rightarrow \frac{1}{50\,\text{ms}} = 20\,\text{Hz}$$

This means a frequency resolution = 20 Hz

Sampling with a flat top—zero order hold

The zero-order hold takes samples of the signal in a stair-case (flat top) form.

The result is a set of rectangular pulses of varying amplitude, with the signal. $x(t)$ is a staircase function.

T second is the hold time.

$$h(t) = \begin{cases} 1 \text{ for } 0 < t \le T \\ 0 \text{ for other values of } t \end{cases}$$

$$H(f) = \int_{-\infty}^{\infty} h(t) e^{-2j\pi f t}\, dt$$

$$= \int_0^T e^{-j2\pi ft}dt = T \frac{\sin \pi fT}{\pi fT} \exp(-j\pi fT)$$

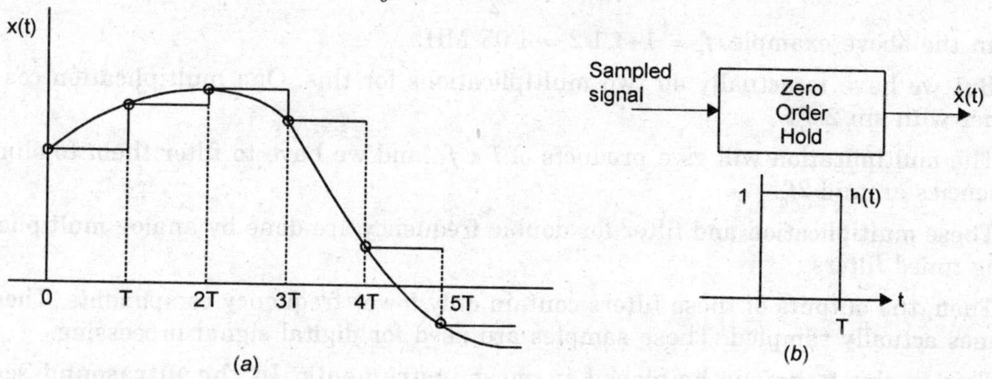

Fig. 9 (*a*) Sampling and zero order holding an analog signal (*b*) Impure response of Z.O.H.

A plot of this function is shown. Such a frequency response is not having a sharp cut-off. In this process, it could introduce undesirable aliasing frequencies beyond $f_s/2$.

So, after a zero order hold, we normally use a further low pass filter which reduces all frequencies above $f_s/2$ to very low levels.

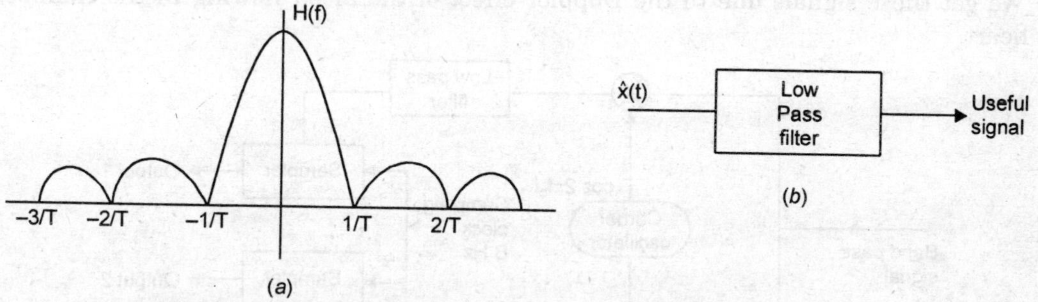

Fig. 10 (*a*) Frequency response of Z.O.H. (*b*) Use of further L.P.F.

Otherwise, another method is to do oversampling and decimation (not in our scope).

BANDPASS SAMPLING

Suppose we have a signal which is a bandpass signal. For example, a signal could have a frequency range 1 MHz to 1.1 MHz, over the band 1–1.1 MHz. The actual signal covers only 0.1 MHz or 100 kHz.

What should be our sampling rate for the signal?

Is it 2.2 MHz or 200 kHz?

The first answer is correct if we sampled the signal as such. That is we take samples of the original signal, which is a bandpass signal.

But we would like to sample at the lower rate of 200 kHz. How could we do it?

Here is where the property of modulation can be useful.

Take the signal; multiply it by a carrier frequency between the upper and lower frequency edges of the signal.

If B_1 and B_2 are these edge frequencies, we can chose a carrier frequency

$$f_c = \frac{B_1 + B_2}{2}$$

In the above example, f_c = 1+1.1/2 = 1.05 MHz.

But we have to actually do two multiplications for this. One multiplication cos $2\pi f_c t$, another with sin $2\pi f_c t$.

The multiplication will give products of $f \pm f_c$, and we have to filter them to eliminate frequencies around $2f_c$.

These multiplication and filter for double frequency are done by analog multiplier and analog tuned filters.

Then, the outputs of these filters contain only lower frequency components. These are the ones actually sampled. These samples are used for digital signal processing.

This is the technique employed in most instruments. In the ultrasound scanner instrument (for medical diagnosis of heart, abdomen etc.), the signal from the ultrasound probe contains the frequencies over and above the crystal frequency of the ultrasound probe. A 2.5 MHz probe, when used to scan the heart chamber, will receive echo signal which would contain frequencies near 2.5 MHz, above and below it by around just 10 kHz.

We get these signals due to the Doppler effect of the blood flowing in the chamber of the heart.

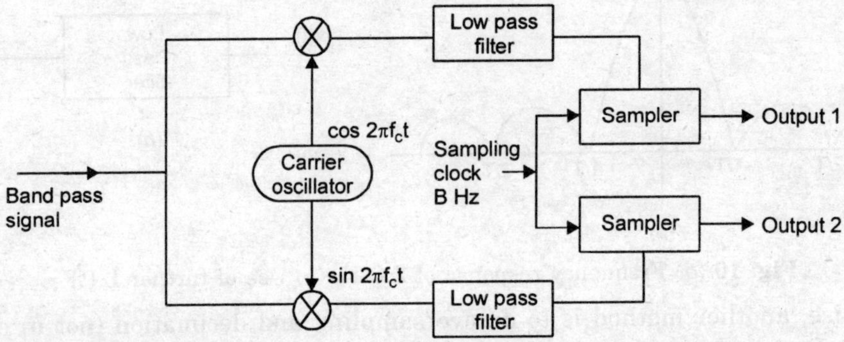

Fig. 11 Quadrature modulation for band pass to low pass signal conversion
The resulting filtered outputs 1 and 2 have a bandwidth B/2 where $B = B_2 - B_1$

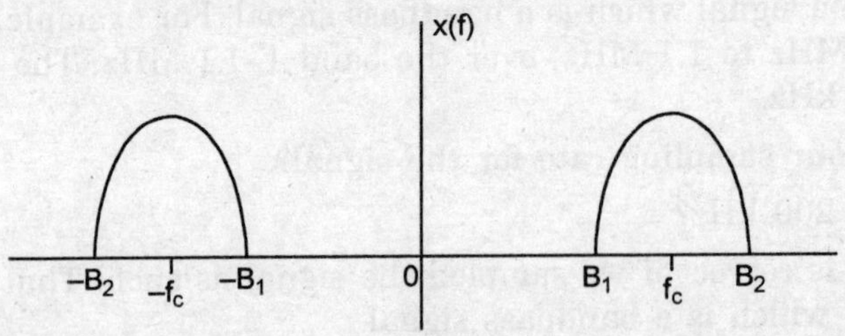

Fig. 12 Band pass signal with frequency components from B_1 to B_2.

Hence the resulting low pass output signal can be represented uniquely by samples taken at the rate of B samples/sec for each of the quadrature components.

Thus the sampling can be done on each of the low pass filter outputs at B samples/sec.

Thus the net rate is 2B samples per second.

But this 2B rate applies only when f_c is such that f_c + B/2 is a multiple of B.

$$f_c + B/2 = kB$$

$$2f_c = (2k - 1)B \text{ or } f_c = (2k - 1)B/2.$$

But it f_c = B/2 or f_c + B/2 $\simeq$ 2B, then the rate of sampling of 2B is not enough; it should be 4B.

In general the sampling rate of the low pass filtered quadrature signal should lie between 2B and 4B.

EXERCISES

1. A signal contains maximum frequencies upto 10 kHz. If it is sampled at 20 kHz, will it aliase in the digital processing further?

   If the frequency of the signal stretches upto 11 kHz, what will be the effect? How will the 11 kHz signal appear to the DSP? [No (as 1 kHz]

2. A Doppler shift signal from an instrument contains upto 20 kHz frequencies.

   We want to sample it and perform a Fourier transform using 1024 samples. This process takes 0.5 ms on the computer.

   We want to present the F.T. spectra one after another taking 1024 samples in each set.

   What is the minimum ADC speed?

   What will be the frequency resolution? [25 μs, 20 Hz]

3. Consider the band pass signal whose spectrum is shown.

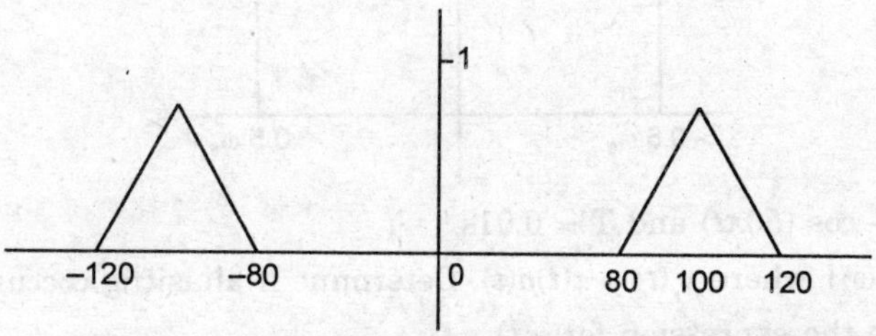

Fig. 13 Determine the min. sampling rate F_s to avoid aliasing.

4. Draw $|Xs(\omega)|$ for the following cases if

$$x_s(t) = x(t)\, p(t) \text{ with sampling period T.}$$

$$p(t) = \sum_{n=-\infty}^{\infty} \delta(t - nT)$$

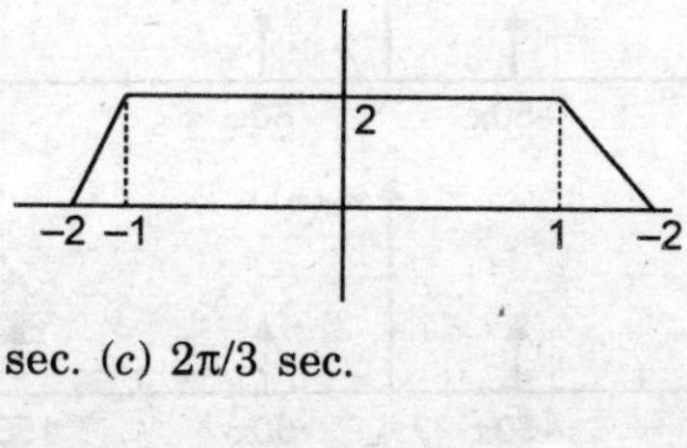

(*a*) T = π/4 sec. (*b*) π/2 sec. (*c*) 2π/3 sec.

Ans.

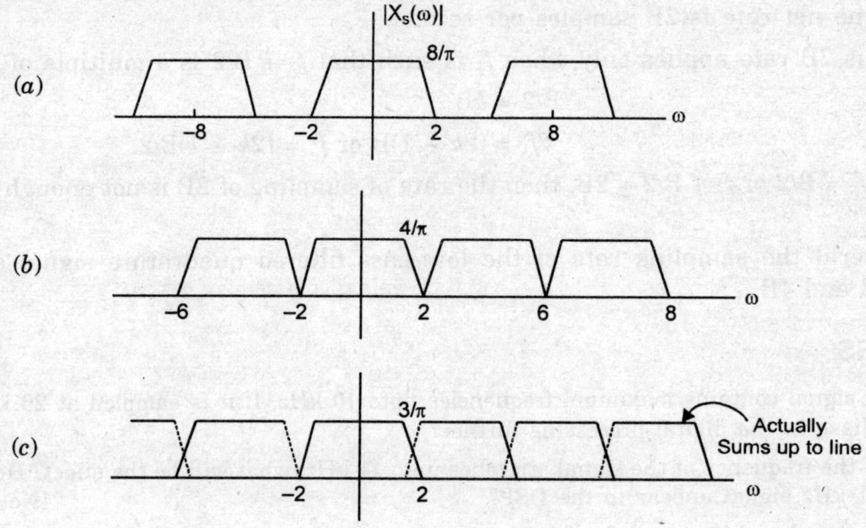

5. **Consider the following sampling and reconstruction.**

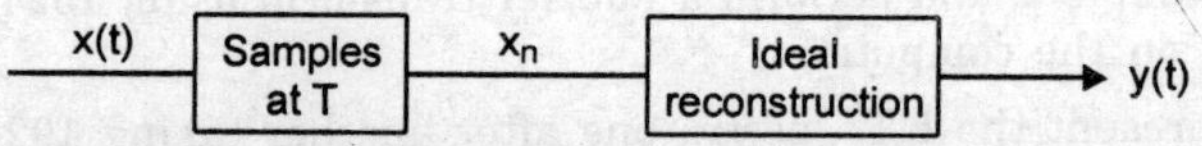

The output $y(t)$ of ideal reconstruction can be found by sending the sampled signal $x_s(t) = x(t)\,p(t)$ through a low pass filter.

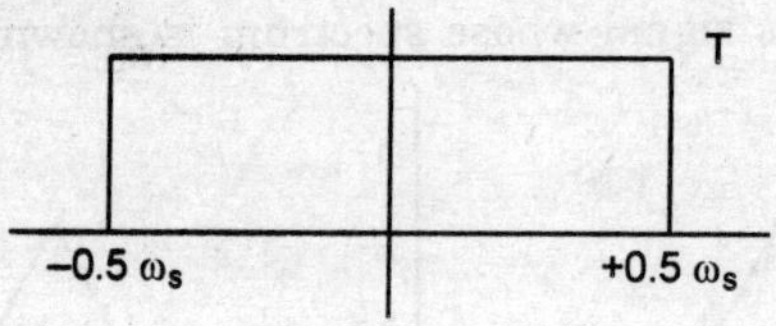

Let $x(t) = 2 + \cos(50\pi t)$ and T = 0.01s.

(*a*) Draw $|x_s(\omega)|$ where $x_s(t) = x(t)p(t)$. Determine if aliasitng coccurs.

(*b*) Determine the expression for $y(t)$

(*c*) Determine an expression for $x[n]$. The *n*th sample *x*.

Ans. (*a*) $x(t) = 2 + \cos(50\pi t)$, $T = 0.01$ sec, $\omega_s = \dfrac{2\pi}{T} = 200\pi$

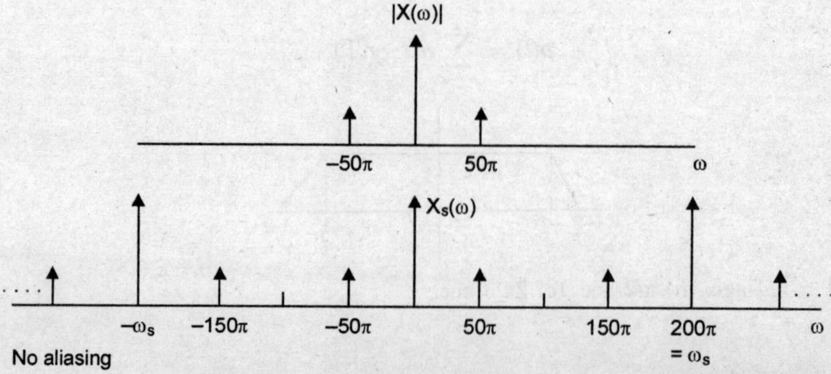

(*b*) After filtering (*i.e.*, reconstruction)

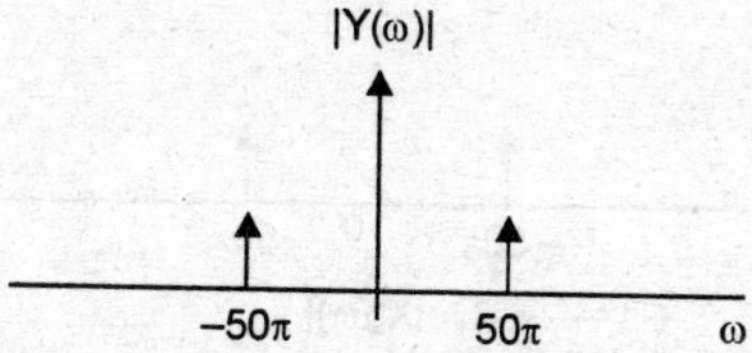

$$y(t) = 2 + \cos(50\pi t)$$

(*c*)

$$x[n] = 2 + \cos(50\pi n\ (0.01))$$

$$x[n] = 2 + \cos(0.5\pi n)$$

6. Repeat the previous problem for $x(t) = 2 + \cos(50\pi t)$ and T = 0.025s.

Ans. (*a*) $x(t) = 2\cos(50\pi t)$, T = 0.025 sec, $\omega_s = \dfrac{2\pi}{T} = 80\pi$

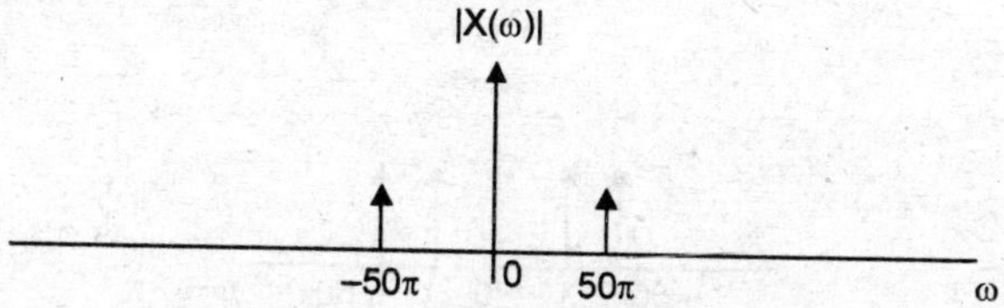

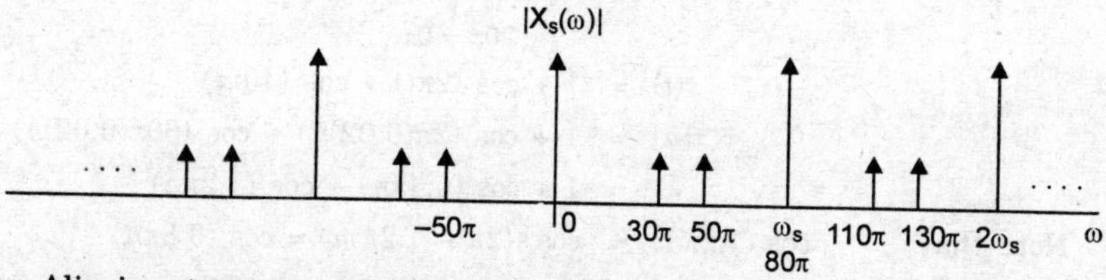

Aliasing, term at ω = 30π came from aliasing.

(*b*) After filtering (reconstruction)

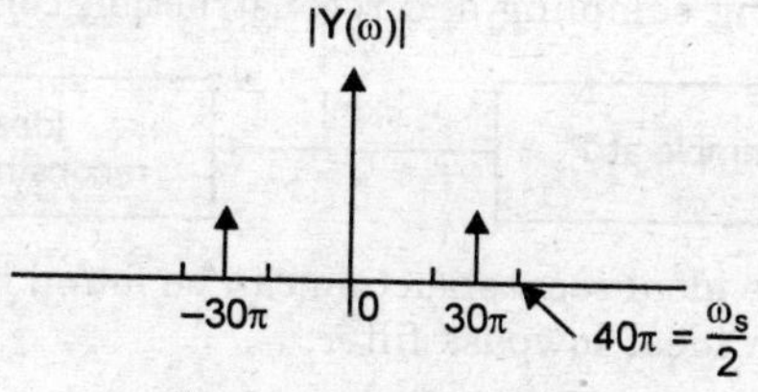

$$y(t) = 2 + \cos(30\pi t)$$

(c)

$$x[n] = 2 + \cos(50\pi(0.025)n) = 2 + \cos(1.25\pi n)$$

Note that $\cos(1.25\pi n) = \cos((2\pi - 1.25\pi)n) = \cos(0.75\pi n)$

$\Rightarrow$ $x[n] = 2 + \cos(0.75\pi n)$

Same as $y[n] = y[nT]$

7. Repeat the above problem for
$x(t) = 1 + \cos(20\pi t) + \cos(60\pi t)$ and T = 0.01s.

Ans. (*a*) $x(t) = 1 + \cos(20\pi t) + \cos(60\pi t)$, T = 0.01 sec, $\omega_s = \dfrac{2\pi}{T} = 100\pi$

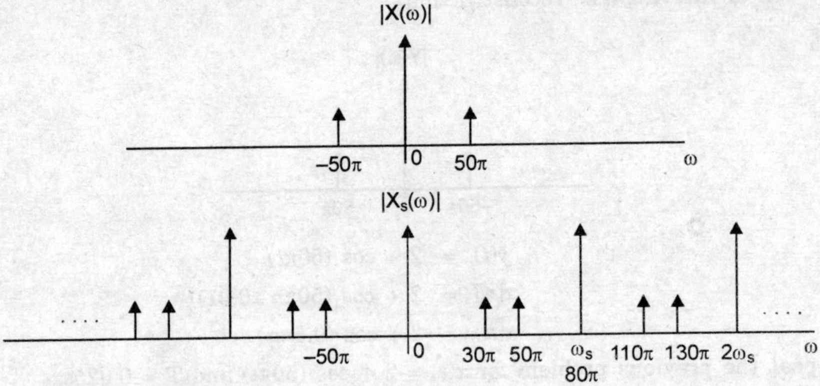

Aliasing occur at $\omega = 40\pi$

(*b*) After reconstruction

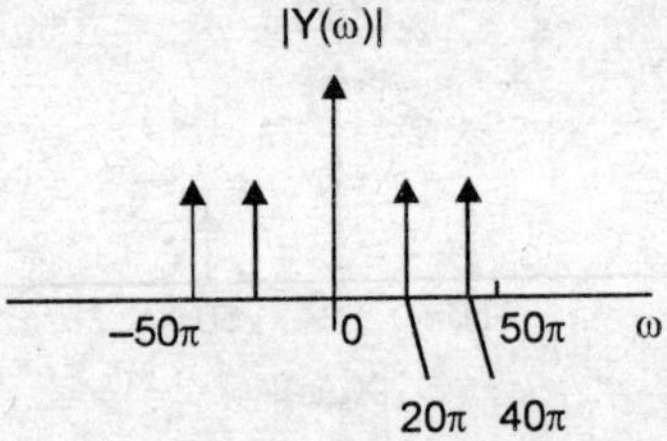

$$y(t) = 1 + \cos(2\pi t) + \cos(40\pi t)$$

(*c*)
$$x[n] = 1 + \cos(2\pi(0.02)n) + \cos(60\pi(0.02)n)$$
$$= 1 + \cos(0.4\pi n) + \cos(1.2\pi n)$$

Note that $\cos(1.2\pi n) = \cos((2\pi - 1.2\pi)n) = \cos(0.8\pi n)$

$$x[n] = 1 + \cos(0.4\pi n) + \cos(0.8\pi n)$$

Same as $y[n] = y(n\text{T})$

8. Consider the following sampling and reconstruction configuration:

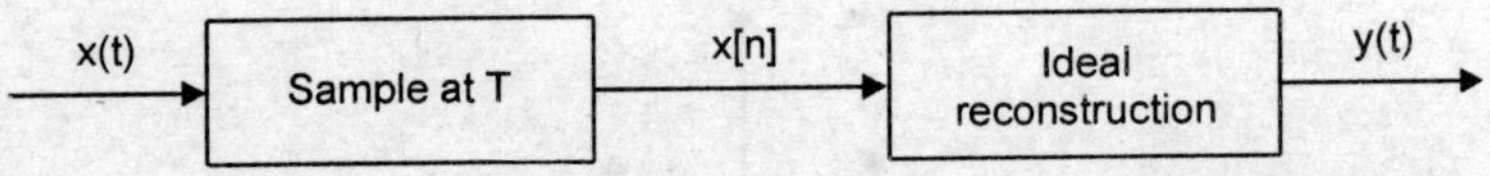

The output $y(t)$ of the ideal reconstruction can be found by sending the sampled signal $x_s(t) = x(t)p(t)$ through an ideal lowpass filter:

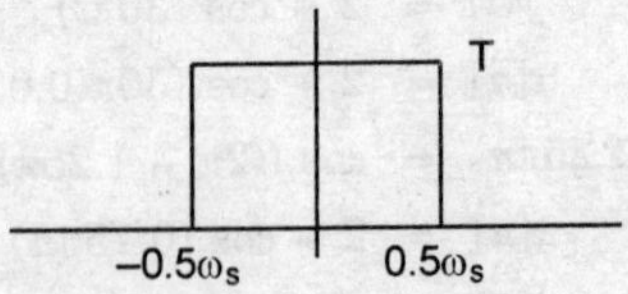

(*a*) Let $x(t) = 1 + \cos(15\pi t)$ and $\text{T} = 0.1$ sec. Draw $|\text{X}_s(\omega)|$ where $x_s(t) = x(t)p(t)$. Determine the epxression for $y(t)$.

(*b*) Let $\text{X}(\omega) = 1/(j\omega + 1)$ and $\text{T} = 1$ sec. Draw $|\text{X}_s(\omega)|$ where $x_s(t) = x(t)p(t)$. Does aliasing occur ? (Justify your answer.)

Ans. 8 (*a*)

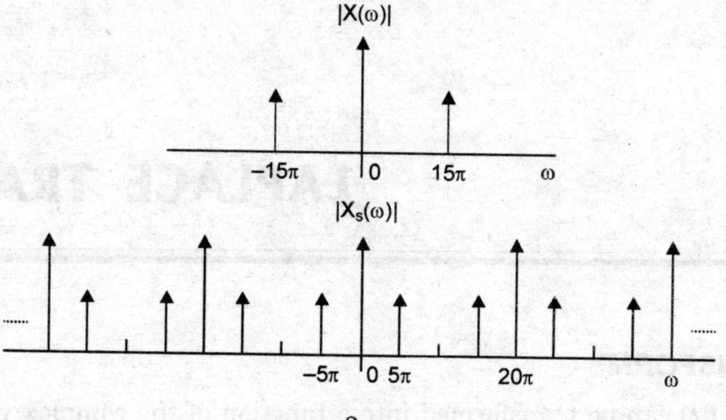

$$\omega_s = \frac{2\pi}{T} = 20\pi$$

(*b*)

$$y(t) = 1 + \cos(5\pi t)$$

$$\omega_s = 2\pi$$

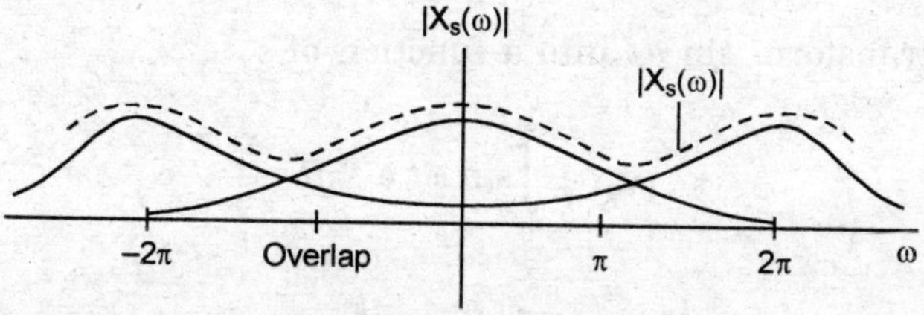

Signal is not band limited so there will be aliasing no matter how large the value of ω_s.

7

LAPLACE TRANSFORM

LAPLACE TRANSFORMS

Any function $f(t)$ can be transformed into a function of the complex variable $s = \sigma + j\omega$ by the Laplace transformation

$$F(s) = \int_0^\infty f(t)\, e^{-st} dt \qquad ...(7.1)$$

As an example transform sin ωt into a function of s.

$$F(s) = \int_0^\infty \sin \omega t\; e^{-st} dt$$

$$= \int \frac{e^{j\omega t} - e^{-j\omega t}}{2j} e^{-st} dt$$

$$= \frac{1}{2j}\left(\frac{1}{s - j\omega} - \frac{1}{s + j\omega}\right) = \frac{\omega}{s^2 + \omega^2}$$

Laplace transform of sin ωt is :

$$L \sin \omega t = \omega / (s^2 + \omega^2)$$

Similarly $\quad L \cos \omega t = s / (s^2 + \omega^2)$

For a unit step function, which is unity for $t > 0$,

$$\int_0^\infty 1e^{-st} dt = \frac{1}{s} \text{ or } L(1) = \frac{1}{s}.$$

For an exponential function e^{-at}.

$$\int_0^\infty e^{-at} e^{-st} dt = \int e^{-(s+a)t} dt = \frac{1}{s + a}$$

Thus a table of Laplace transforms may be written.

$f(t)$	F(s)
1 (step function)	$1/s$
A	A/s
e^{-at}	$1/(s + a)$
$\sin at$	$a/(s^2 + a^2)$
$\cos at$	$s/(s^2 + a^2)$
$e^{-at} \sin bt$	$b/[(s + a)^2 + b^2]$
$e^{-at} \sin bt$	$(s + a)/[(s + a)^2 + b^2]$
t	$1/s^2$
t^n	$n!/s^{n+1}$
te^{-at}	$1/(s + a)^2$
$(t - a)$	e^{-as}/s

SOME THEOREMS ON LAPLACE TRANSFORMS

1. Laplace transform $d/dt\ [f(t)]$ is

$$L\ f'(t) = s\ L\ f(t) - f(0)$$

$$L\ f'(t) = \int e^{-st}(df/dt)dt$$

$$= \left[e^{-st}f(t)\right]_0^\infty + \int se^{-st}f(t)dt$$

$$= sF(s) - f(0) \qquad ...(7.2)$$

since e^{-st} at $t = 0$ is 1 and $t \to \infty$, it is zero.

Likewise, for $f''(t)$ we get:

$$L f''(t) = s^2F(s) - sf(0) - f'(0) \qquad ...(7.3)$$

Relation (7.1) is useful in connection with an inductor, L Henries

$$V_L = L\ di/dt$$

The Laplace transform V_L is got as

$$V_L(s) = L\ s\ I(s) - Li(0) \qquad ...(7.4)$$

where V and I are Laplace transform of $v(t)$ and $i(t)$. (In this equation (7.4) L denotes inductance and not Laplace transform).

2. Laplace transform of $\int f(t)dt$

$$L\ f'(t) = sL\ f(t) - f(0)$$

put $\qquad f'(t) = \theta(t).$

Then $\qquad f(t) = \int \theta(t) = \theta^{-1}(t)$

$\therefore \qquad L\theta(t) = s\ L\theta^{-1}(t) - \theta^{-1}(0)$

$\therefore \qquad L\theta^{-1} = L\theta(t)/s + \theta^{-1}(0)/s$

or
$$\mathrm{L}f^{-1}(t) = \frac{\mathrm{L}f(t)}{s} + \frac{f^{-1}(0)}{s} \qquad \text{...(7.5)}$$

This is useful for the capacitor relation

$$v = 1/\mathrm{C} \int i\,dt$$

$$v(s) = \mathrm{I}(s)/\mathrm{C}s + v(0)/s \qquad \text{...(7.6)}$$

$v(0)$ is the initial voltage on the capacitor C.

3. Initial Value Theorem

$$\lim_{s\to\infty} s\,\mathrm{F}(s) = \lim_{t\to 0} f(t) \qquad \text{...(7.7)}$$

To prove this, consider:

$$\int_0^\infty e^{-st} f'(t)dt = s\mathrm{F}(s) - f(0) \qquad \text{...(7.7a)}$$

Put $s \to \infty$ in the left integral we get zero. So the right side gives the result.

4. Final Value Theorem

$$\lim_{s\to 0} s\,\mathrm{F}(s) = \lim_{t\to\infty} f(t) \qquad \text{...(7.8)}$$

Consider (7.7*a*) again and putting $s = 0$, we get the integral as

$$f(\infty) - f(0)$$

and the right side as

$$\lim_{s\to 0} s\,\mathrm{F}(s) - f(0).$$

Equating both, we get the relation (7.8).

5. Time Translation Theorem

Given $f(t)$, a function of time as shown in Fig. (*a*). We call the function of Fig. (*b*) as $f(t - a)$. The difference between $f(t)$ and $f(t - a)$ is that the latter starts from $t = a$ and is zero for $t < a$.

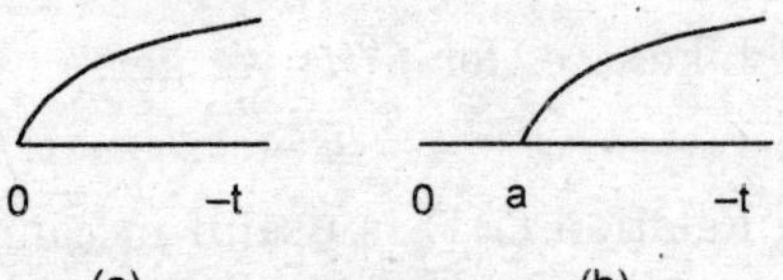

Fig. 1

Now consider the theorem:

$$\mathrm{L}\, f(t - a) = e^{-as}\, \mathrm{L}\, f(t).$$

$$\mathrm{L}\, f(t - a) = \int e^{-st} f(t - a)dt \qquad \text{...(7.9)}$$

Putting $t - a = u$ and integrating, we get the result.

6. Unit Impulse-transform of

The impulse functin $\delta(t)$ is defined as shown. Its value is unity at $t = 0$ and zero at all other t

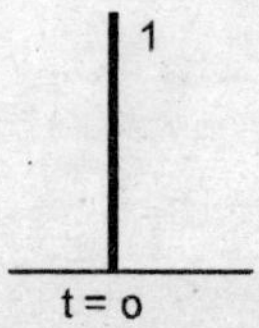

Fig. 2 Impulse function

$$\int e^{-st}\delta(t)dt = [e^{-s0} \times 1] = 1$$

The Laplace transform of $\delta(t)$ is unity. Hence, the inverse, transform of unity is $\delta(t)$, the impulse function.

Region of Convergence (ROC) of Laplace Transforms

We defined F(*s*) in (7.1) from *f*(*t*) simply stating that σ is the real part of *s*, a complex variable.

The condition that the integral

$$\underset{\substack{\in \to 0 \\ T \to \infty}}{\text{Lt}} \int_{\in}^{T} f(t) e^{-st} dt$$

converges in the limits shown, would depend on the value of the variable σ in *s*.

If *f*(*t*) is defined and single-valued for $t > 0$ and if $F(\sigma)$ is absolutely convergent for some real number σ_0, that is

$$F(\sigma) = \int_0^T |f(t)| e^{-\sigma_0 t} dt = \underset{\substack{T \to \infty \\ \in \to 0}}{\text{Lt}} \int_{\in}^{T} |f(t)| e^{-\sigma_0 t} < \infty \text{ for } 0 < \in < T$$

Then the function $f(t)$ has a Laplace transform for it.

The definition of the Inverse Laplace Transform as a Contour Integral

The inverse transform of F(*s*) is defined by

$$L^{-1} F(s) = f(t) = \frac{1}{2\pi j} \int_{c-j\infty}^{c+j\infty} F(s) e^{st} dt$$

where $c > \sigma_0$

The above integral is called Fourier-Mellin inversion integral.

Constraints on ROC for Various Signal Types

1. If *x*(*t*) is of finite duration and is absolutely integrable, then the ROC is the entire *s*-plane.

$$\int_{-\infty}^{\infty} \left| x(t) e^{-st} \right| dt = \int_{-\infty}^{\infty} \left| x(t) e^{-\sigma t} \right| dt < \infty$$

   depends only on $\sigma = \text{Re}\{s\}$

2. The ROC for a right sided signal is a right half plane (spl. case : 'causal signal').
3. The ROC for a left-sided signal is a left half plane.
4. If *x*(*t*) is two-sided, and if the line $\text{Re}\{s\} = \sigma_0$ is in the ROC, then the ROC consists of a strip in the *s*-plane that includes the line $\text{Re}(s) = \sigma_0$.
5. If the ROC of X(*s*) includes the $j\omega$ axis, then the Fourier transform of *x*(*t*) exists.

Example. Plot the ROC of the function

$$H(s) = \frac{s-1}{(s+1)(s-2)}$$

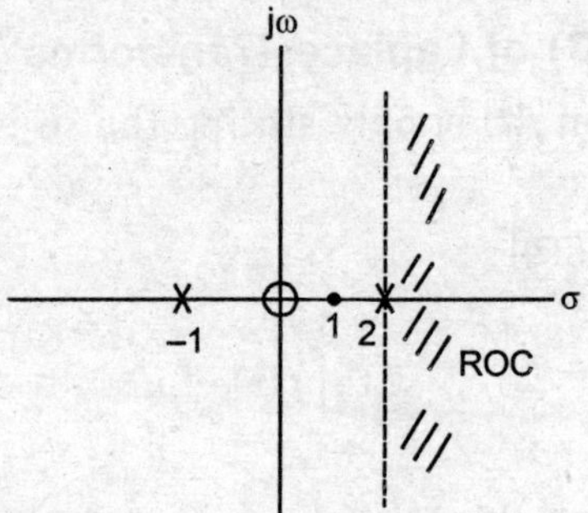

Fig. 3

The ROC comprises of the right half plane upto the vertical line $\sigma = 2$.

Example. If $X(s) = \dfrac{s+3}{(s+1)(s-2)}$

What are the possible ROCs if

(1) $x(t)$ is right sided

(2) $x(t)$ is left sided

(3) $x(t)$ extends for all time.

The possible ROCs are :

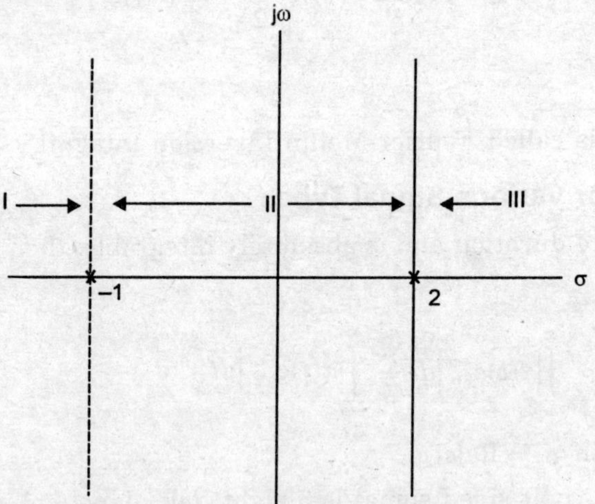

Fig. 4

	ROC	F.T. exists ?
$x(t)$ is right sided →	III	No
$x(t)$ is left sided →	I	No
$x(t)$ extends all time →	II	Yes.

The Fourier transform exists only if the region contains jw axis.

Example. If $x(t)$ is two sided and if the line $\text{Re}(s) = \sigma_0$ is in the ROC, draw the ROC. Explain.

The ROC then consists of a strip in the s-plane that includes the line $\text{Re}(s) = \sigma_0$.

We show three $x(t)$ functions.

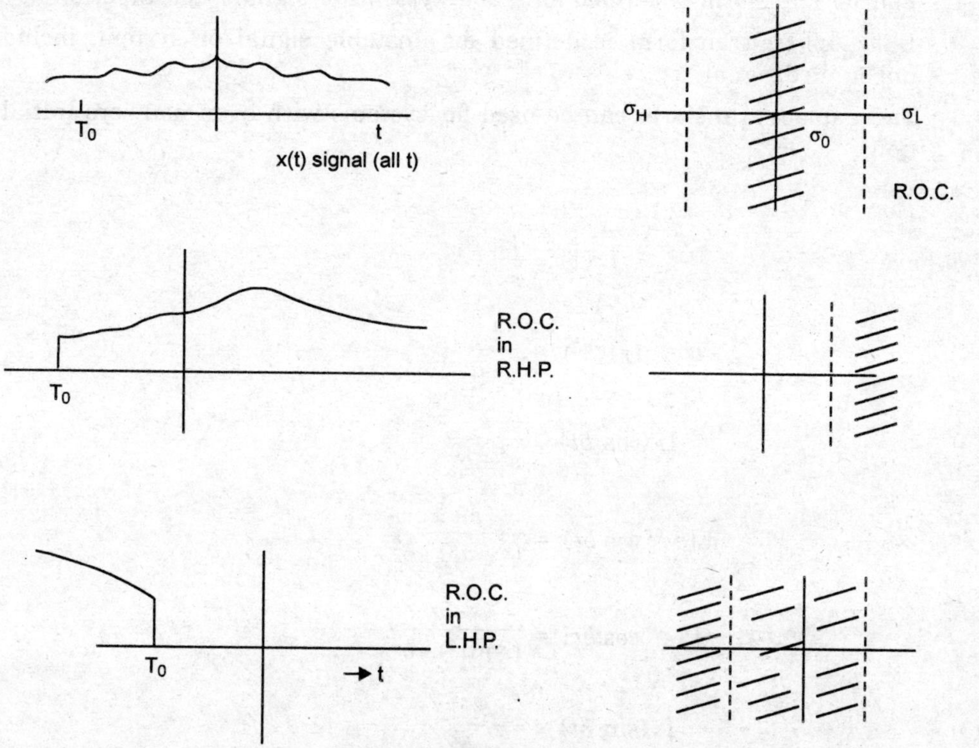

Fig. 5

Example. Find the Laplace transform of

$$x(t) = e^{bt}u(-t) + e^{-bt}u(t)$$

The first term has the transform and ROC as

$$-\frac{1}{s-b}, \ \text{Re}\{s\} < b$$

The second term has the transform and ROC as

$$\frac{1}{s+b}, \ \text{Re}\{s\} > -b$$

So, there will be overlap only if $b > 0$.

If $b < 0$, then there will be no overlap.

Hence there is no Laplace transform.

In the case $b > 0$,

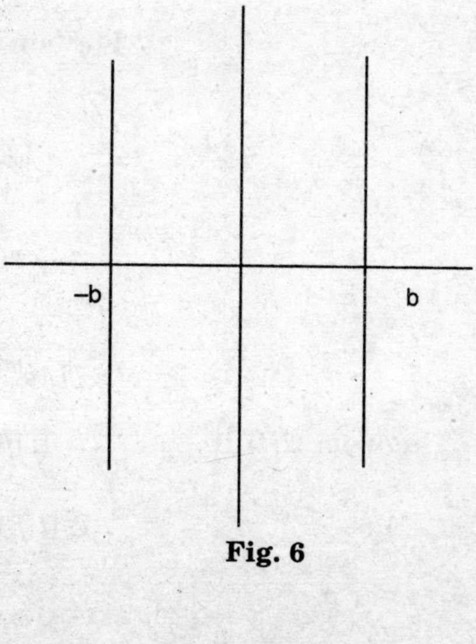

Fig. 6

$$X(s) = \frac{-2b}{s^2 - b^2}$$

Advantage of Laplace Transform Over Fourier Transform

1. Fourier transform is defined for stable systems or signals that taper off at infinity.

   But Laplace transform is defined for unstable signal or system, including an unbounded signal.

2. The Laplace transform can be used for systems with even non-zero initial conditions.

Examples.

We have (1) $$L(1) = \frac{1}{s}$$

$\therefore$ $$L(e^{-at}) = \frac{1}{s+a}$$

(2) $$L(\cos bt) = \frac{s}{s^2+b^2}$$

$\therefore$ $$L(e^{-at}\cos bt) = \frac{s+a}{(s+a^2)+b^2}$$

and $$L(e^{at}\cos bt) = \frac{s-a}{(s+a^2)+b^2}$$

(3) $$L(\sin bt) = \frac{b}{s^2+b^2}$$

$\therefore$ $$L(e^{-at}\sin bt) = \frac{b}{(s+a)^2+b^2}$$

and $$L(e^{at}\sin bt) = \frac{b}{(s-a)^2+b^2}$$

(4) $$L(t^n) = \frac{n!}{s^{n+1}} \text{ if } n \text{ is a positive integer.}$$

$\therefore$ $$L(e^{-at}t^n) = \frac{n!}{(s+a)^{n+1}}$$

$$L(e^{at}t^n) = \frac{n!}{(s-a)^{n+1}}$$

Theorem (*iii*) If $L[f(t)] = F(s)$, then

$$L[t\,f(t)] = -\frac{d}{ds}F(s)$$

$$F(s) = \int_0^\infty e^{-st} f(t)dt$$

$$\frac{d}{ds}F(s) = \frac{d}{ds}\int_0^\infty e^{-st} f(t)dt$$

$$= \int_0^\infty \frac{d}{ds}[e^{-st} f(t)]dt = \int_0^\infty -te^{-st} f(t)dt$$

$$= -\int_0^\infty e^{-st}\, t f(t)dt = -L[t f(t)]$$

$$\therefore \quad L[t f(t)] = -\frac{d}{ds}F(s)$$

Cor. $$L[t^n f(t)] = (-1)^n \frac{d^n}{ds^n}\{L[f(t)]\}$$

We have $$L[t^2 f(t)] = -\frac{d}{ds} L\,[tf(t)]$$

$$= -\frac{d}{ds}\left\{-\frac{d}{ds}L[f(t)]\right\}$$

$$= (-1)^2 \frac{d^2}{ds^2}L[f(t)].$$

Continuing the process, we get the result.

This result can also be written as follows :

If $$L[f(t)] = F(s), \text{ then}$$

$$F'(s) = L[-t f(t)]$$

$$F''(s) = L[(-t)^2 f(t)]$$

$$F^n(s) = L[(-t)^n f(t)].$$

Partial Fraction Expansions

This is necessary when we solve problems using Laplace transform.

Splitting up a ratio of large polynomials into a sum of ratios of small polynomials can be a useful tool, especially for many problems involving Laplace-like transforms. This technique is known as partial fraction expansion. Here's an example of one ratio being split into a sum of three simpler ratios :

$$\frac{8x^2 + 3x - 21}{x^3 - 7x - 6} = \frac{1}{x+2} + \frac{3}{x-3} + \frac{4}{x+1} \qquad ...(1)$$

There are several methods for expanding a rational function via partial fractions. These include the method of clearing fractions, the Heaviside "cover-up" method, and different combinations of these two. For many cases, the Heaviside "cover-up" method is the easiest, and is therefore the method that we will introduce here. For a more complete discussion, see *Signal Processing and Linear Systems* by B. P. Lathi, Berkeley-Cambridge Press, 1998, pp. 24-33. Some of the material below is based upon this book.

No Repeated Roots

Let's say we have a proper function $G(x) = \frac{N(x)}{D(x)}$ (by proper we mean that the degree m of the numerator $N(x)$ is less than the degree p of denominator $D(x)$. In this section we assume that there are no repeated roots of the polynomial $D(x)$.

The first step is to factor the denominator $D(x)$:

$$G(x) = \frac{N(x)}{(x-a_1)(x-a_2)\dots(x-a_p)} \qquad \dots(2)$$

where $a_1 \dots a_p$ are the roots of $D(x)$. We can then rewrite $G(x)$ as a sum of partial fractions :

$$G(x) = \frac{\alpha_1}{x-a_1} + \frac{\alpha_2}{x-a_2} + \dots + \frac{\alpha_p}{x-a_p} \qquad \dots(3)$$

where $a_1 \dots a_p$ are constants. Now, to complete the process, we must determine the values of these α coefficients. Let's look at how to find α_1. If we multiply both sides of the equation of $G(x)$ as a sum of partial fractions (#EQ3) by $x - a_1$ and then let $x = a_1$, all of the terms on the right-hand side will go to zero except for α_1. Therefore, we will be left over with :

$$\alpha_1 = (x-a_1)G(x)\Big|_{x=a_1} \qquad \dots(4)$$

We can easily generalize this to a solution for any one of the unknown coefficients :

$$\alpha_r = (x-a_r)G(x)\Big|_{x=a_r} \qquad \dots(5)$$

This method is called the "cover-up" method because multiplying both sides by $x - a_r$ can be thought of as simply using one's finger to cover up this term in the denominator of $G(x)$. With a finger over the term that would be canceled by the multiplication, you can plug in the value $x = a_r$ and find the solution for α_r.

Example 1. In this example, we'll work through the partial fraction expansion of the ratio of polynomials introduced above. Before doing a partial fraction expansion, we must make sure that the ratio we are expanding is proper. If it is not, we should do long division to turn it into the sum of a proper fraction and polynomial. Once this is done, the first step is to factor the denominator of the function :

$$\frac{8x^2+3x-21}{x^3-7x-6} = \frac{8x^2+3x-21}{(x+2)(x-3)(x+1)} \qquad \dots(6)$$

Now, we set this factored function equal to a sum of smaller fractions, each of which has one of the factored terms for a denominator.

$$\frac{8x^2+3x-21}{(x+2)(x-3)(x+1)} = \frac{\alpha_1}{x+2}+\frac{\alpha_2}{x-3}+\frac{\alpha_3}{x+1} \quad ...(7)$$

To find the alpha terms, we just cover up the corresponding denominator terms in G(*x*) and plug in the root associated with the alpha :

$$\alpha_1 = (x+2)G(x)\Big|_{x=-2}$$

$$= \frac{8x^2+3x-21}{(x-3)(x+1)}\Bigg|_{x=-2} \quad ...(8)$$

$$= 1$$

$$\alpha_2 = (x-3)G(x)\Big|_{x=3} = \frac{8x^2+3x-21}{(x+2)(x+1)}\Bigg|_{x=3} \quad ...(9)$$

$$= 3$$

$$\alpha_3 = (x+3)G(x)\Big|_{x=-1} = \frac{8x^2+3x-21}{(x+2)(x-3)}\Bigg|_{x=-1} \quad ...(10)$$

$$= 4$$

We now have our completed partial fraction expansion :

$$\frac{8x^2+3x-21}{(x+2)(x-3)(x+1)} = \frac{1}{x+2}+\frac{3}{x-3}+\frac{4}{x+1} \quad ...(11)$$

Repeated Roots

When the function G(*x*) has a repeated root in its denominator, as in

$$G(x) = \frac{N(x)}{(x-b)^r(x-a_1)(x-a_2)\ldots(x-a_j)} \quad ...(12)$$

Somewhat more special care must be taken to find the partial fraction expansion. The non-repeated terms are expanded as before, but for the repeated root, an extra fraction is added for each instance of the repeated root :

$$G(x) = \frac{\beta_0}{(x-b)^r}+\frac{\beta_1}{(x-b)^{r-1}}+\ldots+\frac{\beta_{r-1}}{x-b}+\frac{\alpha_1}{x-a_1}+\frac{\alpha_2}{x-a_2}+\ldots+\frac{\alpha_j}{x-a_j} \quad ...(13)$$

All of the alpha consants can be found using the non-repeated roots method above. Finding the beta coefficients (which are due to the repeated root) has the same Heaviside feel to it, except that this time we will add a twist by using the derivative to eliminate some unwanted terms.

Starting off directly with the cover-up method, we can find β_0. By multiplying both sides by $(x-b)^r$, we'll get :

$$(x-b)^r G(x) = \beta_0 + \beta_1(x-b) + \ldots + \beta_{r-1}(x-b)^{r-1} + \alpha_1 \frac{(x-b)^r}{x-a_1} + \alpha_2 \frac{(x-b)^r}{x-a_2} + \ldots + a_j \frac{(x-b)^r}{x-a_j} \qquad \ldots(14)$$

Now that we have "covered up" the $(x-b)^r$ term in the denominator of G(x), we plug in $x = b$ to each side; this cancels every term on the right-hand side except for β_0, leaving the formula

$$\beta_0 = (x-b)^r G(x)\big|_{x=b} \qquad \ldots(15)$$

To find the other values of the beta coefficients, we can take advantage of the derivative. By taking the derivative of the *equation after cover up (Eq. 14)* (with respect to x the right-hand side becomes β_1 plus terms containing an $x - b$ in the numerator. Again, plugging in $x = b$ eliminates everything on the right-hand side except for β_1, leaving us with a formula for β_1 :

$$\beta_1 = \frac{d}{dx}\left((x-b)^r G(x)\right)\Big|_{x=b} \qquad (16)$$

Generalizing over this pattern, we can continue to take derivatives to find the other beta terms. The solution for all beta terms is

$$\beta_k = \frac{1}{k!}\frac{d^k}{dx^k}\left((x-b)^r G(x)\right)\Big|_{x=b} \qquad (17)$$

[**Hint :** To check if you've done the partial fraction expansion correctly, just add all of the partial fractions together to see if their sum equals the original ratio of polynomials.]

Example 1. Find L $(t \sin at)$.

$$L(t \sin at) = -\frac{d}{ds} L(\sin at)$$

$$= -\frac{d}{ds}\frac{a}{s^2+a^2} = \frac{2as}{(s^2+a^2)^2}.$$

Example 2. Find L $(t^2 e^{-3t})$.

$$L(t^2 e^{-3t}) = (-1)^2 \frac{d^2}{ds^2} L(e^{-3t})$$

$$= (-1)^2 \frac{d^2}{ds^2}\left(\frac{1}{s+3}\right) = \frac{2}{(s+3)^3}.$$

Example 3. Find L $(t\,e^{-t}\sin t)$.

$$\text{L}\,(t\,e^{-t}\sin t) = -\frac{d}{ds}\text{L}\,(e^{-t}\sin t)$$

$$= -\frac{d}{ds}\left[\frac{1}{(s+1)^2+1}\right]$$

since

$$\text{L}\,(\sin t) = \frac{1}{s^2+1}.$$

$$\therefore \quad \text{L}(e^{-t}\sin t) = \frac{1}{(s+1)^2+1}.$$

$$\therefore \quad \text{L}(te^{-t}\sin t) = \frac{2(s+1)}{(s^2+2s+2)^2}.$$

Theorem (*iv*) If L$[f\,(t)]$ = F (s) and if $\dfrac{f(t)}{t}$ has a limit as $t \to 0$, then

$$\text{L}\left[\frac{f(t)}{t}\right] = \int_0^{\infty} \text{F}\,(s)\,ds.$$

$$\text{F}(s) = \text{L}\,[f\,(t)] = \int_0^{\infty} e^{-st}\,f\,(t)\,dt$$

$$\therefore \quad \int_s^{\infty} \text{F}\,(s)\,ds = \int_s^{\infty} ds \int_0^{\infty} e^{-st} f\,(t)\,dt$$

$$= \int_0^{\infty} f(t)\left\{\int_s^{\infty} e^{-st}\,ds\right\}dt$$

on interchanging the order of integration

$$= \int_0^{\infty} f(t)\left[\frac{e^{-st}}{-t}\right]_s^{\infty} dt$$

$$= \int_0^{\infty} e^{-st}\,\frac{f(t)}{t}\,dt = \text{L}\left[\frac{f(t)}{t}\right].$$

Example 4. Find L $\left(\dfrac{1-e^t}{t}\right)$.

$$\text{L}\left(\frac{1-e^t}{t}\right) = \int_s^{\infty} \text{L}\,(1-e^t)\,ds$$

$$= \int_s^{\infty} \{L(1) - L(e^t)\} ds$$

$$= \int_s^{\infty} \left(\frac{1}{s} - \frac{1}{s-1}\right) ds$$

$$= \left[\log \frac{s}{s-1}\right]_s^{\infty} = -\log \frac{s}{s-1}$$

$$= \log \frac{s-1}{s}.$$

Example 5. $L^{-1}\left[\dfrac{s}{(s^2+a^2)^2}\right]$.

$$F'(s) = \frac{s}{(s^2+a^2)^2}; \therefore F(s) = \int \frac{s}{(s^2+a^2)^2} ds$$

$$= -\frac{1}{2(s^2+a^2)}$$

$$\therefore \quad L^{-1}\left[\frac{s}{(s^2+a^2)^2}\right] = -t\, L^{-1}\left[-\frac{1}{2(s^2+a^2)}\right]$$

$$= \frac{t}{2} L^{-1}\left(\frac{1}{s^2+a^2}\right)$$

$$= \frac{t}{2a} \sin at.$$

Example 6. $L^{-1}\left[\dfrac{s}{(s^2-1)^2}\right]$.

Here $$F'(s) = \frac{s}{(s^2-1)^2};$$

$$\therefore \quad F(s) = \int \frac{s}{(s^2-1)^2} ds = -\frac{1}{2(s^2-1)}$$

$$\therefore \quad L^{-1}\left[\frac{s}{(s^2-1)^2}\right] = -t\, L^{-1}\left[-\frac{1}{2(s^2-1)}\right]$$

$$= \frac{t}{2} L^{-1}\left(\frac{1}{s^2-1}\right) = \frac{t}{2} \sinh t.$$

Inverse Transforms by Direct Transform Result Modification

We can modify the results we have obtained in finding the Laplace transforms of functions to get the inverse transforms of functions.

(*i*) If L $[f(t)]$ = F (s), then L e^{-at} $[f(t)]$ = F $(s + a)$.

Hence we get the result

$$L^{-1}[F(s+a)] = e^{-t} f(t)$$
$$= e^{-at} L^{-1}[F(s)]$$

Thus for example

1. $$L^{-1}\left[\frac{1}{(s+a)^2}\right] = e^{-at} L^{-1}\left[\frac{1}{s^2}\right]$$
$$= e^{-at} t.$$

2. $$L^{-1}\left[\frac{1}{(s+2)^2+16}\right] = e^{-2t} L^{-1}\left[\frac{1}{s^2+4^2}\right]$$
$$= \frac{e^{-2t}\sin 4t}{4}.$$

3. $$L^{-1}\left[\frac{s-3}{(s-3)^2+4}\right] = e^{3t} L^{-1}\left[\frac{s}{s^2+2^2}\right]$$
$$= e^{3t}\cos 2t.$$

4. $$L^{-1}\left[\frac{1}{(s+a)^n}\right] = e^{-at} L^{-1}\left[\frac{1}{s^n}\right]$$
$$= \frac{e^{-at} t^{n-1}}{(n-1)!}.$$

Theorem (*v*) If L$[f(t)]$ = F (s), then L $[f(at)]$ = $\frac{1}{a}F\left(\frac{s}{a}\right)$.

This result can be written in the form

$$L^{-1}\left[\frac{1}{a}F\left(\frac{s}{a}\right)\right] = f(at), \text{ where } f(t) = L^{-1}[F(s)].$$

Putting $\frac{1}{a} = k$, we have

$$L^{-1}[F(ks)] = \frac{1}{k} f\left(\frac{t}{k}\right), \text{ where } f(t) = L^{-1}[F(s)].$$

Example. 1. Find $L^{-1}\left[\frac{s}{s^2a^2+b^2}\right]$.

$$\frac{s}{s^2a^2+b^2} = \frac{1}{a}\left(\frac{sa}{s^2a^2+b^2}\right) = \frac{1}{a}F(sa).$$

$$\therefore \qquad F(s) = \left(\frac{s}{s^2+b^2}\right).$$

$$\therefore \qquad L^{-1}\left(\frac{s}{s^2a^2+b^2}\right) = \frac{1}{a}L^{-1}\left(\frac{sa}{s^2a^2+b^2}\right)$$

$$= \frac{1}{a}L^{-1}[F(sa)] = \frac{1}{a}\frac{1}{a}.f\left(\frac{t}{a}\right)$$

where

$$f(t) = L^{-1}\left(\frac{s}{s^2+b^2}\right)$$

$$= \cos(bt).$$

$$\therefore \qquad L^{-1}\left(\frac{s}{s^2a^2+b^2}\right) = \frac{1}{a^2}\cos\left(\frac{bt}{a}\right).$$

(*iii*) If $L[f(t)] = F(s)$, then $L[t\,f(t)] = -F'(s)$.

Hence we get the result

$$L^{-1}[F'(s)] = -t\,f(t)$$

$$= -tL^{-1}[F(s)].$$

Example 2. Find $L\left[\frac{\sin at}{t}\right]$.

$$L\left[\frac{\sin at}{t}\right] = \int_s^\infty L(\sin at)\,ds$$

$$= \int_s^\infty \frac{a}{s^2+a^2}\,ds = \left[\tan^{-1}\frac{s}{a}\right]_s^\infty = \tan^{-1}\infty - \tan^{-1}\left(\frac{s}{a}\right)$$

$$= \frac{\pi}{2} - \tan^{-1}\frac{s}{a} = \cot^{-1}\left(\frac{s}{a}\right).$$

Example 3. $L^{-1}\left[\frac{s+2}{(s^2+4s+5)^2}\right]$.

Here

$$F'(s) = \frac{s+2}{(s^2+4s+5)^2}; \quad \therefore F(s) = -\frac{1}{2(s^2+4s+5)}.$$

$$\therefore \qquad L^{-1}\left[\frac{s+2}{(s^2+4s+5)^2}\right] = -t\,L^{-1}\left[\frac{1}{-2(s^2+4s+5)}\right]$$

$$= \frac{t}{2}L^{-1}\frac{1}{(s+2)^2+1^2}$$

$$= \frac{t}{2} e^{-2t} L^{-1} \frac{1}{s^2+1^2} = \frac{t\, e^{-2t} \sin t}{2}.$$

(iv) If L $[f(t)]$ = F (s), then L $[t\, f(t)] = -F'(s)$.

This theorem can be used in the following way to get inverse transforms of certain functions :

Take for example $L^{-1}\left(\log \frac{s+1}{s-1}\right)$.

Let this be equal to $f(t)$.

Then
$$L\,[t\, f(t)] = -\frac{d}{ds} \log\left(\frac{s+1}{s-1}\right)$$

$$= -\frac{d}{ds}[\log(s+1) - \log(s-1)]$$

$$= -\frac{1}{s+1} + \frac{1}{s-1}$$

$\therefore$
$$t\, f(t) = L^{-1}\left(-\frac{1}{s+1} + \frac{1}{s-1}\right)$$

$$= -L^{-1}\frac{1}{s+1} + L^{-1}\frac{1}{s-1}$$

$$= -e^{-t} + e^{t}$$

$$= 2 \sinh t.$$

$\therefore$
$$f(t) = \frac{2 \sinh t}{t}.$$

(v) If L $[f(t)] = s$ F (s) and if $\phi(t)$ is a function such that L $[\phi(t)]$ = F (s) and $\phi(0) = 0$, then $f(t) = \phi'(t)$.

We have
$$L\,[\phi'(t)] = s\, L\,[\phi(t)] - \phi(0)$$

$$= s\, L\,[\phi(t)]$$

$$= s\, F(s)$$

$$= L\,[f(t)]$$

$\therefore$
$$f(t) = \phi'(t).$$

This result for finding inverse transforms can be put in the form

$$L^{-1}[s\, F(s)] = \frac{d}{dt} L^{-1}[F(s)]$$

provided
$$L^{-1}[F(s)] = 0 \text{ when } t = 0.$$

Example 1. Find $L^{-1}\left(\frac{s}{s^2+b^2}\right)$.

$$L^{-1}\left(\frac{s}{s^2+b^2}\right) = \frac{d}{dt}L^{-1}\left(\frac{1}{s^2+b^2}\right)$$

$$= \frac{d}{dt}\left(\frac{1}{b}\sin bt\right)$$

$$= \cos bt.$$

Example 2. Find $L^{-1}\left[\frac{s}{(s+3)^2+4}\right]$.

$$L^{-1}\left[\frac{s}{(s+3)^2+4}\right] = \frac{d}{dt}L^{-1}\left[\frac{1}{(s+3)^2+4}\right]$$

$$= \frac{d}{dt}\left\{e^{-3t}L^{-1}\left(\frac{1}{s^2+4}\right)\right\}$$

$$= \frac{d}{dt}\left(e^{-3t}\frac{\sin 2t}{2}\right)$$

$$= e^{-3t}\cos 2t - \frac{3}{2}e^{-3t}\sin 2t$$

$$= \frac{e^{-3t}}{2}(2\cos 2t - 3\sin 2t).$$

Example 3. Find $L^{-1}\left(\frac{s-3}{s^2+4s+13}\right)$.

$$L^{-1}\left(\frac{s-3}{s^2+4s+13}\right) = L^{-1}\left(\frac{s}{s^2+4s+13}\right) - 3L^{-1}\left(\frac{1}{s^2+4s+13}\right)$$

$$= \frac{d}{dt}L^{-1}\left(\frac{1}{s^2+4s+13}\right) - 3L^{-1}\left(\frac{1}{s^2+4s+13}\right)$$

$$= \frac{d}{dt}L^{-1}\left[\frac{1}{(s+2)^2+3^2}\right] - 3L^{-1}\left[\frac{1}{(s+2)^2+3^2}\right]$$

$$= \frac{d}{dt}\left(\frac{e^{-2t}\sin 3t}{3}\right) - 3.\frac{e^{-2t}\sin 3t}{3}$$

$$= \frac{e^{-2t}}{3}(3\cos 3t - 5\sin 2t).$$

Example 4. Find $L^{-1}\left[\frac{s}{(s+2)^2}\right]$.

$$L^{-1}\left[\frac{s}{(s+2)^2}\right] = \frac{d}{dt} L^{-1}\left[\frac{1}{(s+2)^2}\right]$$

$$= \frac{d}{dt}\left\{e^{-2t} L^{-1}\left(\frac{1}{s^2}\right)\right\}$$

$$= \frac{d}{dt}(e^{-2t}t) = e^{-2t}(1-2t).$$

Example 5. Find $L^{-1}\left[\frac{s^2}{(s-1)^3}\right]$.

$$L^{-1}\left[\frac{s^2}{(s-1)^3}\right] = \frac{d}{dt} L^{-1}\left[\frac{s}{(s-1)^3}\right]$$

$$= \frac{d^2}{dt^2} L^{-1}\left[\frac{1}{(s-1)^3}\right] = \frac{d^2}{dt^2} e^t L^{-1}\left(\frac{1}{s^3}\right) = \frac{d^2}{dt^2}\left(\frac{e^t . t^2}{2}\right)$$

$$= \frac{e^t}{2}(t^2+4t+2).$$

(*vi*)
$$L\left[\int_0^t f(x)\, dx\right] = \frac{1}{s} L[f(t)].$$

Hence
$$L^{-1}\left\{\frac{1}{s} L[f(t)]\right\}_0^t = \int_0^t f(x)\, dx.$$

$\therefore$ If $L[f(t) = \phi(s)$, we have

$$L^{-1}\left[\frac{1}{s}\phi(s)\right] = \int_0^t L^{-1}[\phi(s)]\, dt.$$

Let
$$\int_0^t f(x)\, dx = F(t)$$

Then
$$F(0) = \int_0^0 f(x)\, dx = 0.$$

Also
$$F'(t) = f(t)$$

Hence $$L[F'(t)] = L[f(t)].$$

But
$$L[F'(t)] = s\,L[F(t)] - F(0)$$
$$= s\,L[F(t)] \text{ as } F(0) = 0.$$

$\therefore$
$$L[f(t)] = s\,L[F(t)]$$
$$= s\,L\left[\int_0^t f(x)\,dx\right]$$

$\therefore$
$$L^{-1}\left\{\frac{1}{s}\,L[f(t)]\right\} = \int_0^t f(x)\,dx.$$

$\therefore$ If $L[f(t)] = \phi(s)$, we have

$$L^{-1}\left[\frac{1}{s}\,\phi(s)\right] = \int_0^t L^{-1}[\phi(s)]\,dt.$$

Example 1. Find $L^{-1}\left[\dfrac{1}{s(s+a)}\right]$.

$$L^{-1}\left[\frac{1}{s(s+a)}\right] = \int_0^t L^{-1}\left[\frac{1}{s+a}\right]dt$$

$$= \int_0^t e^{-at}dt = \left[\frac{e^{-at}}{-a}\right]_0^t = \frac{1}{a}(1-e^{-at}).$$

Example 2. Find $L^{-1}\left[\dfrac{1}{s(s^2+a^2)}\right]$.

$$L^{-1}\left[\frac{1}{s(s^2+a^2)}\right] = \int_0^t L^{-1}\left(\frac{1}{s^2+a^2}\right)dt$$

$$= \int_0^t \frac{\sin at}{a}\,dt = \frac{1}{a}\left[\frac{-\cos at}{a}\right]_0^t$$

$$= \frac{1}{a^2}(1-\cos at).$$

Example 3. Find $L^{-1}\left[\dfrac{1}{(s^2+a^2)^2}\right]$.

$$L^{-1}\left[\frac{1}{(s^2+a^2)^2}\right] = L^{-1}\left[\frac{1}{s}\,\frac{s}{(s^2+a^2)^2}\right]$$

$$= \int_0^t L^{-1}\left[\frac{s}{(s^2+a^2)^2}\right] dt$$

$$= \int_0^t \frac{t \sin at}{2a} dt$$

$$= \frac{1}{2a}\left[\frac{-t \cos at}{a} + \frac{\sin at}{a^2}\right]_0^t$$

$$= \frac{1}{2a^3}(\sin at - at \cos at).$$

(*vii*) The method of partial fractions can be used to find the inverse transforms of certain functions.

The method is illustrated in the following examples :

Example 1. $L^{-1}\left[\frac{1}{s(s+1)(s+2)}\right]$.

We can split $\frac{1}{s(s+1)(s+2)}$ into partial fractions and we have

$$\frac{1}{s(s+1)(s+2)} = \frac{1}{2}\frac{1}{s} - \frac{1}{s+1} + \frac{1}{2}\frac{1}{s+2}.$$

$$\therefore \quad L^{-1}\left\{\frac{1}{s\ (s+1)\ (s+2)}\right\} = \frac{1}{2}L^{-1}\left(\frac{1}{s}\right) - L^{-1}\left(\frac{1}{s+1}\right) + \frac{1}{2}L^{-1}\left(\frac{1}{s+2}\right)$$

$$= \frac{1}{2} - e^{-t} + \frac{1}{2}e^{-2t}.$$

Example 2. $L^{-1}\left[\frac{1}{(s+1)(s^2+2s+2)}\right]$.

Splitting into partial fractions we have

$$\frac{1}{(s+1)(s^2+2s+2)} = \frac{1}{s+1} - \frac{s+1}{s^2+2s+2}.$$

$$\therefore \quad L^{-1}\left[\frac{1}{(s+1)\ (s^2+2s+2)}\right] = L^{-1}\left(\frac{1}{s+1}\right) - L^{-1}\left(\frac{s+1}{s^2+2s+2}\right)$$

$$= e^{-t} - L^{-1}\left\{\frac{s+1}{(s+1)^2+1^2}\right\}$$

$$= e^{-t} - e^{-t}L^{-1}\left(\frac{s}{s^2+1^2}\right)$$

$$= e^{-t} - e^{-t} \cos t$$

$$= e^{-t}(1 - \cos t).$$

Example 3. $L^{-1}\left[\dfrac{1+2s}{(s+2)^2 (s-1)^2}\right]$.

$$\frac{1+2s}{(s+2)^2 (s-1)^2} = \frac{1}{3}\,\frac{(s+2)^2 - (s-1)^2}{(s+2)^2 (s-1)^2}$$

$$= \frac{1}{3}\left[\frac{1}{(s-1)^2} - \frac{1}{(s+2)^2}\right]$$

Hence $$L^{-1}\left[\frac{1+2s}{(s+2)^2 (s-1)^2}\right] = \frac{1}{3} L^{-1}\left[\frac{1}{(s-1)^2}\right] - \frac{1}{3} L^{-1}\left[\frac{1}{(s+2)^2}\right]$$

$$= \frac{1}{3} t\, e^{t} - \frac{1}{3} t\, e^{-2t} = \frac{t}{3}(e^{t} - e^{-2t}).$$

Use of Laplace Transforms for Transient Problems

Consider the differential equation of the L – R transient

$$L\, di/dt + R\, i = E$$

Transforming i into its Laplace transform and denoting the transform as $I(s)$,

$$Ls\, I(s) - L\, i(0) + R\, I(s) = E/s.$$

$\because$ di/dt transforms into $sI(s) - i(0)$

E transforms into E/s for step input at $t = 0$.

$$\therefore \quad I(s) = \frac{E}{s(Ls+R)} + \frac{L\, i(0)}{L s + R}$$

Since $i(0) = 0$, $I(s) = E/s(Ls + R)$

Let $$I(s) = \frac{E}{s(Ls+R)} = \frac{A}{s} + \frac{B}{sL + R} \qquad \text{...(7.10)}$$

The above expansion is called *partial fraction* expansion.

A and B are to be found. Cross multiply by s $(Ls + R)$.

$$E = A\,(Ls + R) + Bs$$

Put $s = 0$ in this identity. We get $A = E/R$.

Put $s = -R/L$ and then $B = -LE/R$.

The relation (7.10) is a function of s. To transform into the function of time, we find the inverse tranform of $I(s)$. From the table of transforms, the inverse of $1/s$ is unity; that for $1/(sL + R)$ is $(1/L)\exp(-R/L)t$. So

$$i(t) = A + B/L\, e^{-Rt/L}$$

$$= E/R\,(1 - e^{-Rt/L})$$

Solving Differential Equations Using L.T. Additional Problems

1. *Solve the following differential equations using the Laplace Transform method:*

(a) $\dot{y} + 4y = 3x;\ x(t) = \sin(2t),\ y(0) = 1$

(b) $\ddot{y} + 4\dot{y} + 20y = 2\dot{x} - x;\ x(t) = u(t),\ y(0) = 0,\ \dot{y}(0) = 1$

(c) $\ddot{y} + 7\dot{y} + 12y = 6x;\ x(t) = u(t),\ y(0) = 0,\ \dot{y}(0) = -2$

(d) $\ddot{y} + 9\dot{y} + 20y = x(t),\ y(0) = 1,\ \dot{y}(0) = -2,\ x(t) = 2u(t)$

Ans. 1 (*a*) $\dot{y} + 4y = 3x;\ x = \sin(2t),\ y(0) = 1$

$$sY(s) - y(0) + 4Y(s) = 3X(s)$$

$$(s + 4)\,Y(s) = \frac{6}{s^2 + 4} + 1$$

$$Y(s) = \frac{6}{(s^2 + 4)(s + 4)} + \frac{1}{s + 4}$$

$$= \frac{C_1 + C_2}{s^2 + 4} + \frac{C_3}{s + 4} + \frac{1}{s + 4}$$

$$C_3 = 3/10$$

$$6 = (C_1 s + C_2)(s + 4) + 3/10(s^2 + 4)$$

$$\Rightarrow \quad C_1 = -3/10$$

$$6 = 4C_2 + 6/5$$

$$6/5 = C_2$$

$$Y(s) = \frac{-3/10s + 6/5}{s^2 + 4} + \frac{13/10}{s + 4}$$

$$= \frac{-3/10s}{s^2 + 4} + \frac{6}{10}\frac{2}{s^2 + 4} + \frac{13/10}{s + 4}$$

$$y(t) = \left(-3/10\cos(2t) + 6/10\sin(2t) + 13/10e^{-4t}\right)u(t)$$

(*b*) $\ddot{y} + 4\dot{y} + 20y = 2\dot{x} - x,\ x(t) = u(t),\ y(0) = 0,\ \dot{y}(0) = 1$

$$s^2Y(s) - 5y(0) - \dot{y}(0) + 4sY(s) - 4y(0) + 20Y(s) = 2sX(s) - 2x(0) - X(s)$$

$$(s^2 + 4s + 20)\,Y(s) = 1 + 2 - 1/s \qquad (x(0) = 0)$$

$$Y(s) = \frac{3}{(s + 2)^2 + 16} - \frac{1}{((s + 2)^2 + 16)s}$$

$$= \frac{3}{(s + 2)^2 + 16} + \frac{C_1 s + C_2}{(s + 2)^2 + 16} + \frac{C_3}{s}$$

$$C_3 = 1/20,\ C_1 = 1/20,\ C_2 = 1/5$$

$$Y(s) = \frac{3}{4}\frac{4}{(s+2)^2+16} + \frac{1/20s + 1/5}{(s+2)^2+16} - \frac{1/20}{s}$$

$$\frac{1/20(s+2)}{(s+2)^2+16} + \frac{1/10}{(s+2)^2+16}$$

$$y(t) = \frac{3}{4}e^{-2t}\sin(4t) + \frac{1}{20}e^{-2t}\cos(4t) + \frac{1}{40}e^{-2t}\sin(4t)$$

for $t \geq 0$

(c) $\ddot{y} + 7\dot{y} + 12y = 6x,\ x = u(t),\ y(0) = 0,\ \dot{y}(0) = -2$

$$s^2Y(s) - sy(0) - \dot{y}(0) + 7(sY(s) - y(0)) + 12Y(s) = \frac{6}{s}$$

$$(s^2 + 7s + 12)\ Y(s) = \frac{6}{5} - 2$$

$$Y(s) = \frac{-2s+6}{s(s^2+7s+12)} = \frac{C_1}{s} + \frac{C_2}{s+4} + \frac{C_3}{s+3}$$

$$C_1 = 1/2,$$

$$C_2 = \left.\frac{-2s+6}{s(s+3)}\right|_{s=-4} = -\frac{1}{2}$$

$$C_3 = \left.\frac{-2s+6}{s(s+4)}\right|_{s=-3} = -4$$

$$Y(s) = \frac{1/2}{s} - \frac{1/2}{s+4} - \frac{4}{s+3}$$

$$y(t) = 1/2 - 1/2e^{-4t} - 4e^{-3t},\ t \geq 0$$

(d) $\dfrac{d^2y}{dt^2} + 9\dfrac{dy}{dt} + 20y = x(t),\ y(0) = 1,\ \dot{y}(0) = -2,\ x(t) = 2u(t)$

$$s^2Y(s) - sy(0) - \dot{y}(0) + 9(sY(s) - y(0)) + 20Y(s) = X(s)$$

$$(s^2 + 9s + 20)\ Y(s) = s - 2 + 9 + \frac{2}{s}$$

$$Y(s) = \frac{s^2 + 7s + 2}{s(s^2 + 9s + 20)} = \frac{s^2 + 7s + 2}{s(s+4)(s+5)}$$

$$Y(s) = \frac{C_1}{s} + \frac{C_2}{s+4} + \frac{C_3}{s+5}$$

$$= \frac{1/10}{s} + \frac{5/2}{s+4} + \frac{-8/5}{s+5}$$

$$y(t) = \frac{1}{10} + \frac{5}{2}e^{-4t} - \frac{8}{5}e^{-5t},\ t \geq 0$$

Advantages of Laplace Transform for Transients

(1) Initial conditions of transient problems can be easily incorporated.

(2) The particular integral and the complimentary function (the steady-state and transient solutions) are obtained together simultaneously.

(3) The writing of differential equations for meshy circuits is difficult without Laplace transforms.

As an example suppose the voltage across R_2 is required.

Method (*a*) $$R_1 i_1 = 1/C \int i_2 \, dt$$

$$E = R_1 i_1 + R_2 (i_1 + i_2)$$

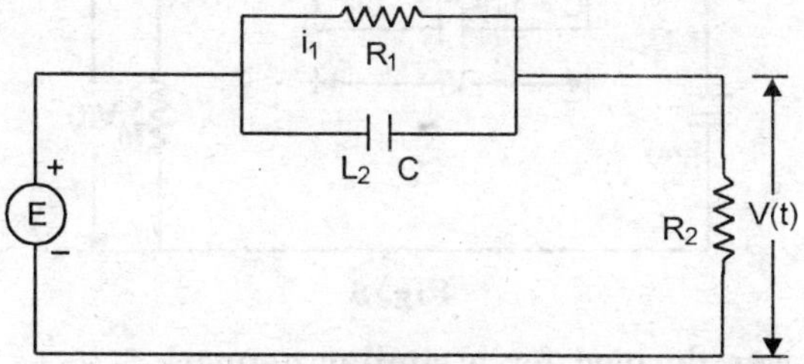

Fig. 7

$$= \frac{R_1 + R_2}{R_1 C} \int_0^t i_2 \, dt + R_2 i_2 \qquad ...(a)$$

The required voltage across R_2 is

$$v = R_2(i_1 + i_2) = (R_2 / R_1)(1/C) \int i_2 \, dt + R_2 \, i_2 \qquad ...(b)$$

Transform relations (*a*) and (*b*).

$$E\,(s) = \frac{R_1 + R_2}{R_1 C} \frac{I_2(s)}{s} + R_2 I_2(s) \qquad ...(a')$$

The above assumes that there is no initial charge on C.

$$v(s) = \frac{R_2}{R_1 C} \frac{I_2\,(s)}{s} + R_2 I_2(s) \qquad ...(b')$$

Eliminate I_2 from a' and b' to get

$$v(s) = \frac{R_2(1 + R_1 \, C \, s)}{R_1 + R_2 + R_1 R_2 \, Cs} E(s) \qquad ... (7.11)$$

This can be transformed back to get $v(t)$, the transient solution.

Method (*b*). Suppose we find the impedance of the whole circuit using the symbolic method for a frequency ω and compute $v(t)$ for that input.

$$\text{Impedance } Z = R_2 + \frac{R_1(1/j\omega C)}{R_1 + (1/j\omega C)}$$

The voltage $$V = [E/Z] \times R_2$$

$$= \frac{E\,R_2\,[R_1 + (1/j\omega C)]}{R_2\,[R_1 + 1/j\omega C] + R_1/j\omega C}$$

Now, in this, if we replace $j\omega$ by s we find that we get relation (7.11) above. So we see this method as simpler. In method *b*, we are generalising the result which we know as true for a frequency ω to be true also for a complex frequency $s = \sigma + j\omega$. We discuss this further in the next article.

Example 1. Find the transient currents i_1 and i_2 and voltages v_1, v_2 after closure of switch. The capacitor is initially uncharged.

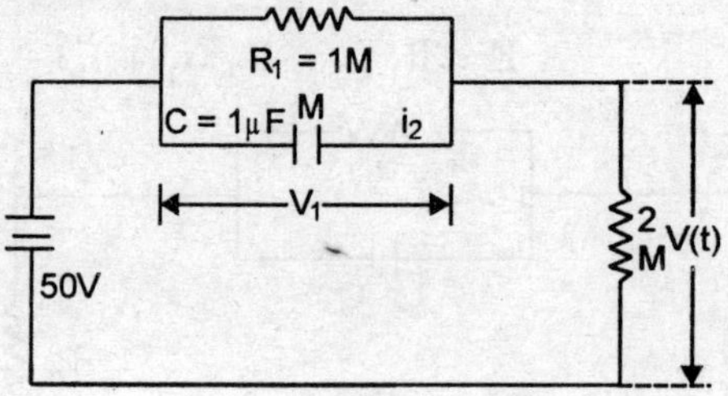

Fig. 8

From (7.11), which was derived for a similar network,

$$V(s) = \frac{R_2\,(1 + R_1\,C_1\,s)}{R_1 + R_2 + R_1\,R_2\,C\,s}\,E(s)$$

For a step input (*i.e.*, switch suddenly closing at $t = 0$), $E(s) = E/s$.

$$V(s) = \frac{E}{s}\,\frac{(1 + R_1\,C_1\,s)}{(s + 1/\tau)R_1\,C};\ \text{where } \tau = \frac{R_1\,R_2\,C}{R_1 + R_2} \qquad ...(a)$$

$$= \frac{A}{s} + \frac{B}{(s + 1/\tau)} \qquad \left\{ = \frac{1 \times 2 \times 1}{1 + 2} = \frac{2}{3} \right.$$

To get A, put $s = 0$ in expression (*a*), omitting the denominator terms.

$$A = \frac{E\,(1 + R_1\,C_1 \times 0)}{(10 + 1/\tau)\,R_1\,C} = \frac{E\tau}{R_1 C} = \frac{50 \times 2/3}{1 \times 1} = \frac{100}{3}$$

To get B, put $s = -1/\tau$ in expression (*a*), omitting $(s + 1/\tau)$.

$$B = \frac{E\,[1 + R_1\,C_1(-1/\tau)]}{R_1\,C(-1/\tau)} = \frac{50[1 - (3/2)]}{-3/2} = \frac{50}{3}$$

Taking the inverse Laplace transform of V(*s*), we get

$$v(t) = A + B\,e^{-t/\tau}$$

$$= 50/3(2 + e^{-3t/2})$$

Hence $\qquad v(t) + v_1(t) = 50$

$$v_1(t) = 50 - v(t) = 50/3\ (1 - e^{-3t/2}).$$

Example 2. An impulse generator circuit, similar to Fig. 9 is used to produce high voltage inpulse wave, akin to voltages developed due to lightning surges. Develop an expression for the voltage across C_2. The sparkgap S breaks down and starts conducting at t = 0.

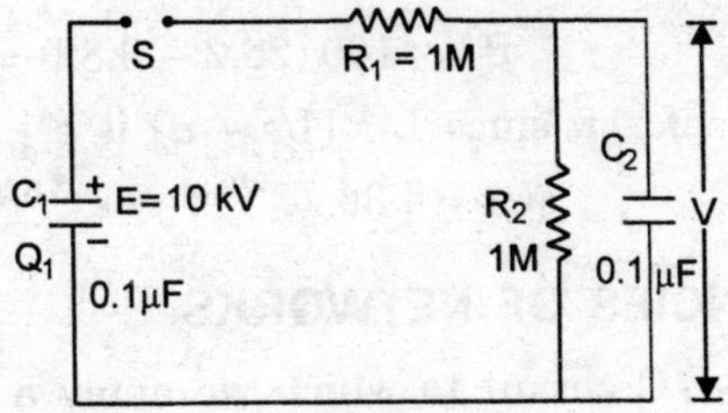

Fig. 9

The impedance function of the R_2 C_2 branch is

$$Z_2 = \frac{R_2 / s\, C_2}{R_2 + (1/s\, C_2)} = \frac{R}{1 + s\, C_2\, R_2}$$

The capacitor C_1 is already charged to voltage E.

The Kirchoff's law equation for the circuit is

$$\frac{1}{C_1} \int i\ dt + R_1 i + Z_2\ i = 0$$

Transforming, $[1/C_1\, s] + (E/s) + (R_1 + Z_2)\, I = 0.$

$$I = -\frac{E}{s} \frac{1}{R_1 + Z_2 + (1/C_1 s)}$$

Substituting for Z_2 and simplifying,

$$I(s) = \frac{-E\, C_1\, (1 + sC_2R_2)}{1 + s\,(R_1C_1 + R_2C_2 + R_2C_1) + s^2R_1C_1R_2C_2}$$

The voltage v is got as I Z_2

$$V(s) = \frac{-E\, C_1\, R_2}{1 + s\,(R_1C_1 + R_2C_2 + R_2C_1) + s^2R_1C_1R_2C_2}$$

$$= \frac{-E\, R_2\, C_1}{1 + as + bs^2}$$

where $a = (R_1 + R_2)\, C_1 + R_2C_2 = 2 \times 0.1 + 0.1 = 0.3$

$$b = R_1\, R_2\, C_1\, C_2 = 1 \times 1 \times 0.1 \times 0.1 = 0.01$$

$$E\, R_2\, C_1 = 10\ \text{kV} \times 1 \times 0.1 = 1.0$$

$$V = \frac{-1.0}{0.01\, s^2 + 0.3\, s + 1} = \frac{-1}{0.01\,(s + 26.2)\,(s + 3.82)}$$

$$= \frac{A}{s + 26.2} + \frac{B}{s + 3.82}$$

where
$$A = -100/(-26.2 + 3.82) = 4.35$$
$$B = -100/(26.2 - 3.82) = -4.35.$$

Taking the inverse transform, since L^{-1} $[1/s + a]$ is e_{-at},

$$v(t) = 4.35\ (e^{-26.2t} - e^{-3.82t})\ \text{kV}.$$

THE NATURAL FREQUENCIES OF NETWORKS

Suppose we have an R-L-C circuit to which we apply a step voltage and observe the transient current that passes. Let R be a variable resistance. Let us vary R from zero upto a high value. The current will first be oscillatory for low values of R, of the form e^{-at} sin $(bt + \theta)$. Then at a critical R value, the response becomes non-oscillatory when $a = b$. Further increase of R makes the transient current exponential in nature.

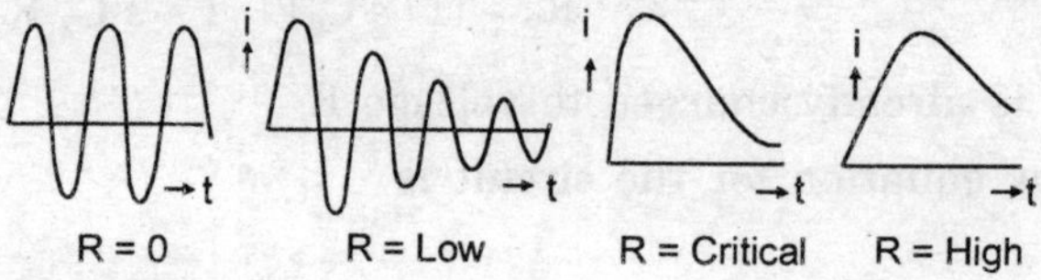

Fig. 10 Natural frequencies in R-L-C series circuit.

The natural frequencies are the *roots* of the characteristic equation of the R-L-C circuit differential equation.

$$L\,d^2i/dt^2 + R\,di/dt + i/C = 0 \qquad \text{...(7.12)}$$

viz.,
$$(Ls^2 + Rs + 1/C) = 0 \qquad \text{...(7.13)}$$

The roots may be real or imaginary; the solution of (7.12) is got as

$$A \exp(s_1 t) + B \exp(s_2 t) \qquad \text{...(7.14)}$$

by Differential equation theory. s_1 and s_2 may be real or imaginary. They are natural frequencies, which are, in general complex numbers. Thus the roots s_1, s_2 may be written is general as

$$s = \sigma + j\omega \qquad \text{...(7.15)}$$

where σ is the real part and ω the imaginary part.

If the response is oscillatory, ω will be the frequency of oscillation. If the response is exponential, the imaginary part will be zero. This too can be considered as an oscillation, if we generalise the frequency variable to s, a complex quantity.

System Response : Additional Problem

1. *Sketch the response of each of the systems below to a step input.*

(a) $H(s) = \dfrac{10}{s+2}$

(b) $H(s) = \dfrac{0.2}{s+0.2}$

2. *Given, the two step responses shown below, the first one is a first order system and the second one is a second order system. Determine the transfer functions for both systems.*

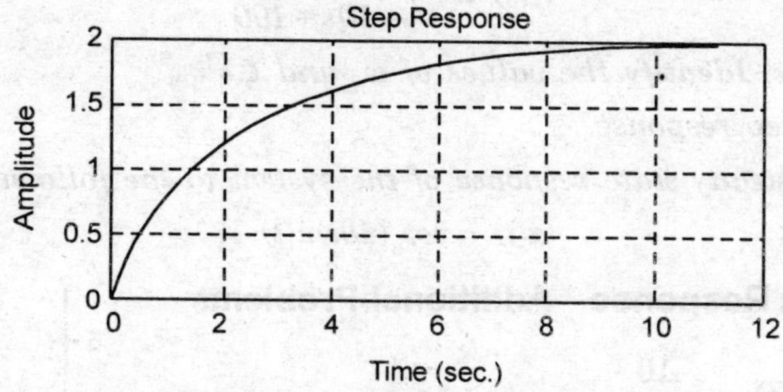

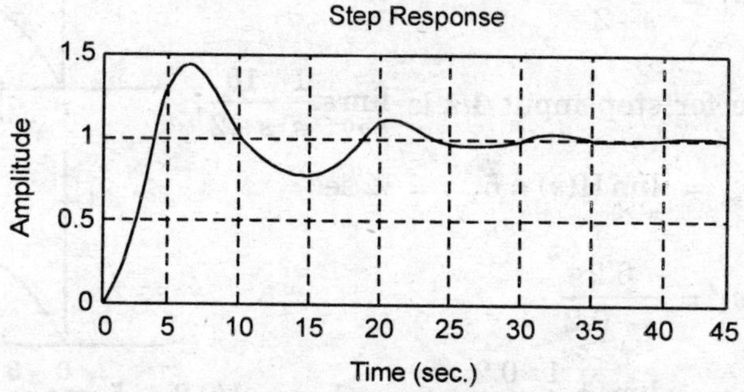

3. *Plot the pole positions for each of the following systems, determine the values for* ζ *and* ω_n *for the stable second order systems with complex poles.*

(a) $H(s) = \dfrac{1}{s+4}$

(b) $H(s) = \dfrac{1}{s+10}$

(c) $H(s) = \dfrac{1}{s-2}$

(d) $H(s) = \dfrac{1}{s^2+4s+16}$

(e) $H(s) = \dfrac{1}{s^2+4s+3}$

(f) $H(s) = \dfrac{1}{s^2+4s+2}$

(g) $H(s) = \dfrac{1}{s^2-4s+16}$

4. *Give the general form of the response of the systems in Problem 3 to a step input.*

5. *Determine the steady-state response of the systems in Problem 3(a), (d), and (f) to an input of* $x(t) = 2\cos(4t - 20°)u(t)$.

6. *Given the following system:*

$$H(s) = \frac{10}{s^2 + 10s + 100}$$

(a) Plot the poles. Identify the values of ω_n *and* ζ*.*

(b) Sketch the step response.

(c) What is the steady-state response of the system to the following input?

$$x(t) = \cos(10t)u(t)$$

Answers to System Response—Additional-Problems

1. (*a*) $\quad H(s) = \dfrac{10}{s+2}$

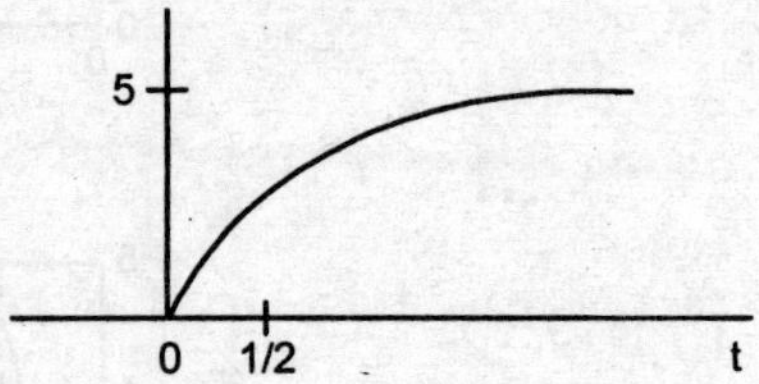

Stedy state value for step input 1/s is $\lim\limits_{s\to 0} s.\dfrac{1}{s}.\dfrac{10}{s+.2}$;

$$y_{ss} = \lim_{s\to 0} H(s) = 5,\ \tau = ½ \text{ sec}$$

(*b*) $\quad H(s) = \dfrac{0.2}{s+0.2}$

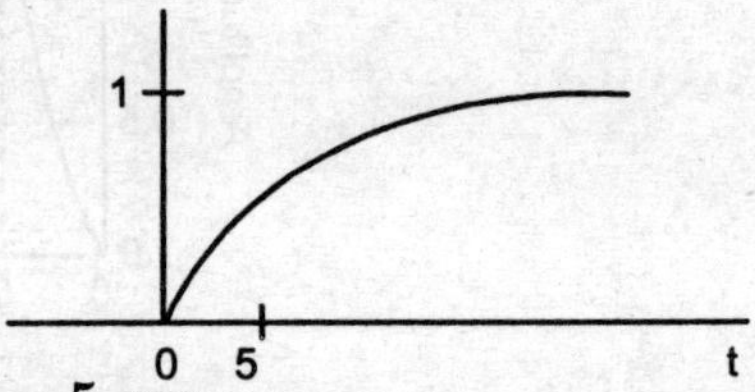

$$y_{ss} = \lim_{s\to 0} s.\frac{1}{s}.\frac{0.2}{s+.2};\ y_{ss} = 1,\ \tau = 1/0.2 = 5 \text{ sec}$$

2.

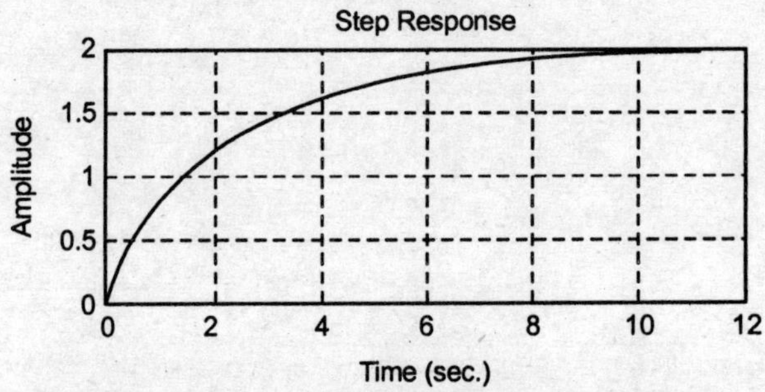

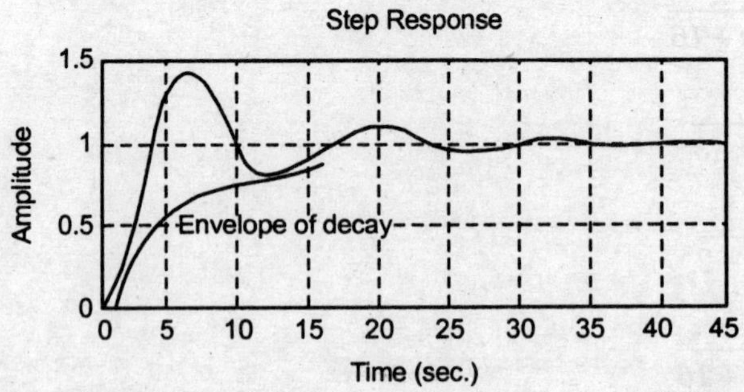

First order system (top plot) has general form $H(s) = \dfrac{k}{s+a}$. The time-constant is the time that the response is equal to $2(1 - e^{-1}) = 63\%$ of $2 = 1.26$, so $\tau \approx 2$ sec. $a = 1/\tau = 0.5$.

The steady-state value (due to a unit step input) is $H(0) = k/a = k/0.5$. From the plot $2 = k/0.5$ so $k = 1$.

Final answer: $$H(s) = \frac{1}{s+0.5}$$

Second order system (bottom plot) has general form $H(s) = \frac{k}{s^2 + 2\zeta\omega_n s + \omega_n^2}$ or

$H(s) = \frac{k}{(s+\zeta\omega_n)^2 + \omega_d^2}$ where the real part of the pole is at $-\zeta\omega_n$ is the real part of the pole (it governs the envelope of decay) and ω_d is the imaginary part of the pole (it governs the frequency of the oscillations, $\omega_d = 2\pi/T$).

From the plot, $T \approx 12$ sec, so $\omega_d = 2\pi/13$. The time constant of the envelope of decay is about $\tau \approx 7$ sec, so $\zeta\omega_n = 1/7$. k is found from the steady-state value $1 = H(0) = \frac{k}{(\zeta\omega_n)^2 + \omega_d^2}$. Solving for k yields $k \approx 0.254$.

Final answer : $$H(s) = \frac{0.254}{s^2 + 0.286s + 0.254}.$$

$$H(s) = \frac{k}{s^2 + 2\zeta\omega_n s + \omega_n^2}$$

$$\omega_d = \frac{2\pi}{T} = 9.7$$

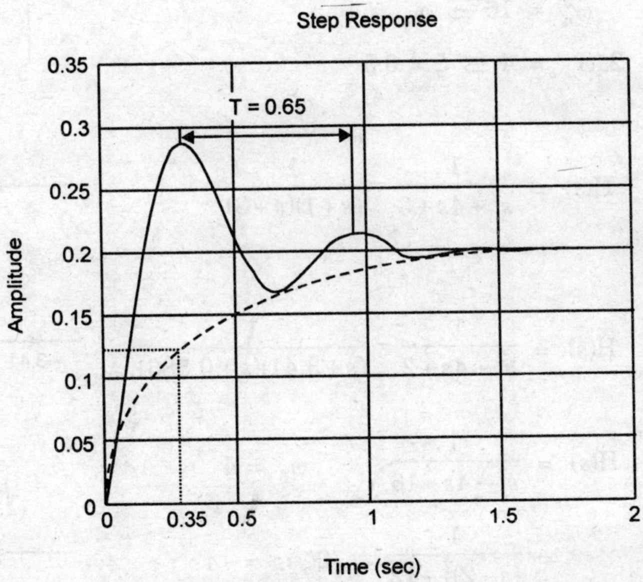

$$\tau \approx .35 \text{ sec} \Rightarrow \frac{1}{\zeta\omega_n} = \tau \Rightarrow \zeta\omega_n = 2.9$$

$$H(s) = \frac{k}{(s+2.9)^2 + 9.7^2}, \; H(0) = 0.2 \Rightarrow k = 20.5$$

$$H(s) = \frac{20.5}{(s+2.9)^2 + 9.7^2} \text{ from plot}$$

Actual is $$H(s) = \frac{22}{(s+3)^2 + 10^2}$$

3. (*a*) $$H(s) = \frac{1}{s+4}$$

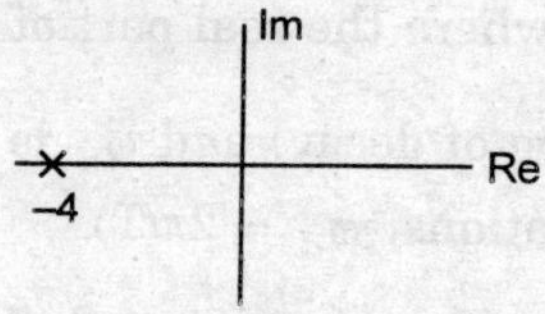

(*b*) $$H(s) = \frac{1}{s+10}$$

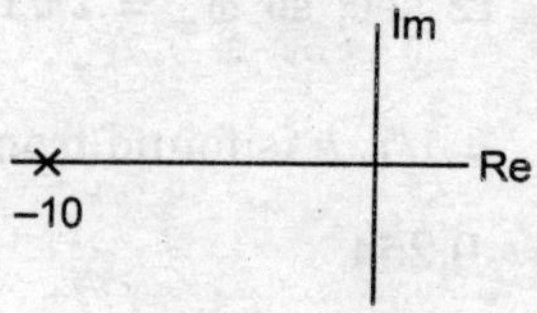

(*c*) $$H(s) = \frac{1}{s-2}$$

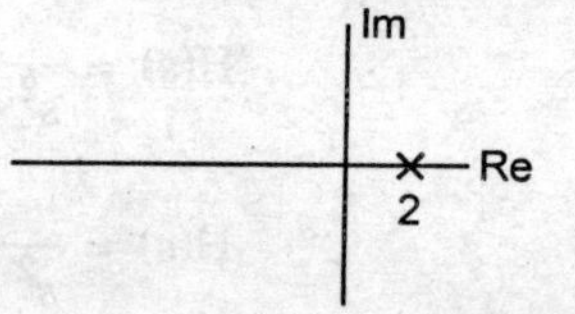

(*d*) $$H(s) = \frac{1}{s^2 + 4s + 16} = \frac{1}{(s+2)^2 + 12}$$

$$\omega_n^2 = 16 \Rightarrow \omega_n = 4$$

$$2\zeta\omega_n = 4 \Rightarrow \zeta = 0.5$$

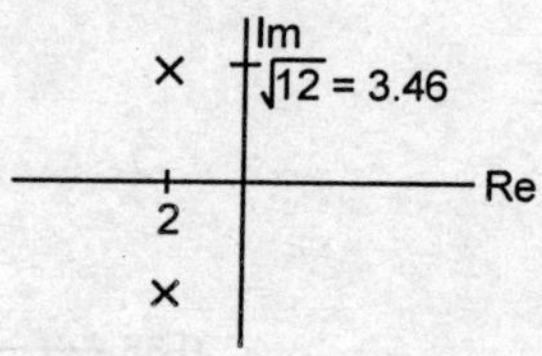

(*e*) $$H(s) = \frac{1}{s^2 + 4s + 3} = \frac{1}{(s+1)(s+3)}$$

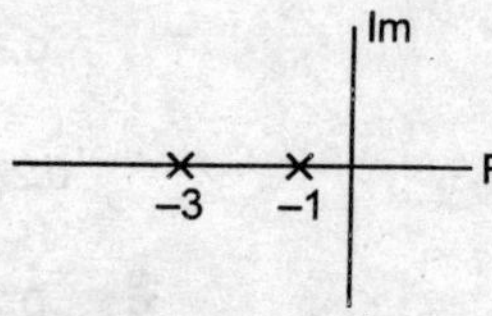

(*f*) $$H(s) = \frac{1}{s^2 + 4s + 2} = \frac{1}{(s+3.41)(s+0.586)}$$

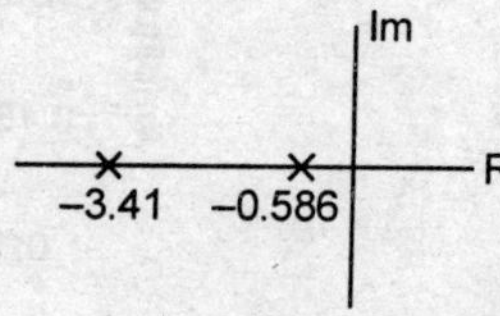

(*g*) $$H(s) = \frac{1}{s^2 - 4s + 16}, \qquad \omega_n = 4$$

$$= \frac{1}{(s-2)^2 + 12} \qquad 2\zeta\omega_n = -4$$

So $\zeta = -0.5$

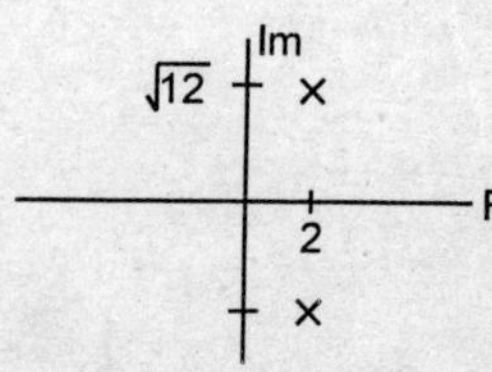

4. (*a*) $$y(t) = k_1 + k_2 e^{-4t}, \ t \geq 0$$

$\left(\text{Note, } k_1 \text{ is found from steady-state } = \text{H}(0) = \frac{1}{4}\right)$

(*b*) $$y(t) = k_1 + k_2 e^{-10t}, \ t \geq 0$$

(*c*) $$y(t) = k_1 + k_2 e^{2t}, \ t \geq 0$$

(*d*) $$y(t) = k_1 + k_2 e^{-2t} \cos\left(\sqrt{12t} + \theta\right), \ t \geq 0$$

(*e*) $$y(t) = k_1 + k_2 e^{-t} - k_3 e^{-3t}, \ t \geq 0$$

(*f*) $$y(t) = k_1 + k_2 e^{-3.4t} + k_3 e^{-0.586t}, \ t \geq 0$$

(*g*) $$y(t) = k_1 + k_2 e^{2t} \cos\left(\sqrt{12t} + \theta\right), \ t \geq 0$$

in all cases, $k_1 = \text{H}(0)$

5. $\text{A} \cos(\omega t + \theta) u(t) \rightarrow y_t(t) + y_{ss}(t)$

$\text{A}|\text{H}(\omega)| \cos(\omega t + \theta + \angle \text{H}(\omega))$

let $x(t) = 2 \cos(4t - 20°)\, u(t)$

(*a*) $$\text{H}(s) = \frac{1}{s+4} \text{ so } \text{H}(\omega) = \frac{1}{j\omega + 4}$$

$$\text{H}(4) = \frac{1}{4j+4} = 0.176 e^{-45° j}$$

$$y_{ss}(t) = 0.352 \cos(4t - 20° - 45°)$$
$$= 0.352 \cos(4t - 65°)$$

(*b*) $$\text{H}(s) = \frac{1}{s^2 + 4s + 16}, \ \text{H}(\omega) = \frac{1}{(16 - \omega^2) + 4j\omega}$$

$$\text{H}(4) = \frac{1}{16j} = 0.0625 e^{-j90°}.$$

$$y_{ss}(t) = 0.125 \cos(4t - 110°)$$

(*c*) $$\text{H}(s) = \frac{1}{s^2 + 4s + 2}, \ \text{H}(\omega) = \frac{1}{(2 - \omega^2) + 4j\omega}$$

$$\text{H}(4) = \frac{1}{-14 + 16j} = 0.047 e^{-j131°}.$$

$$y_{ss}(t) = 0.094 \cos(4t - 151°)$$

6. (*a*) $$\omega_n = 10, \ 2\zeta\omega_n = 10 \rightarrow \zeta = 0.5$$

$$\text{P}_1, \text{P}_2 = -5 \pm 8.66j$$

(*b*) $\tau = \dfrac{1}{\zeta\omega_n} = \dfrac{1}{5}$, $T = \dfrac{2\pi}{\omega d} = 0.726$

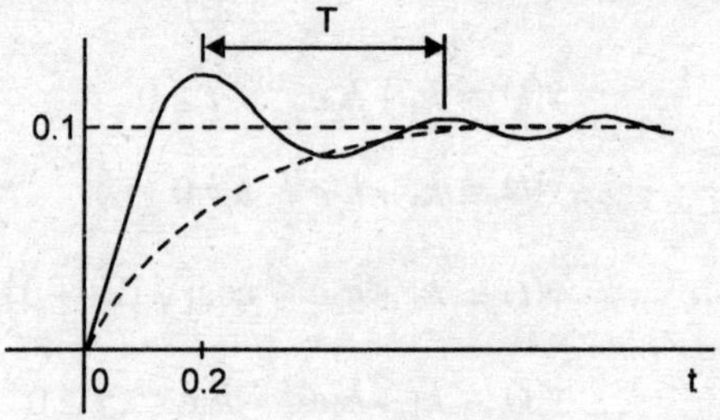

$$y_{ss} = H(0) = 0.1$$

(*c*)
$$H(10) = \frac{10}{(j10)^2 + 10j\omega + 100} = \frac{10}{100j} = 0.1e^{-j\pi/2}$$

$$y_{ss}(t) = 0.1 \cos(10t - \pi/2)$$

NETWORK FUNCTIONS AND LAPLACE TRANSFORMS

The response of a network to a complex frequency can be found in the same manner as we did in chapter 3 for real frequencies, *viz.*, voltages of the form V $e^{j\omega t}$.

$$L\, di/dt + Ri + 1/C \int i\, dt = e_i(t) = E e^{st} \qquad \text{...(7.16)}$$

is the R-L-C circuit differential equation, and the input is $e_i(t) = Ee^{st}$. The particular integral, as we know, gives the steady state current. Let us assume that the steady state current is Ie^{st}. Substituting in (7.16),

$$[L\,s + R + 1/Cs]\, Ie^{st} = Ee^{st}$$

giving
$$E/I = Ls + R + (1/Cs) \qquad \text{...(7.17)}$$

In chapter III, when we considered the input of $e^{j\omega t}$, we got the impedance Z as

$$Z = Lj\omega + R + (1/Cj\omega) \qquad \text{...(7.18)}$$

(7.17) is a general impedance function for a complex excitation exp (st) while (7.18) is the same obtained for the case of $s = j\omega$.

if we take the Laplace transform of (7.5), we get,

$$LsI(s) + RI(s) + (1/Cs)\, I(s) = E_t(s) \qquad \text{...(7.19)}$$

where it is assumed that initial values are zero and $E_t(s)$ is the transform of $e_i\,(t)$. From (7.19) we get

$$E(s)/I(s) = Ls + R + (1/Cs)$$

which is the same as (7:17). Thus we see that the Laplace transform when *s* is substituted for $j\omega$ gives the already familiar relation (7.18) for a frequency ω. Functions such the input impedance, transfer function etc., of networks expressed in terms of *s* are called network functions in terms of the complex frequency *s*. Such functions are said to be the frequency domain representations of the circuits. The corresponding time functions can be obtained by the aid of the Inverse Laplace transform.

Example 1. Find the current in the network after the switch is closed.

The circuit differential equation

$$4\,di/dt + 3i = 10e^{-2t}$$

can be transformed to give

$$4sI(s) - 4i(0) + 3I(s) = 10/(s+2)$$

If $i(0) = 0$,

$$I(s) = \frac{10}{(s+2)(4s+3)} = \frac{-2}{s+2} + \frac{8}{4s+3}$$

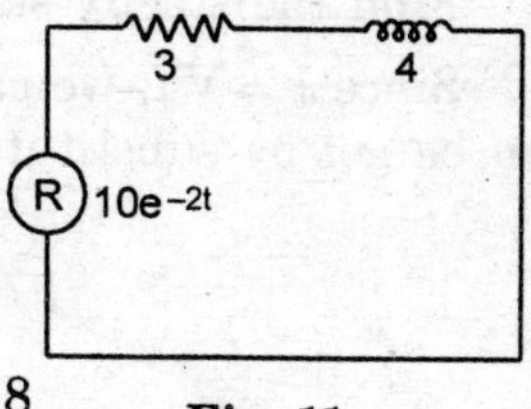

Fig. 11

The inverse transform may be found as

$$i(t) = -2e^{-2t} + 2e^{-(3/4)t}.$$

Example 2. The Laplace transform I(s) of a certain time-varying current $i(t)$ is given by

$$I(s) = \frac{3s^2+1}{(s^2+4)(2s+3)}$$

Find the value of the current at $t = 0$. Find the voltage and charge acquired by a 2 farad capacitor at the end of a long period of time, if this $i(t)$ passes through it. Express the current as a function of time.

By the theorem of initial value (7.7a),

$$\lim_{s\to\infty} s\,I(s) = i(0)$$

$$= \lim_{s\to\infty}[s(3s^2+1)/(s^2+4)(2s+3)]$$

$$= 3/2 \text{ as } s \to \infty$$

By the final value theorem,

$$\lim_{s\to 0} s\,F(s) = f(t) \text{ for } t \to \infty$$

The capacitor voltage will be $I(s)/sC = I(s)/2s = F(s)$

Hence we get $\lim sF(s) \to \frac{1}{2}(1/12) = 1/24$ volts.

The current can be expressed as a function of time by taking the inverse transform. Expand in partial fractions :

$$\frac{3s^2+1}{(s^2+4)(2s+3)} = \frac{As+B}{s^2+4} + \frac{C}{2s+3} \quad \text{...(7.20)}$$

$$3s^2+1 = (As+B)(2s+3) + C(s^2+4)$$

$$A = -23/13,\ B = -69/26,\ C = 31/13.$$

Then

$$i(t) = A\cos 2t + \tfrac{1}{2}B\sin 2t + \tfrac{1}{2}Ce^{-(3/2)t}. \quad \text{...(7.21)}$$

Example 3. A certain impedance has the following admittance function :

$$Y(s) = \frac{4s+8}{3s^2+12s+8}$$

Find the steady state current for a sinusoidal input of 10 sin $2t$.

Since I = VY, we can find the current. The admittance for the given frequency $\omega = 2$ can be got by substituting $s = j\omega$ in the above.

$$Y(j\omega) = (4j\omega+8) \div [(3j\omega)^2 + 12(j\omega)+8]$$

$$= (8j+8) \div [-24j+24j+8] = (j+1) = \sqrt{2}\angle 45°.$$

Hence current is got from A.C. theory as

$$I = 10\sqrt{2}\angle 45°$$

or in time function as $$i(t) = 10\sqrt{2}\sin(2t+\pi/4).$$

Example 4. A passive one-port (two terminal) network has the following impedance Z(*s*).

$$Z(s) = \frac{2s^2+2\times 10^5 s+10^4}{8s^2+0.1s+10^4}$$

Find the frequency at which this one-port behaves as a pure reactance. Find the value of this reactance.

We have seen that by putting $s = j\omega$, we obtain the symbolic form of the impedance at frequency ω.

$$Z(j\omega) = \frac{2(j\omega)^2+2.10^5 j\omega+10^4}{8(j\omega)^2+0.1j\omega+10^4} = \frac{(2\omega^2+10^4)+2.10^5 j\omega}{(8\omega^2+10^4)+0.1j\omega}$$

$$= \frac{[(10^4-2\omega^2)+2.10^5 j\omega][(10^4-8\omega^2)-0.1j\omega]}{(10^4-8\omega^2)^2+(0.1\omega)^2}$$

The frequency at which it is a pure reactance implies that the real part of $Z(j\omega) = 0$.

$$\text{Re}[Z(j\omega)] = 0$$

$$\therefore (10^4-2\omega^2)(10^4-8\omega^2)+0.2\times 10^5\omega^2 = 0$$

$$16\omega^4 - 8\omega^2.10^4 + 10^8 = 0$$

giving ω = 500 rad/sec upon solving the above equation.

Transfer Functions : Additional Problems

1. *Find the transfer functions of the following systems :*

(*a*) $$\dot{y}+4y = 3x$$

(*b*) $$\ddot{y}+4\dot{y}+20y = 2\dot{x}-x$$

(*c*) $$\dddot{y}-3\ddot{y}+4\dot{y}+8y = 4\ddot{x}-2\dot{x}+x$$

2. *Find the transfer function of*

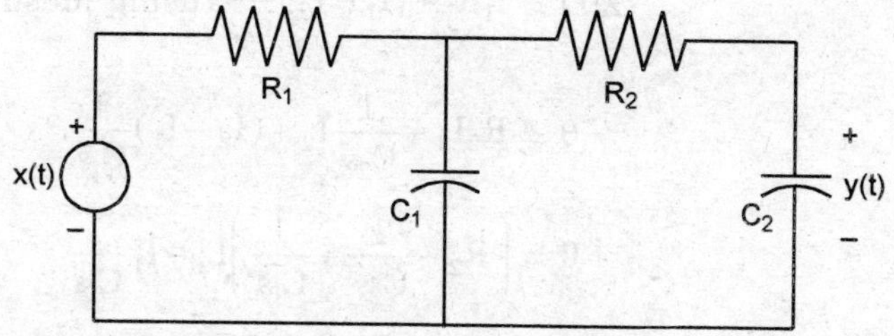

Give the result for $C_1 = C_2 = 100\mu f$, $R_1 = R_2 = 2000\ \Omega$.

3. *Find the transfer function of the following circuit where* $R_1 = R_2 = 1000\ \Omega$ *and* $C = 100\mu f$.

(*a*)

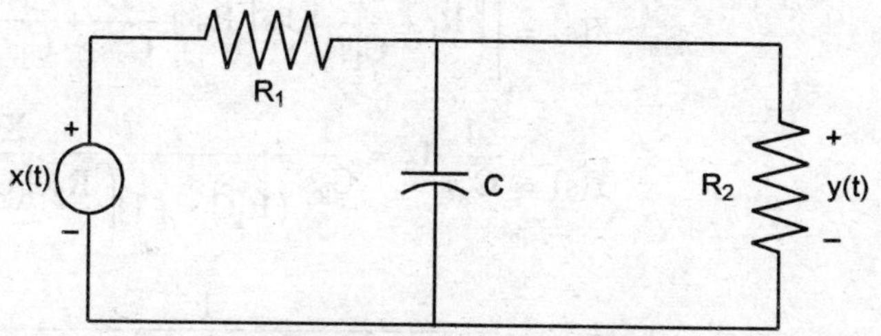

Answers to Transfer Functions – Additional Problems

1. (*a*)

$$\dot{y} + 4y = 3x$$

$$(s + 4)\ Y(s) = 3X(s)$$

$$H(s) = \frac{3}{s+4}$$

(*b*)

$$\ddot{y} + 4\dot{y} + 20y = 2\dot{x} - x$$

$$(s^2 + 4s + 20)\ Y(s) = (2s - 1)\ X(s)$$

$$H(s) = \frac{2s-1}{s^2+4s+20}$$

(*c*)

$$\dddot{y} - 3\ddot{y} + 4\dot{y} + 8y = 4\ddot{x} - 2\dot{x} + x$$

$$(s^3 - 3s^2 + 4s + 8)\ Y(s) = (4s^2 - 2s + 1)\ X(s)$$

$$H(s) = \frac{4s^2 - 2s + 1}{s^3 - 3s^2 + 4s + 8}$$

2.

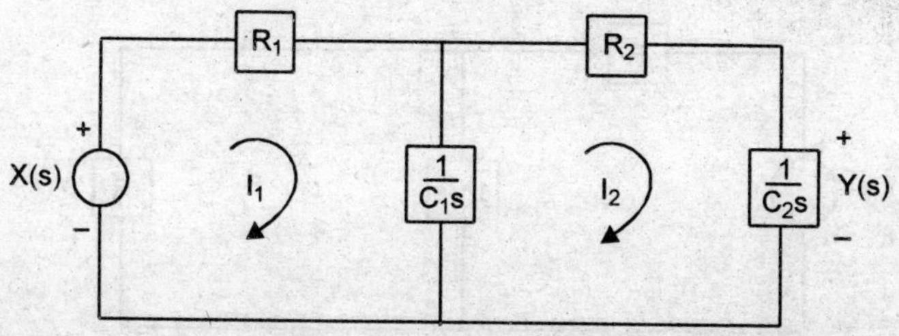

(1) $$X(s) = I_1R_1 + (I_1 - I_2)\frac{1}{C_1 s} \text{ using Mesh and analysis}$$

(2) $$0 = R_2I_2 + \frac{1}{C_2 s}I_2 + (I_2 - I_1)\frac{1}{C_1 s}$$

$$0 = \left(R_2 + \frac{1}{C_2 s} + \frac{1}{C_1 s}\right)I_2 - I_1\frac{1}{C_1 s}$$

$\Rightarrow$ $$I_1 = C_1 s\left(R_2 + \frac{1}{C_2 s} + \frac{1}{C_1 s}\right)I_2$$

into (1) : $$X(s) = \left[\left(R_1 + \frac{1}{C_1 s}\right)C_1 s\left(R_2 + \frac{1}{C_2 s} + \frac{1}{C_1 s}\right) - \frac{1}{C_1 s}\right]I_2$$

$\Rightarrow$ $$Y(s) = \frac{1}{C_2 s}I_2 = \frac{1}{C_2 s}\,\frac{X(s)}{(R_1C_1 s + 1)\left(\dfrac{R_2C_1C_2 s + C_1 + C_2}{C_1C_2 s}\right) - \dfrac{1}{C_1 s}}$$

$$H(s) = \frac{1}{(R_1C_1 s + 1)\left(\dfrac{R_2C_1C_2 s + C_1 + C_2}{C_1}\right) - \dfrac{C_2}{C_1}}$$

$$= \frac{C_1}{(R_1C_1 s + 1)(R_2C_1C_2 s + C_1 + C_2) - (C_2)}$$

If $C_1 = C_2 = 100\mu f$ and $R_1 = R_2 = 2000\Omega$

$$H(s) = \frac{0.0001}{(.2s + 1)(.00002s + .0002) - .0001}$$

$$H(s) = \frac{25}{s^2 + 15s + 50}$$

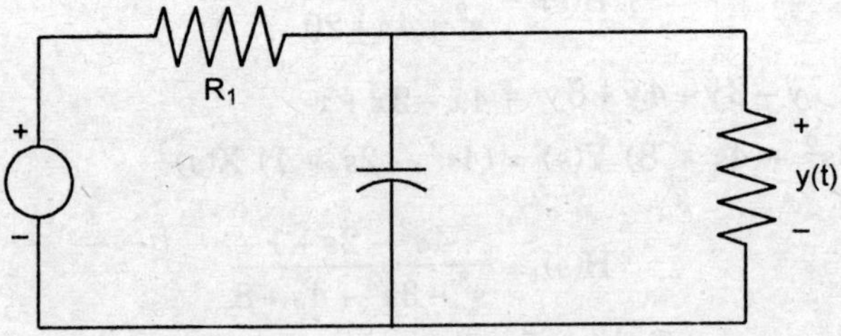

3.

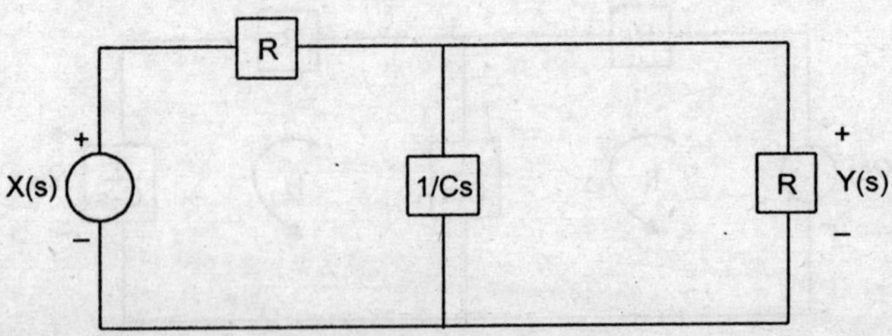

$$R \| \frac{1}{Cs} = \frac{1}{1/R + Cs} = \frac{R}{1 + RCs}$$

$$1/(s) = X(s)\frac{\dfrac{R}{1+RCs}}{R + \dfrac{R}{1+RCs}} \text{ by voltage divider law}$$

$$H(s) = \frac{R}{R + R^2Cs + R}$$

$$= \frac{1000}{2000 + 100s} = \frac{10}{s + 20}$$

Laplace Transforms

4. *Compute the Laplace transforms of the following functions :*

(a) $x(t) = 4\sin(100t)u(t)$

(b) $x(t) = 4\sin(100t - 10)u(t - 0.1)$

(c) $x(t) = 2u(t) + \delta(t-4) - \cos(5t)u(t)$

(d) $x(t) = tu(t) - 2(t-2)u(t-2) + (t-3)u(t-3)$

(e) $x(t) = u(t) - e^{-2t}\cos(10t)u(t)$

5. *Compute the inverse Laplace Transforms of the following functions :*

(a) $X(s) = \dfrac{10(s+1)}{s^2 + 4s + 3}$

(b) $X(s) = \dfrac{10(s+1)}{s^2 + 4s + 8}$

(c) $X(s) = \dfrac{2s + 100}{(s+1)(s+8)(s+10)}$

(d) $X(s) = \dfrac{10(s+1)}{s^2 + 4s + 3}e^{-2s}$

(e) $X(s) = \dfrac{20}{s(s^2 + 10s + 16)}$

(f) $X(s) = \dfrac{10(s+1)}{(s^2 + 4s + 8)s}$

6. *Find the limit as $t \to \infty$ of x(t) (if the limit exists)*

(a) $X(s) = \dfrac{10(s+1)}{s(s^2 + 4s + 3)}$

(b) $X(s) = \dfrac{10(s+1)}{s(s^2 + 4s + 8)}$

(c) $X(s) = \dfrac{10(s+1)}{s(s^2 + 2s - 3)}$

7. *Give the general form of x(t) (do not solve for the coefficients explicity).*

(a) $$X(s) = \frac{2s+100}{(s+2)(s+6)(s+10)}$$

(b) $$X(s) = \frac{2s+100}{s(s+1)(s+8)(s-4)}$$

(c) $$X(s) = \frac{s-40}{(s+1)(s+8)(s+10)}$$

(d) $$X(s) = \frac{10(s+1)}{s(s^2+4s+3)}$$

(e) $$X(s) = \frac{10(s+1)}{s(s^2+4s+8)}$$

(f) $$X(s) = \frac{s+1}{s(s^2+4)(s+8)}$$

(g) $$X(s) = \frac{20(s+1)}{(s^2+16)((s+4)^2+25)(s+1)}$$

Answers

4. (a) $$4\sin(100t)u(t) \leftrightarrow \frac{400}{s^2+100^2}$$

(b) $$4\sin(100t-10)u(t-0.1) = 4\sin(100(t-0.1))$$

$$4\sin(100t) \leftrightarrow \frac{400}{s^2+100^2}$$

$$4\sin(100(t-.1)u(t-.1) \leftrightarrow \frac{400}{s^2+100^2}e^{-0.1s}$$

(c) $$2u(t)+(t-4)-\cos\ (5t)\ u(t) \leftrightarrow \frac{2}{s}+e^{-4s}-\frac{s}{s^2+25}$$

(d) $$tu(t)-2(t-2)u(t-2)+(t-3)u(t-3) \leftrightarrow \frac{1}{s^2}-\frac{2e^{-2s}}{s^2}+\frac{1}{s^2}e^{-3s}$$

(e) $$u(t)-e^{-2t}\cos(10t) \leftrightarrow \frac{1}{s}-\frac{s+2}{(s+2)^2+10^2}$$

5. (a) $$X(s) = \frac{10(s+1)}{s^2+4s+3} = \frac{10(s+1)}{(s+1)(s+3)} = \frac{10}{s+3}$$

$$X(t) = 10e^{-3t}\,u(t)$$

(b) $$X(s) = \frac{10(s+1)}{s^2+4s+8} = \frac{10(s+1)}{(s+2)^2+4} = \frac{10(s+2)}{(s+2)^2+4} - \frac{10}{(s+2)^2+4}$$

$$X(t) = (10e^{-2t}\cos(4t) - 5e^{-2t}\sin(4t))\,u(t)$$

(c) $$X(s) = \frac{2s+100}{(s+1)(s+8)(s+10)} = \frac{C_1}{s+1} + \frac{C_2}{s+8} + \frac{C_3}{s+10}$$

$$C_1 = X(s)(s+1)\big|_{s=-1} = 1.555$$

$$C_2 = -4.667$$

$$C_3 = 4.44$$

$$X(t) = (1.555e^{-t} - 4.667e^{-8t} + 4.44e^{-10t})\,u(t)$$

(d) $$X(s) = \frac{10(s+1)}{s^2+4s+3}e^{-2s} = X_1(s)e^{-2s}$$

From (a) $$X_1(t) = 10e^{-3t}\,u(t)$$

so $$x(t) = 10e^{-3(t-2)}\,u(t-2)$$

(e) $$X(s) = \frac{20}{s(s^2+10s+16)} = \frac{C_1}{s} + \frac{C_2}{s+2} + \frac{C_3}{s+8}$$

$$C_1 = \frac{5}{4},$$

$$C_2 = \frac{20}{s(s+8)}\bigg|_{s=-2} = \frac{-20}{12} = \frac{-5}{3}, \quad C_3 = \frac{20}{s(s+2)}\bigg|_{s=-8} = \frac{5}{12}$$

$$X(t) = \left(\frac{5}{4} - \frac{5}{3}e^{-2t} + \frac{5}{12}e^{-8t}\right)u(t)$$

(f) $$X(s) = \frac{10(s+1)}{s(s^2+4s+8)} = \frac{10(s+1)}{s\left((s+2)^2+4\right)} = \frac{C_1}{s} + \frac{C_2 s + C_3}{(s+2)^2+4}$$

$$C_1 = 5/4$$

$$10(s+1) = \frac{5}{4}(s^2+4s+8) + C_2 s^2 + C_3 s \Rightarrow C_2 = \frac{-5}{4}, C_3 = 5$$

$$X(s) = \frac{5/4}{5} + \frac{-5/4s+5}{(5+2)^2+4} = \frac{5/4}{s} - \frac{5/4(s+2)}{(s+2)^2+4} + \frac{15/2}{(s+2)^2+4}$$

$$X(t) = \frac{5}{4}u(t) - \frac{5}{4}e^{-2t}\cos(2t)\,u(t) + \frac{15}{4}e^{-2t}\sin(2t)\,u(t)$$

6. (a) $$X(s) = \frac{10(s+1)}{s(s^2+4s+3)}$$

poles of $sX(s)$ are -3, $-1 \Rightarrow$ limit exists

$$\lim_{t\to\infty} x(t) = \lim_{s\to 0} sX(s) = \frac{10}{3}$$

(*b*) $$X(s) = \frac{10(s+1)}{s(s^2+4s+8)}$$

poles of $sX(s)$ are $-2 \pm 2j$

$< 0 \Rightarrow$ limit exists

$$\lim_{t\to\infty} x(t) = \lim_{s\to 0} sX(s) = \frac{5}{4}$$

(*c*) $$X(s) = \frac{10(s+1)}{s(s^2+2s-3)}$$

poles of $sX(s)$ are 3, –1 ; 3 > 0 so no limit exists

7. (*a*) $$X(s) = \frac{2s+100}{(s+2)(s+6)(s+10)}$$

$$x(t) = C_1 e^{-2t} + C_2 e^{-6t} + C_3 e^{-10t},\ t \geq 0$$

(*b*) $$X(s) = \frac{2s+100}{s(s+1)(s+8)(s-4)}$$

$$x(t) = C_1 + C_2 e^{-t} + C_3 e^{-8t} + C_4 e^{4t},\ t \geq 0$$

(*c*) $$X(s) = \frac{s-40}{(s+1)(s+8)(s+10)},$$

$$x(t) = C_1 e^{-t} + C_2 e^{-8t} + C_3 e^{-10t},\ t \geq 0$$

(*d*) $$X(s) = \frac{10(s+1)}{s(s^2+4s+3)} = \frac{10(s+1)}{s(s+1)(s+3)}$$

$$x(t) = C_1 + C_2 e^{-t} + C_3 e^{-3t},\ t \geq 0$$

(*e*) $$X(s) = \frac{10(s+1)}{s(s^2+4s+8)} = \frac{10(s+1)}{s\left((s+2)^2+4\right)}$$

$$x(t) = C_1 + C_2 e^{-2t}\cos(2t+\theta),\ t \geq 0$$

(*f*) $$X(s) = \frac{s+1}{s(s^2+4)(s+8)},$$

$$x(t) = C_1 + C_2\cos(2t+\theta) + C_3 e^{-8t},\ t \geq 0$$

(*g*) $$X(s) = \frac{20(s+1)}{(s^2+16)(s+4)^2+25)(s+1)},$$

$$x(t) = C_1\cos(4t+\theta_1) + C_2 e^{-4t}\cos(5t+\theta_2) + C_3 e^{-t},\ t \geq 0$$

EXERCISES

1. Find the Laplace transforms of the following functions :

1. $t\,e^{-5t}$. **[Ans.** $(1/(s + 5)^2]$

2. $t^2\,e^{3t}$. **[Ans.** $2/(s - 3)^2]$

3. $t \cos 2t$. **[Ans.** $(s^2 - 4)/(s^2 + 4)^2]$

4. $t \cos^2 t$. **[Ans.** $1/2s^2 + 2s/\ (s^2 + 4)^2]$

5. $t^2 \sin 2t$. **[Ans.** $12s^2 - 16/(s^2 + 4)^3]$

6. $t^2\,e^{-3t}$. **[Ans.** $2/(s + 3)^2]$

7. $t \sinh at$. **[Ans.** $2as/(s^2 - a^2)^2]$

8. $t \cosh at$. **[Ans.** $s^2 + a^2/(s^2 - a^2)^2]$

9. $\sin at - at \cos at$. **[Ans.** $2a^3/(s^2 + a^2)^2]$

10. $t\,e^{-t} \cos t$. **[Ans.** $s^2 + 2s/(s^2 + 2s + 2)^2]$

2. Find the inverse transforms :

1. $\dfrac{1}{s^2\,(s^2 + a^2)}$.

2. $\dfrac{s^2}{(s^2 + a^2)^2}$. $\left[\textbf{Ans. } \dfrac{1}{2a}\sin at + \dfrac{t}{2}\cos at\right]$

3. $\dfrac{s}{(s^2 + 4)^2}$. **[Ans.** $t/4 \sin 2t]$

4. $\dfrac{1}{(s^2 + 9)^2}$.

5. $\dfrac{3a^2}{s^3 + a^3}$.

6. $\dfrac{s^2 - a^2}{(s^2 + a^2)^2}$.

7. $\dfrac{1}{s^2(s^2 + 1)(s^2 + 9)}$.

8. $\dfrac{1}{s^4 - a^4}$.

9. $\dfrac{6s + 3}{(s - 1)\,(s^2 + 2s + 5)}$.

10. $\dfrac{s + 1}{s^2 + 2s}$. **[Ans.** $(1/2 + 1/2e^{-2t})]$

11. $\log (s+1)$. [**Ans.** e^{-t}/t]

12. $\log (s-1)$. [**Ans.** e^{t}/t]

13. $\log \left(\dfrac{1+s}{s^2}\right)$.

14. $\dfrac{1}{s\,(s+2)^3}$.

15. $\dfrac{s}{(s^2+4)\,(s^2+1)}$. $\left[\textbf{Ans. } \dfrac{1}{3}(\cos t - \cos 2t)\right]$

16. $\dfrac{s^2-s+2}{s\,(s-3)\,(s+2)}$. $\left[\textbf{Ans. } \dfrac{1}{3} + \dfrac{8}{15}e^{3t} + \dfrac{4}{5}e^{-2t}\right]$

17. $\dfrac{2s-1}{s^2\,(s-1)^2}$. [**Ans.** $t(e^t - 1)$]

18. $\dfrac{1}{s(s^2-2s+5)}$. $\left[\textbf{Ans. } \dfrac{1}{5}\left[1 - e^t \cos 2t + \dfrac{1}{2}e^t \sin 2t\right]\right]$

3. Find the region of convergence of the transfer function :

(1) $\dfrac{s-5}{(s+2)\,(s-6)}$. [**Ans.** between 2 and 6]

(2) $\dfrac{s}{(s+1)\,(s-3)}$. [**Ans.** between 1 and 3]

4. For the system given below,

$$\ddot{y} + 8\dot{y} + 116y = 116x$$

(*a*) Find the transfer function.

(*b*) Give the poles and zeros.

(*c*) Give the general form of the response $y(t)$ to a step input (do not solve explicitly).

[**Ans.** (*a*) $116/(s^2 + 8s + 116)$,

(*b*) Zeros → none, poles $-4 \pm 10j$

(*c*) $y(t) = C_1 + C_2 e^{-4t} \cos(10t + \theta),\ t \geq 0$]

5. Repeat Problem 4 for the system given below. In additon, compare the types of poles of this system to those in Problem 4 and use this to explain the resulting behaviour in the step response plots.

$$\ddot{y} + 8\dot{y} + 116y = 116x$$

[**Ans.** (*a*) $12/s^2 + 8s + 12$,

(*b*) No zeros Poles at -2, -6,

(*c*) $y = C_1 + C_2 \exp(-6t) + C_3 \exp(-2t)$ for $t \geq 0$

(No oscillations)]

8

Z-TRANSFORMS

INTRODUCTION TO DISCRETE SIGNALS

Signals are generated in the real world as information relating to events of various causes. Whether signals are generated by us for a purpose, such as in a broadcast station, or they are gleaned from an external source, such as the ECG of the physiological origin, they need to be processed in some way or other. A simple tuned filter in a radio receiver processes the received signal by selectively tuning to the desired broadcast frequency. It does this all the time the signal is in the air and is a real-time signal processor. Consisting as it does of a simple coil and a tuning capacitor, it filters out the wanted frequency and hands it over to the detector circuit for continuous listening of the broadcast at the wanted frequency. It does the wonderful job of picking the signal of that chosen frequency from the spectrum of the various frequencies that may be picked by the receiver antenna. It is an analog signal processor which does the job of a tunable band-pass filter.

Using passive components of resistors, capacitors and inductors, several filters can be fabricated and that is the realm of analog filters. This subject was developed during the fifties and sixties of this century. Subsequently, when operational amplifiers and active circuitry on a chip were developed during the late sixties, many active filter circuits were evolved to provide improved performance and to meet strict design specifications of filters. For example, with only passive components, it is not easy to design, say, a low pass filter, with sharp cut-off at the low-pass edge or with a clean flat response without any droop in the pass-band. The use of OPAs and active filter theory enabled the development of such precise filters. Those were needed in the improved communication techniques that were developed in the 1970's in line communication.

Signal processing comprises generally a certain number of functions or operations which could be made on the signals to improve the desired energy components and discriminate against noise or interference which could have corrupted the signals. The main processing operations are: filtering estimation, transformation, coding and identification.

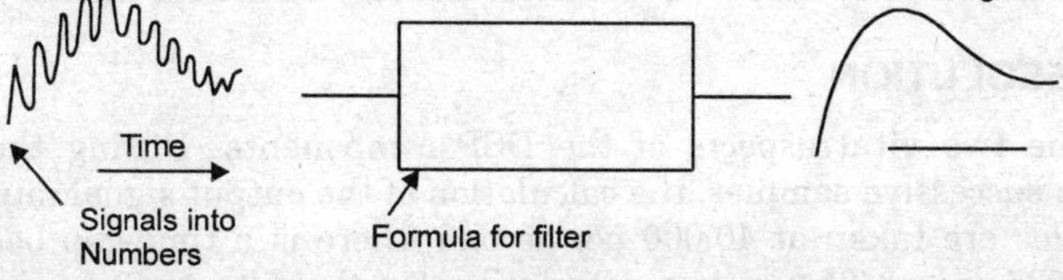

Fig. 1. DSP deals with numbers that are samples of the signals

Today, the several methods of line and radio communication that are employed to squeeze information of various kinds (voice, data, image signals etc.) and transmit all over the globe and into space, have stringent requirements in the retrieval of such signals that are processed by the equipment of the receiving end. First, signals need to be processed in specific ways particular to the type of the signal. For example, a quadrature-amplitude-modulated phase-shift-keyed (QAM-PSK) signal from a telephone modem that sends data has to be processed in circuitry that are specific to the QAM technique. Filters which have to deal with these signals are not easy to design using merely active and passive components.

In digital communication, cellular telephony and satellite based mobile telephony, DSP is used. In the transmitter, by a suitable choice of the modulation, the speech signal is encoded and filtered to satisfy the low transmit bit rate in the transmit channel. At the receiver, the signal is again digitally filtered to reduce noise and adjacent-channel interferences. The digital CD (Compact Disc), the Video-phone which senses the coded images over telephony channel together with compressed voice signals, and the answering machine in digital telephone (tapeless), semiconductor memory storage etc. are all typical present day DSP consumer applications in communication.

Can we not numerically evaluate the signal for processing it in the designed manner ? If the input signal is available in numbers and if the operation of filtering or whatever processing can be written down into a mathematical formula (or algorithm), then surely one could evaluate the numbers that represent that output signal after such processing. What would this process need ? Essentially, it will involve:

1. The signal at input must be **converted into numbers.** This must be done by taking samples of the analog signals at instants in time that are equally spaced. The more the samples that are taken per second, the better it would be, because even fine and fast variations of input signal can only then be taken care of.
2. To do the formula for finding the numeric value of the output signal, a numerical calculation has to be done. That would need a computer or similar microprocessor.
3. The numbers that are **evaluated** have to be done fast enough; as fast as the input samples are being collected. If the evaluation takes time, then it is required to reduce the time of sampling. That means, fast or high frequency components of a signal will be missed.
4. The evaluated signals in numbers have to be tranformed back into analog signal. This needs a DAC. It takes numbers and forms a voltage level in an analog fashion that represents the number. The essentials of such digital signal processing is thus shown in Fig. 2. The analog-to-digital converted (ADC) is at the front-end. The digital-to-analog converter (DAC) is at the rear end. The numbers represent signal samples from the data which are processed using a microprocessor. It needs the associated components or program memory and data memory.

SPEED AND RESOLUTION

These are the two vital aspects of the DSP components. During the time that is available between successive samples, the calculation of the output signal must be done. For example, if samples are taken at 40,000 per second, there is a time-gap between samples of 1/40,000 or 0.025 ms or 25 μs. Any calculation for the filtering or processing that is intended will have to be completed well before this 25 μs. The calculations may involve several multiplications and additions depending on the type and nature of the filter operation. Hence, unless the unit time for one multiply and add operation is itself very small (less than 1 μs), it is not possible to perform real-time filtering.

Another aspect is that of the digitization of the signal, or converting the signal sample into numbers. The ADC itself takes a finite time for conversion. A fast device may typically take 1 μs per conversion. A sample and hold circuit is usually employed so that the signal is held constant during the time of conversion.

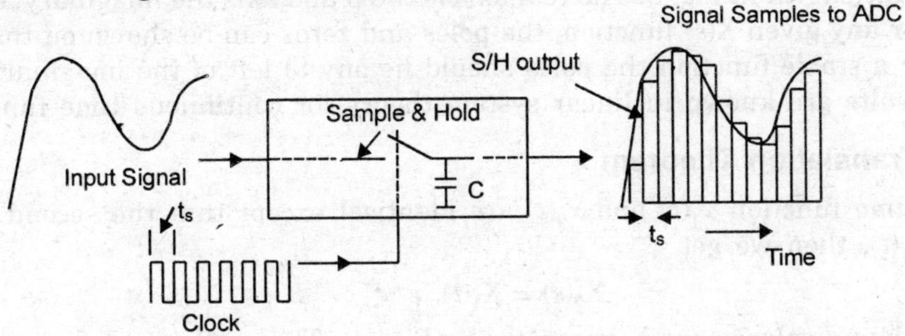

Fig. 2. A sample and hold circuit keeps the chosen sample constant for the ADC to convert correctly.

LAPLACE AND Z-TRANSFORMS

From the Laplace transform of the time function $x(t)$, given by

$$X(s) = \int_0^{\infty} x(t)e^{-st}dt \qquad ...(8.1)$$

we can get the frequency response of the continuous time function $x(t)$ at any frequency ω by substituting $s = j\omega$.

Here s is complex variable given by

$$s = \sigma + \omega j \qquad ...(8.2)$$

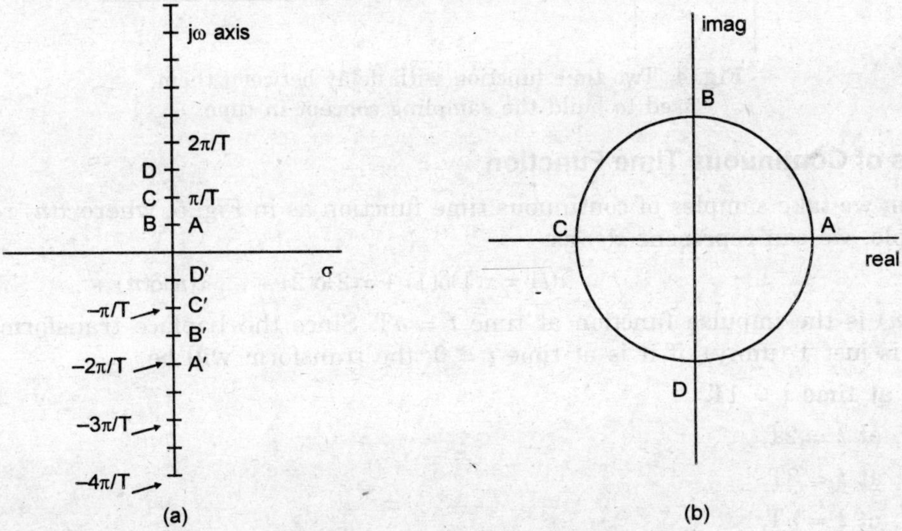

Fig. 3. (*a*) *s*-plane (*b*) *z*-plane.

X(s) is a complex frequency function. The analogy with the Fourier transform is noted, but the two differences are:

1. Instead of just $j\omega t$, we have in (8.1), $\sigma + j\omega t$. The factor $e^{-\alpha t}$ is a convergent factor to permit the integral to converge.
2. The limits of integration starts at 0 instead of $-\infty$. The function X(s) is the frequency domain representation of the time function $x(t)$. The s-plane is the complex plane shown in Fig. 3. The real axis is the σ and axis the imaginary axis is $j\omega$ axis. For any given X(s) function, the poles and zeros can be shown on ths s-plane and for a stable function the poles should lie ony to left of the imaginary axis. These results are known in linear system theory for continuous time functions.

The Time Translation Theorem

If two time function $x_1(t)$ and $x_2(t)$ are identical except that the second is a delayed version of $x_1(t)$, then we get

$$X_2(s) = X_1(t).\ e^{-sT}, \quad ...(8.3)$$

where T is time displacement between them (Fig. 4). Thus e^{-sT} is the factor representing the time delay T.

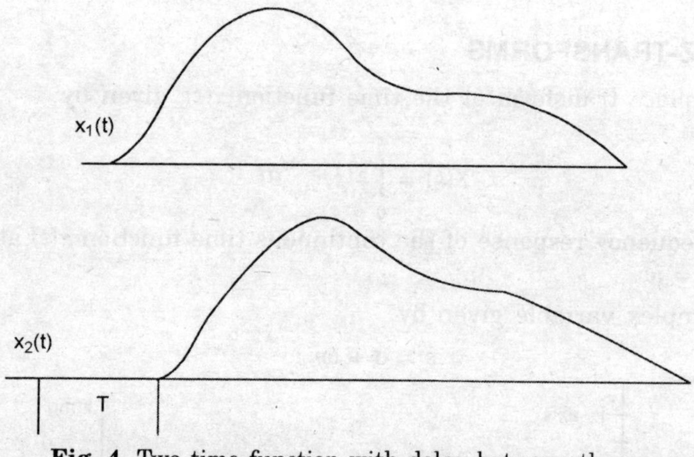

Fig. 4. Two time function with delay between them, used to build the sampling concept in time.

Samples of Continuous Time Function

When we take samples of continuous time function as in Fig. 5, where $x(n)$ represents nth sample, we can represent $x(t)$ as

$$x(t) = x(1)\delta(1) + x(2)\delta(2) + ... x(n)\delta(n) + ... \quad ...(8.4)$$

where $\delta(n)$ is the impulse function at time $t = n$T. Since the Laplace transform of the δ function is just 1 (unity), if it is at time $t = 0$, the transform will be

e^{-sT}, at time t = 1T,

e^{-2sT}, at t = 2T,

e^{-3sT}, at t = 3T

and e^{-nsT}, at t = nT,

we can write this as

$$X(s) = \sum_{n=0}^{\infty} x(n)z^{-n} \qquad ...(8.5)$$

where $z = e^{sT}$. The same equation can therefore be written in terms of the transformed variable z as

$$X(z) = \sum_{n=0}^{n=\infty} x(n)z^{-n} \qquad ...(8.6)$$

The above is the z-transform of the time sequence $x(n)$ petaining to the continuous time function $x(t)$.

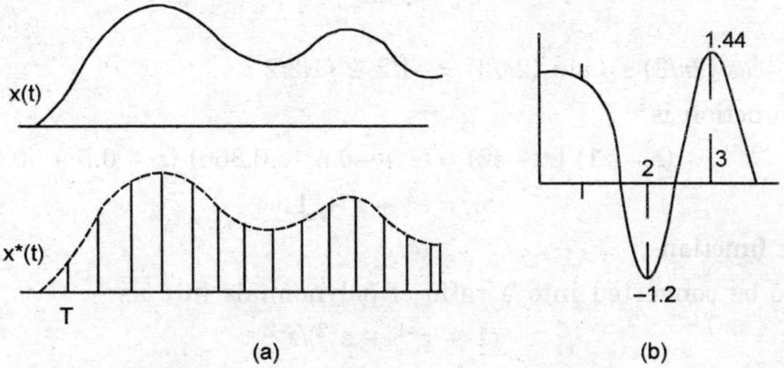

Fig. 5 (*a*) Samples of continuous time function in steps of T give a sequence of impulses. (*b*) Example of sequence for problem 8.6.

The $z = e^{sT}$ is the complex transformation, which converts the s plane into the z-plane as in Fig. 3(*b*). When $s = j\omega$, s varies on the imaginary axis. As s takes values from $-\infty$ to $+\infty$, the z function varies only by a limited amount.

When $s = j\omega = 2\pi j/T$, $z = \exp(j2\pi) = 1 + j0$.

When $s = -2\pi j/T$, it is again $z = \exp(-j2\pi) = 1 + j0$.

Between the frequencies $-1/T$ to $+1/T$, the z-function varies on a unit circle. For example at the frequency f_s,

$z = \exp(j2\pi f_s T) = \exp(j2\pi.1/2T.T) = \exp(j\pi)$ which is equal to $-1 + j0$. Thus, at the sampling frequency, the z function is negative and of value -1. So, it is clear that the $j\omega$ axis transforms into the unit circle shown in Fig. 1(*b*).

Let $$f_s = 1/T \text{ and } f_0 = 1/2T. \qquad ...(8.7)$$

$\exp(j2\pi fT) = \exp(j2\pi\{f + 1/T\}T)$ since $\exp(j.2\pi\{1/T\}.T) = \exp(j.2\pi)$, the z-function is periodic in $f_0 = 1/T$. So, the entire frequency region of the imaginary axis from $j\omega = 0$ to $j\omega = \infty$ repeats several times on the unit circle in the z-plane. The same applies for negative frequencies $\omega = -\infty$ to 0 as well.

In the s-plane, the real part σ is negative in the left half plane and these are mapped into the inside of the unit circle in the z-plane. Positive value of σ correspond to the right of the s-plane which map to points outside of the unit circle.

In the s-plane, poles to the right of $j\omega$ axis will lead to growing oscillations or instability.

When the s function is on the left hand side of the imaginary axis, it is known that the function gives rise to stable responses. On the right hand side, it is unstable. Likewise, in the z-plane, points inside the circle are stable and those outside are not. Also poles on the unit circle can be stable only if they are simple poles (not multiple ones).

Examp'ɜ 1. Determine the z transform for a notch filter which is to filter a 50 Hz signal, if the sampling rate is 150 Hz.

Sampling frequency = 150, T = 1/150

Required zero of the z-function (notch) = 50 Hz

The zero is therefore at a angle on unit circle given by

$$2\pi.(50/150) = 2\pi/3$$

There are two zeros corresponding to $\pm j\omega$, as shown at $\pm 2\pi/3$. The zeros are having the co-ordinates.

$$-\cos(2\pi/3) \pm j \sin(2\pi/3) = -1/2 \pm j\sqrt{3}/2$$

Thus the function is

$$(z - z1)(z - z2) = (z + -0.5 - j0.866)(z - 0.5 + j0.866)$$

$$= z^2 + z + 1$$

the notch filter function.

This has to be converted into a ratio of polynomials in z as

$$(1 + z^{-1} + z^{-2})/z^{-2}$$

The above can be approximated to $1 + z^{-1} + z^{-2}$, neglecting the double pole at $z = 0$.

The discrete time function difference equation results from the this as

$$Y(z) = x(z)\{1 + z^{-1} + z^{-2}\}$$

or $$y(t) = x(t) + x(t - T) + x(t - 2T)$$

or $$y_n = x_n + x_{n-1} + x_{n-2}$$

where n, n–1, n–2 refer to sample numbers.

Example 2. Find the z-transform of the exponentially decaying discrete time function given by

t	0	1	2	3	4	5
$x(t)$	1	.8	.64	.51	.41	.33

The z transform equation is just the sum of the discrete values multiplied by the exponent function.

$$X(z) = 1 + z^{-1}(0.8) + z^{-2}\ .64 + z^{-3}\ .51 + ...$$

$$= 1 + (0.8z^{-1}) + (0.8z^{-1})^2 + (0.8z^{-1})^3 + ...$$

Bionomial theorem is used to sum this infinte series to

$$= z/(z - 0.8)$$

Suppose the function had no more samples than n (not infinite), the sum would become

$$= z(1 - 0.8^n)/(z - 0.8)$$

These are the z-transforms.

Basic Definition of the z-Transform

The **z-transform** of a sequence is defined as

$$X(z) = \sum_{n=0}^{\infty} (x[n]z^{-n}) \qquad ...(8.8)$$

Sometimes this equation is referred to as the **bilateral z-transform.** At times the z-transform is defined as

$$X(z) = \sum_{n=0}^{\infty} (x[n]z^{-n}) \qquad ...(8.9)$$

which is known as the **unilateral z-transform.**

There is a close relationship between the z-transform and the **Fourier transform** of a discrete time signal, which is defined as

$$X(e^{j\omega}) = \sum_{n=0}^{\infty} (x[n]e^{-(j\omega n)}) \qquad ...(8.10)$$

Notice that when the z^{-n} is replaced with $e^{-(j\omega n)}$ the z-transform reduces to the Fourier transform. When the Fourier transform exists, $z = e^{j\omega}$, which gives the magnitude of z equal to unity.

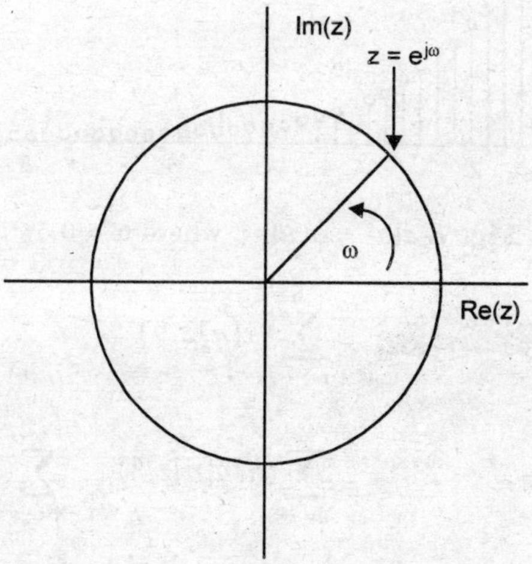

Fig. 6

The Z-plane is a complex plane with an imaginary and real axis referring to the complex-valued variable z. The position on the complex plane is given by $re^{j\omega}$, and the angle from the positive, real axis around the plane is denoted by ω. X(z) is defined everywhere on this plane. X($e^{j\omega}$) on the other hand is defined only where $|z| = 1$, which is referred to as the unit circle. So for example, $\omega = 0$ at $z = 1 + j0$ and $\omega = \pi$ at $z = -1$. This is useful because, by representing the Fourier tranform as the z-transform on the unit circle, the periodicity of Fourier transform is easily seen.

Region of Convergence

The region of convergence, known as the **ROC**, is important to understand because it defines the region where the z-transform exists. The ROC for a given $x[n]$, is defined as the range of z for which the z-transform converges. Since the z-transform is a **power series,** it converges when $x[n]z^{-n}$ is absolutely summable. Stated differently,

$$\sum_{n=\infty}^{\infty}\left(\left|x[n]z^{-n}\right|\right) < \infty \qquad ...(8.11)$$

must be satisfied for convergence. This is best illustrated looking at the different ROC's of the z-tranfoms of $\alpha^n u[n]$ and $\alpha^n u[n-1]$.

Example. 1 For $$x[n] = \alpha^n u[n] \qquad ...(8.12)$$

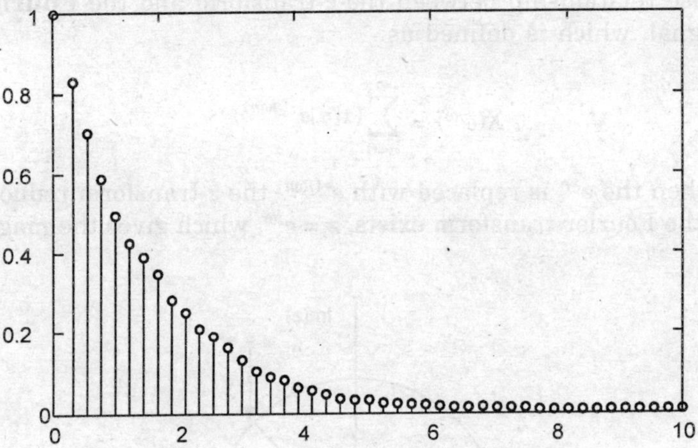

Fig. 7. $x[n] = \alpha^n u[n]$, where $\alpha = 0.5$

$$X(z) = \sum_{n=-\infty}^{\infty}(x[n]z^{-n})$$

$$= \sum_{n=-\infty}^{\infty}(\alpha^n u[n]z^{-n}) = \sum_{n=-\infty}^{\infty}(\alpha^n z^{-n}) \qquad ...(8.13)$$

$$= \sum_{n=-\infty}^{\infty}((\alpha z^{-1})^n)$$

This sequence is an example of a right-sided exponential sequence because it is non-zero for $n \geq 0$. It only converges $\left|\alpha z^{-1}\right| < 1$. When it converges,

$$X(z) = \frac{1}{1-\alpha z^{-1}} = \frac{z}{z-\alpha} \qquad ...(8.14)$$

If $\left|\alpha z^{-1}\right| \geq 1$, then the series, $\sum_{n=0}^{\infty} (\alpha z^{-1})^n$ does not converge. Thus the ROC is the range of values where

$$\left|az^{-1}\right| < 1 \qquad ...(8.15)$$

or, equivalently,

$$|z| > |\alpha| \qquad ...(8.16)$$

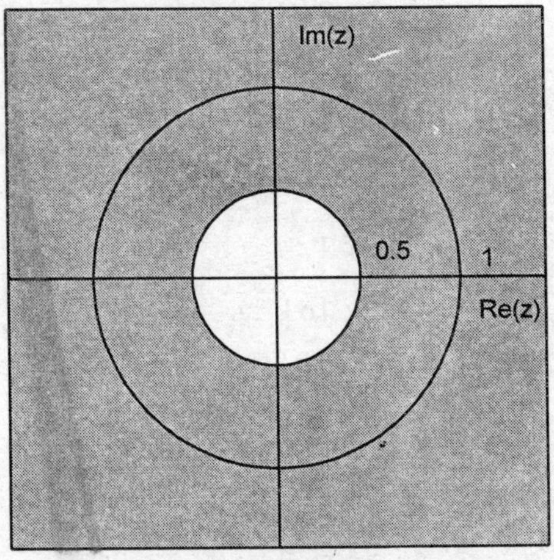

Fig. 8. ROC for $x[n] = \alpha^n u[n]$ where $\alpha = 0.5$

Example. 2 For

$$x[n] = (-(\alpha^n))u[-n-1] \qquad ...(8.17)$$

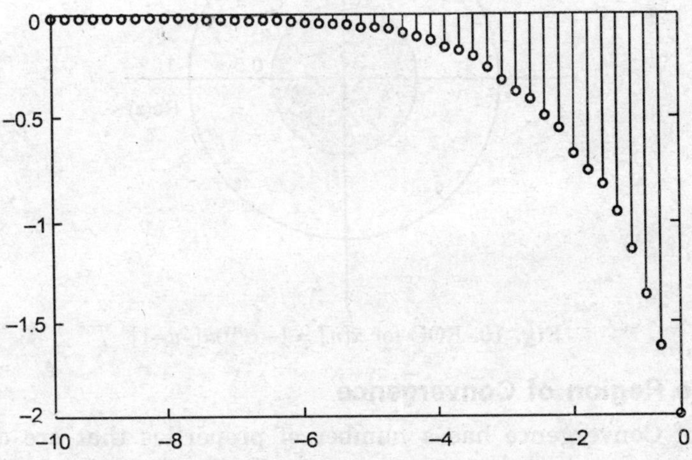

Fig. 9 $x[n] = (-(\alpha^n))u[-n-1]$ where $\alpha = 0.5$.

$$X(z) = \sum_{n=-\infty}^{\infty} (x[n]z^{-n})$$

$$= \sum_{n=-\infty}^{\infty} ((-(\alpha^n))u[-n-1]z^{-n})$$

$$= -\left(\sum_{n=-\infty}^{-1}(\alpha^n z^{-n})\right) = -\left(\sum_{n=-\infty}^{-1}((\alpha^{-1}z)^{-n})\right) \qquad ...(8.18)$$

$$= -\left(\sum_{n=-1}^{\infty}((\alpha^{-1}z)^n)\right)$$

$$= 1-\sum_{n=0}^{\infty}((\alpha^{-1}z)^n)$$

The ROC in this case is the range of values where

$$\left|\alpha^{-1}z\right| < 1 \qquad ...(8.19)$$

or, equivalently

$$|z| < |\alpha| \qquad ..(8.20)$$

If the ROC is satisfied, then

$$X(z) = 1 - \frac{1}{1-\alpha^{-1}z} = \frac{z}{z-\alpha} \qquad ...(8.21)$$

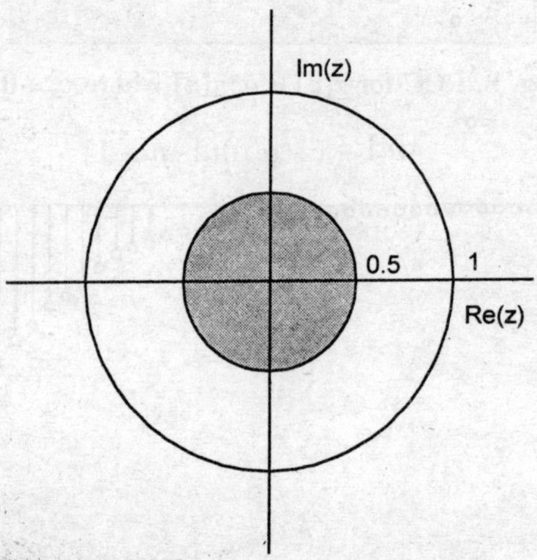

Fig. 10. ROC for $x[n] = (-(\alpha^n))u[-n-1]$

Properties of the Region of Convergence

The Region of Convergence has a number of properties that are dependent on the characteristics of the signal, $x[n]$.

- *The ROC cannot contain any* ***poles.*** By definition a pole is a where $X(z)$ is infinite. Since $X(z)$ must be finite for all z for convergence, there cannot be a pole in the ROC.
- *If x[n] is a finite-duration sequence, then the ROC is the entire z-plane, except possibly $z = 0$ or $|z| = \infty$.* A **finite-duration sequence** is a sequence that is nonzero in a finite interval $n_1 \leq n \leq n_2$. As long as each value of $x[n]$ is finite then the sequence will be absolutely summable. When $n_2 > 0$ there will be a z^{-1} term

and thus the ROC will not include $z = 0$. When $n_1 < 0$ then the sum will be infinite and thus the ROC will not include $|z| = \infty$. On the other hand, when $n_2 \le 0$ then the ROC will include $z = 0$, and when $n_1 \ge 0$ the ROC will include $|z| = \infty$. With these constraints, the only signal, then, whose ROC is the entire z-plane $x[n] = c\ \delta[n]$.

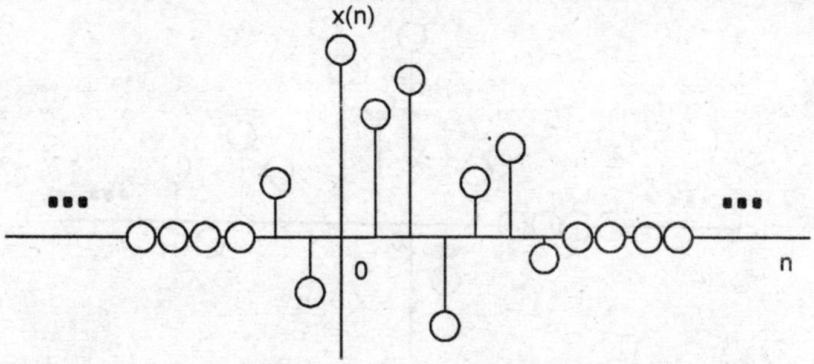

Fig. 11. An example of a finite duration sequence.

The next properties apply to infinite duration sequences. As noted above, the z-transform converges when $|X(z)| < \infty$. So we can write

$$|X(z)| = \left|\sum_{n=-\infty}^{\infty}(x[n]\ z^{-n})\right| \le \sum_{n=-\infty}^{\infty}\left(\left|x[n]\ z^{-n}\right|\right)$$

$$= \sum_{n=-\infty}^{\infty}\left(\left|x[n]\right| (|z|^{-n})\right) \quad ...(8.22)$$

We can then split the infinite sum into positive-time and negative-time portions. So

$$|X(z)| \le N(z) + P(z) \quad ...(8.23)$$

where
$$N(z) = \sum_{n=-\infty}^{-1}\left(\left|x[n]\right| (|z|)^{-n}\right) \quad ...(8.24)$$

and
$$P(z) = \sum_{n=0}^{\infty}\left(\left|x[n]\right| (|z|)^{-n}\right) \quad ...(8.25)$$

In order for $|X(z)|$ to be finite, $|x[n]|$ must be bounded. Let us then set

$$|x(n)| \le C_1 r_1^n \quad ...(8.26)$$

for
$$n < 0$$

and
$$|x(n)| \le C_2 r_2^n \quad ...(8.27)$$

for
$$n \ge 0$$

From this some further properties can be derived:

If x[n] is a right-sided sequence, then the ROC extends outward from the outermost pole in X(z). A **right-sided sequence** is a sequence where $x[n] = 0$ for $n < n_1 < \infty$. Looking at the positive-time portion from the above derivation, it follows that

$$P(z) \le C_2 \sum_{n=0}^{\infty}(r_2^{\,n}(|z|)^{-n}) = C_2 \sum_{n=0}^{\infty}\left(\left(\frac{r_2}{|z|}\right)^n\right) \quad ...(8.28)$$

Thus in order for this sum to converge, $|z| > r_2$, and therefore the ROC of a right-sided sequence is of the form $|z| > r_2$

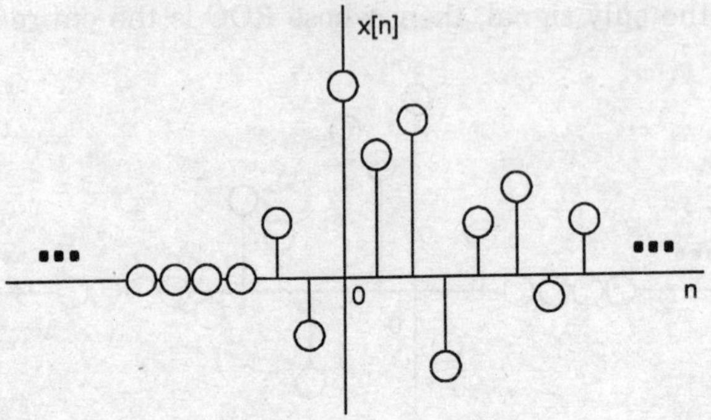

Fig. 12. A right-sided sequence.

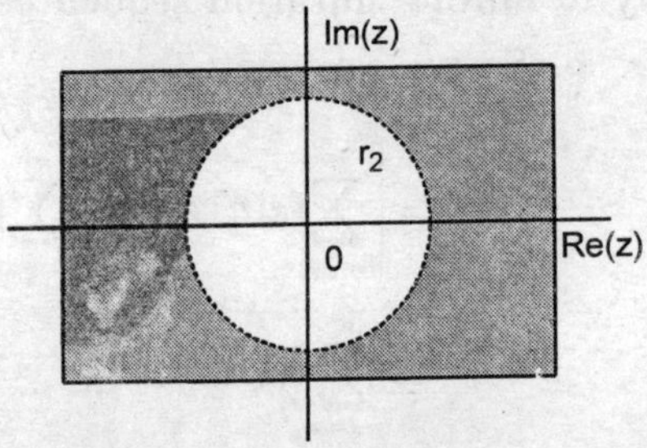

Fig. 13. The ROC of a right-sided sequence.

- *If x[n] is a left-sided sequence, then the ROC extends inward from the innermost pole in X(z).* A **left-sided sequence** is a sequence where $x[n] = 0$ for $n > n_2 > -\infty$. Looking at the negative-time portion from the above derivation, it follows that

$$N(z) \le C_1 \sum_{n=-\infty}^{-1} (r_1{}^n (|z|)^{-n}) = C_1 \sum_{n=-\infty}^{-1} \left(\left(\frac{r_1}{|z|} \right)^n \right) = C_1 \sum_{k=1}^{\infty} \left(\left(\frac{|z|}{r_1} \right)^k \right) \qquad \text{...(8.29)}$$

- Thus in order for this sum to converge, $|z| < r_1$, and therefore the ROC of a left-sided sequence is of the form $|z| < r_1$.

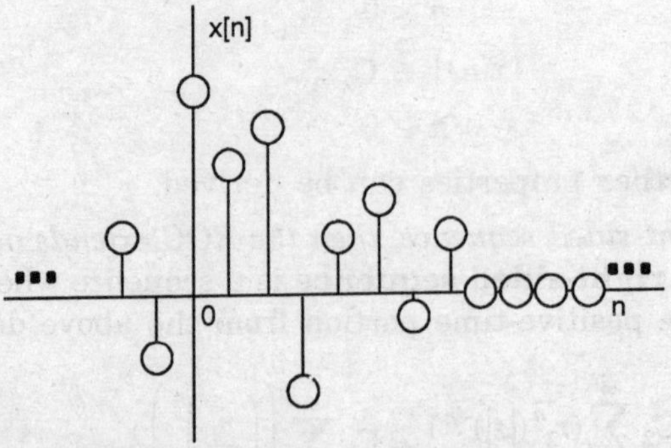

Fig. 14. A left-sided sequence.

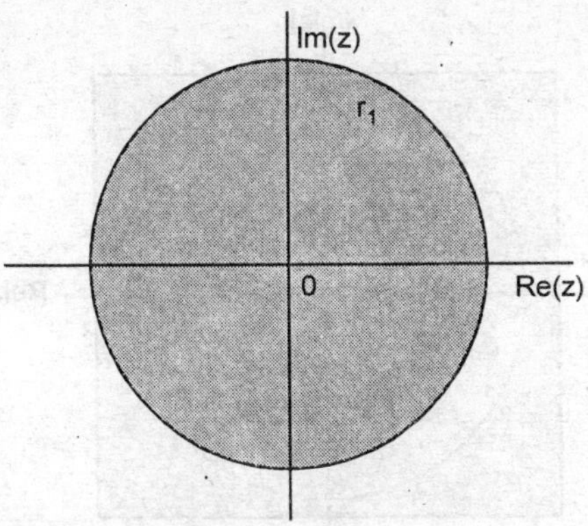

Fig. 14. The ROC of a left-sided sequence.

If $x[n]$ is a two-sided sequence, the ROC will be a ring in the z-plane that is bounded on the interior and exterior by a pole. A **two-sided sequence** is an sequence with infinite duration in the positive and negative directions. From the derivation of the above two properties, it follows that if $r_2 < |z| < r_1$ converges, then both the positive-time and negative-time portions converge and thus $X(z)$ converges as well. Therefore the ROC of a two-sided sequence is of the form $r_2 < |z| < r_1$.

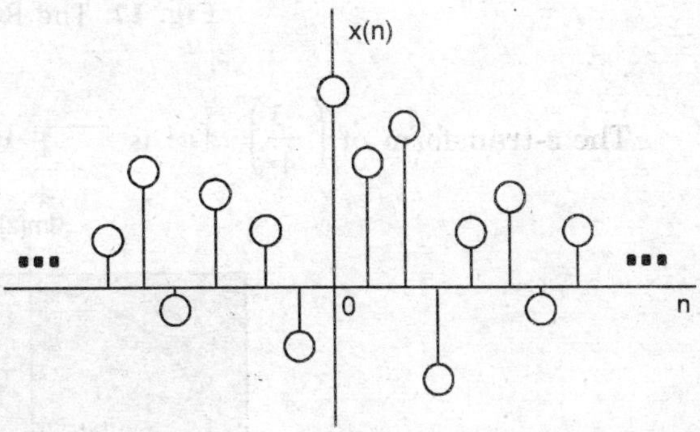

Fig. 15. A two-sided sequence.

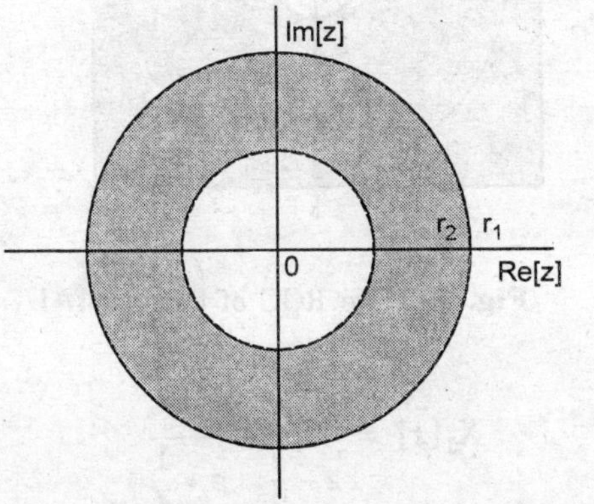

Fig. 16. The ROC of a two-sided sequence.

Example 1.

Lets take
$$x_1[n] = \left(\frac{1}{2}\right)^n u[n] + \left(\frac{1}{4}\right)^n u[n] \quad \text{...(8.30)}$$

The z-transform of $\left(\frac{1}{2}\right)^n u[n]$ is $\dfrac{z}{z-\frac{1}{2}}$ with an ROC at $|z| > \frac{1}{2}$.

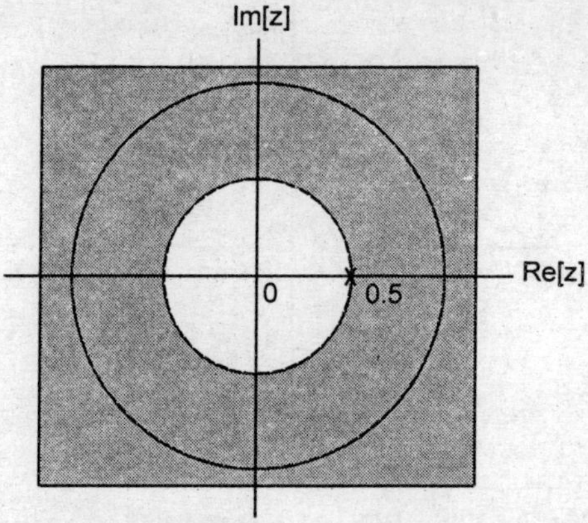

Fig. 17. The ROC of $\left(\frac{1}{2}\right)^n u[n]$

The z-transform of $\left(\frac{-1}{4}\right)^n u[n]$ is $\frac{z}{z+\frac{1}{4}}$ with an ROC at $|z| > \frac{-1}{4}$.

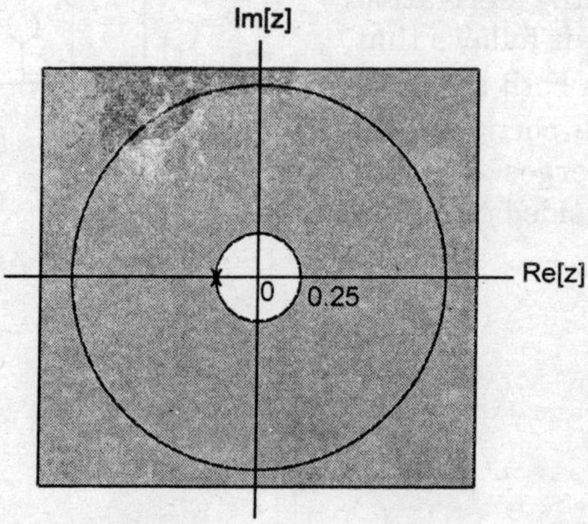

Fig. 18. The ROC of $\left(\frac{-1}{4}\right)^n u[n]$

Due to linearity,

$$X_1[z] = \frac{z}{z-\frac{1}{2}} + \frac{z}{z+\frac{1}{4}}$$

$$= \frac{2z\left(z-\frac{1}{8}\right)}{\left(z-\frac{1}{2}\right)\left(z+\frac{1}{4}\right)} \qquad \text{...(8.31)}$$

By observation it is clear that there are two zeros, at 0 and $\frac{1}{8}$, and two poles, at $\frac{1}{2}$, and $\frac{-1}{4}$. Following the above properties, the ROC is $|z| > \frac{1}{2}$.

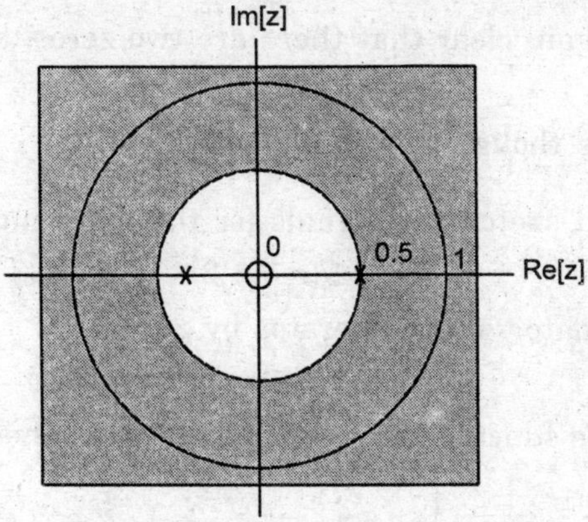

Fig. 19. The ROC of $x_1[n] = \left[\frac{1}{2}\right]^n u[n] + \left[\frac{-1}{4}\right]^n u[n]$

$$x_2[n] = \left(\frac{-1}{4}\right)^n u[n] - \left(\frac{1}{2}\right)^n u[-n-1] \qquad \text{...(8.32)}$$

The z-transform and ROC of $\left(\frac{-1}{4}\right)^n u[n]$ was shown in the example above. The z-transorm of $\left(-\left(\left(\frac{1}{2}\right)^n\right)\right)u[-n-1]$ is $\frac{z}{z-\frac{1}{2}}$ with an ROC at $|z| > \frac{1}{2}$.

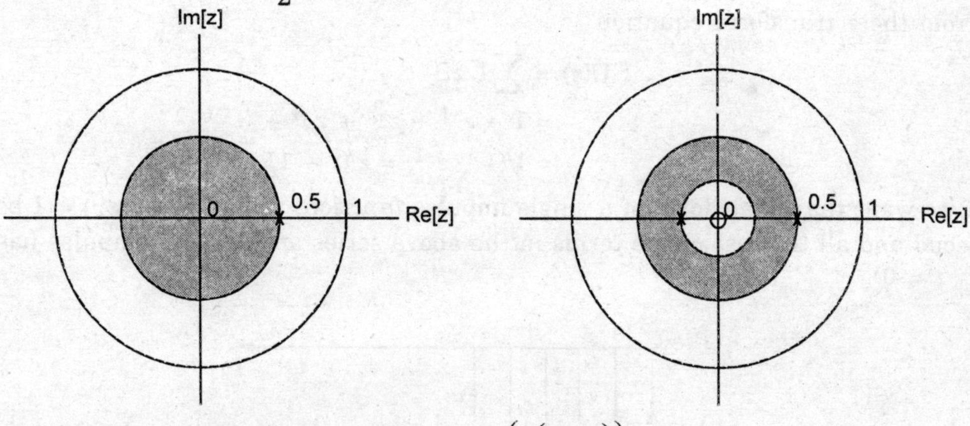

Fig. 20 : The ROC of $\left(-\left(\left(\frac{1}{2}\right)^n\right)\right)u[-n-1]$

Once again, by linearity,

$$X_2[z] = \frac{z}{z+\frac{1}{4}} + \frac{z}{z-\frac{1}{2}} = \frac{z\left(2z-\frac{1}{8}\right)}{\left(z+\frac{1}{4}\right)\left(z-\frac{1}{2}\right)} \qquad \text{...(8.33)}$$

By observation it is again clear that there are two zeros, at 0 and $\frac{1}{16}$, and two poles, at $\frac{1}{2}$, and $\frac{-1}{4}$. In this case though, the ROC is $|z| < \frac{1}{2}$.

Example. 2. Find and sketch the signals for the following function of (z).

$$X(z) = 1/[z + 1.2]$$

Multiply both denominator and numerator by z^{-1},

$$X(z) = z^{-1}/(1 + 1.2z^{-1})$$

Using expansion of the function $1/(1 + x)$ by Binomial theorem,

$$\frac{1}{1+x} = 1 - x + x^2 - x^3 \ldots$$

$$= z^{-1} [1 - 1.2\ z^{-1} + \{1.2\ z^{-1}\}^2 + \ldots$$

$$= z^{-1} - 1.2\ z^{-2} + 1.44\ z^{-3} \ldots$$

The coefficients of the z powers represent the amplitudes of the discrete time samples at times 0, T, 2T ... and are shown in Fig. 5. The continuous time function can be imagined to pass through these points.

Example. 3. Find the z-transform of the discrete time function of a unit step continuous time function.

All points at any $t = n$T $(0 < n < \infty)$ have 1 as the value, for unit step and for $t < 0$, as zero. (Fig. 21).

From the z transform equation

$$U(z) = \sum 1.\ z^{-n}$$

$$= 1 + z^{-1} + z^{-2} + z^{-3} + \ldots$$

$$= 1/(1 - z^{-1}) = z/(z - 1).$$

If we want the z-transform of a single impulse function, we get it as $\delta(z) = 1$ because, the second and all the rest of the terms in the above series are absent. (Impulse has value at only $t = 0$).

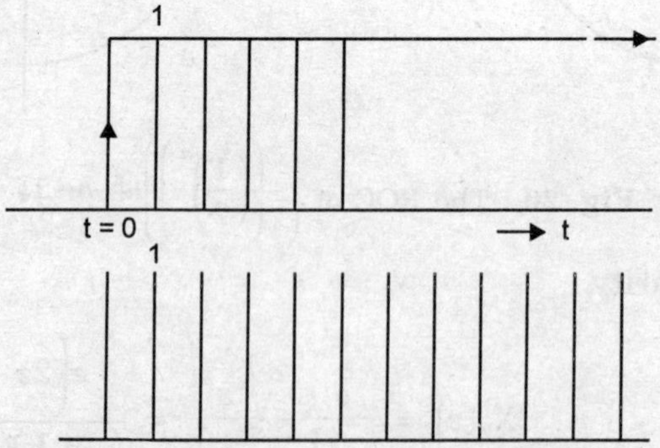

Fig. 21. Unit step and its impulse response

Example : Find z-transform of $x(n) = \left(\frac{1}{2}\right)^n$ for $n \geq 5$, and zero for $n < 5$.

$$X(z) = \sum_{5}^{\infty} (1/2)^n z^{-n}$$

$$= \sum_{0}^{\infty} \left(\frac{1}{2z}\right)^n - \left(1 + \frac{1}{2z} + \frac{1}{(2z)^2} + \frac{1}{8z^3} + \frac{1}{16z^4}\right)$$

The first geometric series upto ∞ is $\frac{1}{1 - 0.5z^{-1}}$.

The second 5 term geometric series is $\frac{1 - (0.5z^{-1})^4}{1 - 0.5z^{-1}}$, giving $\frac{(0.5z^{-1})^4}{1 - 0.5z^{-1}}$ as the result.

Impulse Response of Discrete Time Systems

Example 1 : Find the impuse response for the following discrete time systems:

(*a*) $y(n) + 0.2\, y(n-1) = x(n) - x(n-1)$.

(*b*) $y(n) = 0.24[x(n) + x(n-1) + x(n-2) + x(n-3)$.

Solution : (*a*) $y(n) + 0.2\, y(n-1) = x(n) - x(n-1)$.

For impulse response, the function $x(n)$ is the impulse function $\delta(t)$. Then $y(n)$ is $h(n)$.

$$\therefore \quad h(n) + 0.2\, h(n-1) = \delta(n) - \delta(n-1)$$

$$h(0) = -0.2\, h(-1) + \delta(0) - \delta(-1) = 0 + 1 - 0 = 1$$

$$h(1) = -0.2\, h(0) + \delta(1) - \delta(0) = -0.2 - 1 = -1.2$$

$$h(2) = -0.2\, h(1) = 0.24$$

$$h(3) = -0.2\, h(2) = -0.048$$

$$\therefore \quad h(n) = (-0.2)^{n-1}(-1.2) \text{ for } n \geq 1$$

(*b*)

$$y(n) = 0.24\, (x(n) + x(n-1) + x(n-2) + x(n-3)$$

$$h(n) = 0.24\, (\delta(n) + \delta(n-1) + \delta(n-2)\, \delta(n-3))$$

$$\left.\begin{aligned} &= 0.24 \\ &= 0 \end{aligned}\right\} \begin{aligned} &0 \leq n \leq 3 \\ &\text{otherwise.} \end{aligned}$$

Criterion of Functions which are Linear and Time Invariant

We repeat these two principles here again:

1. Linearity implies superposition property, and hence a sequence $x(n)$ is linear if $y(n)$ is known for a given $x(n)$, then for a sequence which is $ax(n)$ where a is constant, the y sequence is also a times $x(n)$.

   In the same way, two sequences $x_1(n)$ and $x_2(n)$ are giving outputs $y_1(n)$ and $y_2(n)$, by a function that operates on the input to give the output $y(n)$ will also operate on the outputs accordingly :

$$F\{a_1\, x_1(n) + a_2\, x_2(n)\} = a_1\, y_1(n) + a_2\, y_2(n)$$

Example : Check if the function $y(n) = [x(n)]^2$ is linear.

Consider the sequence x as split into two : $x_1(n)$ and $x_2(n)$.

Then $$\{x(n)\}^2 = x_1(n)^2 + x_2(n)^2 + 2\, x_1(n)\, x_2(n)$$

Since this is not equal to $x_1(n)^2 + x_2(n)^2$, which is $y_1(n) + y_2(n)$, the system is not linear.

2. **Time invariance :** A sequence which is $x(n)$ produces its output $y(n)$ through a (filter) function F.

   If the sequence is shifted by k samples, then the output is also shifted by k samples of $y(n)$. Then it is time invariant.

Example : If $y(n) = x(-n)$ then show that the system is not time invariant.

Let x_n be assumed as $a_1, a_2, a_3, a_4....$ Then $x(-n)$ is given by

$$x(n) = \{a_1, a_2, a_3, a_4....,\}$$

$$x(-n) = \{....a_4, a_3, a_2, a_1\}$$

If we introduce a delay of one sample in x, that will give

$\{...0, a_4, a_3, a_2, a_1\}$ ...(8.34)

A delay of the output will give as

$$y_{del}(n) = \{0, ...a_4, a_3, a_2, a_1\}$$

Since this is not equal to the Eq. (8.34) above, the system is not time invariant. However, most physical systems are time invariant. This example is a peculiar case.

Z-transform Tables

The following table of Laplace and Z-transforms is fit to be memorised.

$x(n)$	X(s)	X(z)
$\delta(n)$	1	1
U(n) (Step)	$1/s$	$z/z-1$
nT (ramp)	$1/s^2$	$Tz/(z-1)^2$
$e^{-na}T$	$1/(s+a)$	$z/\{z - e^{-aT}\}$
a^n	$1/[s - ln\ a/T]$	$z/z-a$
sin a nT	$a/(s^2 + a^2)$	$z \sin aT/[z^2 - 2z \cos aT + 1]$
cos a nT	$s/(s^2 + a^2)$	$\{z^2 - z \cos aT\}/[z^2 - 2z \cos aT + 1]$

Additionally, the following operations with Z-transforms are to be understood.

Sum	$ax1(n) + bx2(n)$	$aX1(z) + bX2(z)$
Time translation	$x(n - m)$	z^{-m} X(z)
Power multiplication	$a^{-n}x(n)$	X(az)
Differentiation	$n\ x(n)$	$-z\ \{d/dz\{X(z)\}$
Initial value	$x(0)$	$\lim_{Z\to\infty} X(z)$
Final value	$x(\infty)$	$\lim_{Z\to 1} (z-1)/z.\ X(z)$
Convolution	$\sum_{m=0}^{n} x(m)h(n-m)$	X(z)H(z)

Steady State Response from Z-Transform

Since the definition of the complex variable z is given by $z = e^{sT}$, when we want the steady state response to a sinusoidal input, just as we did substitute $s = j\omega$ in the Laplace continuous transform, so we do here, giving

$$z = \exp(j\omega T)$$

Since $T = 1/f_s = 1/2f_0$, (where f_0 is the folding frequency $f_s/2$)

$$z = \exp(j\,2\pi\,f\,/\,2f_0) = \exp(j\pi\,f\,/\,f_0) \quad ...(8.35)$$

This $f\,/\,f_0$ is called the normalised frequency variable in digital signal processing.

Some books use f/f_s as the normalised frequency. It is better to know how one defines it.

Example 1 : Evaluate the amplitude and phase response of a system whose output $y(t)$ is given in terms of the input $x(t)$ and past values of $x(t)$ samples, as

$$y(n) = 0.5\,y(n-1) + x(n) + x(n-1)$$

If the sampling frequency is 1 kHz, determine the steady state output for a sine wave input of 100 Hz at amplitude 10 units.

Rearranging the time discrete difference equation above, and putting z^{-1} for one delay of any sample (x or y),

$$H(z) = Y(z)/X(z)$$

$$= \frac{1+z^{-1}}{1-0.5z^{-1}}$$

The folding frequency $f_0 = 1000/2 = 500z$, hence the ratio f/f_0 becomes $f/500$. Let it be θ.

$$H(e) = \frac{1+e^{-j\pi\theta}}{1-0.5e^{-j\pi\theta}}$$

$$= \frac{1+\cos\pi\theta - j\sin\pi\theta}{1-0.5\cos\pi\theta + j0.5\sin\pi\theta}$$

The amplitude and phase can be found at any frequency ratio from this complex fraction.

As example for 100 Hz, $\theta = 100/500 = 0.2$

Substitution and complex algebra gives

Amplitude = 2.864

Phase = –0.772 radians.

Hence the steady state output for sine input of 10 amplitude and frequency 100 is

$$y(n) = 28.64\sin\{(.2\pi n) - 0.772\}$$

for any sample number n.

Phase Response and Group Delay

By the time translation theorem, a delay of k samples is z^{-k} in the z-domain and that means a linear phase change (lag or negative) of $k\omega$.

A system whose phase response $\theta(\omega)$ satisfies.

$$-\theta(\omega) = k\,\omega$$

for constant k is termed linear phase. The effect of such a response is to delay the input sequence by k sampling intervals. For a sine wave signal, this may be an angle which varies proportional to the frequency ω. Thus, if a 100 Hz sine wave is delayed by 45°, the 200 Hz sine wave signal is delayed by 2 × 45° = 90°. Thus, a signal which contains a 100 Hz and 200 Hz signal in combination, will preserve its waveshape and will not suffer distortion.

Filter in digital signal processing invariably lead to delays in the output and will have a varying phase component $k\omega$ in the phase function of the filter.

The rate of change of phase with frequency given by

$$-d\theta / d\omega = T'_G \qquad \text{...(8.36)}$$

where T_G is called the Group delay.

ROC for Z-transform–General

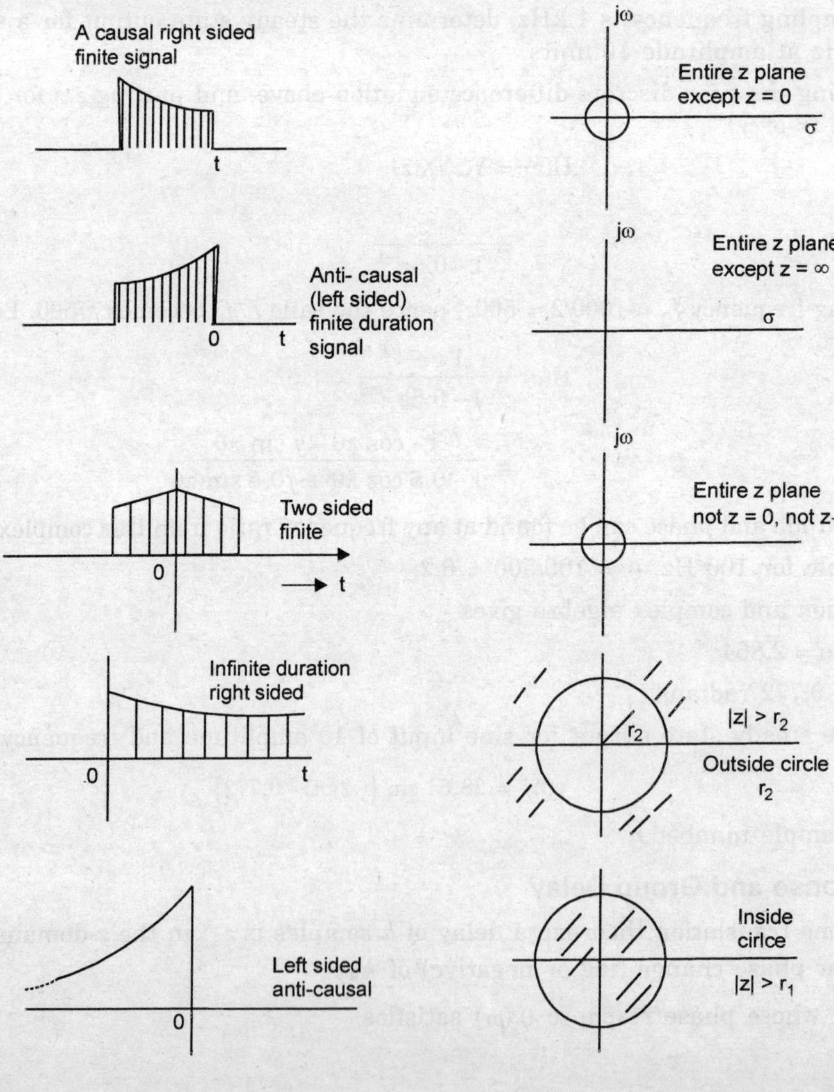

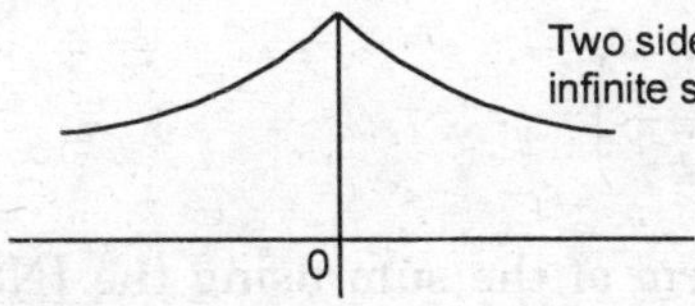

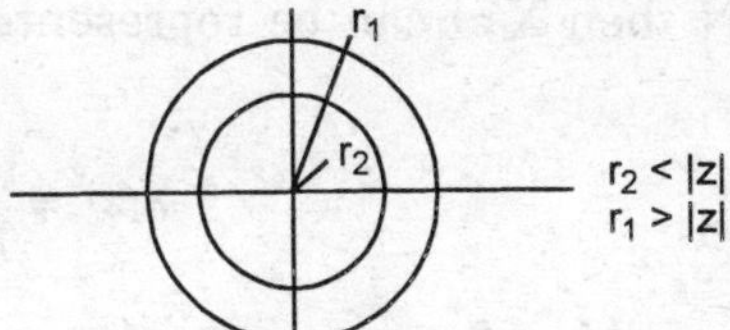

Z-Transform Calculation

When using the **Z-TRANSFORM**

$$X(z) = \sum_{n=-\infty}^{\infty} (x[n]z^{-n})$$

it is often useful to be able to find $x[n]$ given $X(z)$. There are at least 4 different methods to do this :

1. INSPECTION
2. PARTIAL-FRACTION EXPANSION
3. POWER SERIES EXPANSION
4. CONTOUR INTEGRATION

Inspection Method

This "method" is to basically become familiar with the **Z-TRANSFORM PAIR TABLES**

Example 1. When given

$$X(z) = \frac{z}{z-\alpha}$$

with an ROC of

$$|z| > \alpha$$

we could determine "by inspection" that

$$x[n] = \alpha^n u[n]$$

Partial-Fraction Expansion Method

When dealing with **linear time-invariant systems** the z-transform often is in the form

$$X(z) = \frac{B(z)}{A(z)} = \frac{\sum_{k=0}^{M} (b_k z^{-k})}{\sum_{k=0}^{N} (a_k z^{-k})} \qquad ...(8.37)$$

This can also expressed as

$$X(z) = \frac{a_0}{b_0} \frac{\prod_{k=1}^{M} (1 - c_k z^{-1})}{\prod_{k=1}^{N} (1 - d_k z^{-1})} \qquad ...(8.38)$$

where c_k represents the nonzero zeros of $X(z)$ and d_k represents the nonzero poles.

If M < N then $X(z)$ can be represented as

$$X(z) = \sum_{k=1}^{N}\left(\frac{A_k}{1-d_k z^{-1}}\right) \qquad ...(8.39)$$

This form allows for easy inversions of each term of the sum using the INSPECTION METHOD and the TRANSFORM TABLE. Thus if the numerator is a polynomial then it is necessary to use PARTIAL-FRACTION EXPANSION to put $X(z)$ in the above form. If $M \geq N$ then $X(z)$ can be expressed as

$$X(z) = \sum_{r=0}^{M-N}(B_r z^{-r}) + \frac{\sum_{k=0}^{N-1}(b'_k z^{-k})}{\sum_{k=0}^{N}(a_k z^{-k})} \qquad ...(8.40)$$

Example 2. Find the inverse *z*-transform of

$$X(z) = \frac{1+2z^{-1}+z^{-2}}{1+(-3z^{-1})+2z^{-2}}$$

where the ROC is $|z| > 2$. In this case M = N = 2, so we have to use long division to get

$$X(z) = \frac{1}{2} + \frac{\frac{1}{2}+\frac{7}{2}z^{-1}}{1+(-3z^{-1})+2z^{-2}}$$

Next factor the denominator..

$$X(z) = \frac{1}{2} + \frac{0.5+3.5z^{-1}}{(1-2z^{-1})(1-z^{-1})}$$

Now do partial-fraction expansion.

$$X(z) = \frac{1}{2} + \frac{A_1}{1-2z^{-1}} + \frac{A_2}{1-z^{-1}} = \frac{1}{2} + \frac{\frac{9}{2}}{1-2z^{-1}} + \frac{-4}{1-z^{-1}}$$

Now each term can be inverted using the inspection method and the z-transform table. Thus, since the ROC is $|z| > 2$,

$$x[n] = \frac{1}{2}\delta[n] + \frac{9}{2}2^n u[n] + (-4u[n])$$

Power Series Expansion Method

When the z-transform is defined as a power series in the form

$$X(z) = \sum_{n=-\infty}^{\infty}(x[n]z^{-n}) \qquad ...(8.41)$$

then each term of the sequence x[*n*] can be determined by looking at the coefficients of the respective power of z^{-n}.

Example 3. Now look at the z-transform of a **finite-length sequence.**

$$X(z) = z^2(1+2z^{-1})\left(1-\frac{1}{2}z^{-1}\right)(1+z^{-1})$$

$$= z^2+\frac{5}{2}z+\frac{1}{2}+\left(-(z^{-1})\right) \quad \text{...(8.42)}$$

In this case, since there were no poles, we multiplied the factors of $X(z)$. Now, by inspection, it is clear that

$$x[n] = \delta[n+2]+\frac{5}{2}\delta[n+1]+\frac{1}{2}\delta[n]+(-(\delta[n-1])) \quad \text{...(8.42a)}$$

One of the advantages of the power series expansion method is that many functions encountered in engineering problems have their power series' tabulated. Thus functions such as log, sin, exponent, sinh, etc, can be easily inverted.

Example 4. Suppose

$$X(z) = \log_n(1+\alpha z^{-1})$$

Noting that $$\log_n(1+x) = \sum_{n=1}^{\infty}\left(\frac{-1^{n+1}x^n}{n}\right)$$

Then $$X(z) = \sum_{n=1}^{\infty}\left(\frac{-1^{n+1}\alpha^n z^{-n}}{\cdot n}\right)$$

Therefore coefficients $$x(n) = \begin{cases}\dfrac{-1^{n+1}\alpha^n}{n} & \text{if } n \geq 1\\ 0 & \text{if } n \leq 0\end{cases}$$

Contour Integration Method

Without going into much detail

$$x[n] = \frac{1}{2\pi j}\oint_r X(z)z^{n-1}dz \quad \text{...(8.43)}$$

where r is a counter-clockwise contour in the ROC of $X(z)$ encircling the origin of the z-plane. To further expand on this method of finding the inverse requires the knowledge of complex variable theory and thus will not be addressed in this module.

Difference Equation

One of the most important concepts of DSP is to be able to properly represent the input/output relationship to a given LTI system. A linear constant-coefficient **difference equation** (LCCDE) serves as a way to express just this relationship in a discrete-time system. Writing the sequence of inputs and outputs, which represent the characteristics of the LTI system, as a difference equation help in understanding and manipulating a system.

DEFINITION 1 : Difference Equation

An equation that shows the relationship between consecutive values of a sequence and the differences among them. They are often rearranged as a recursive formula so that a systems output can be computed from the input signal and past outputs.

Example

$$y[n]+7y[n-1]+2y[n-2] = x[n]-4x[n-1] \qquad ...(8.44)$$

General Formulas from the Difference Equation

As stated briefly in the definition above, a difference equation is a very useful tool in describing and calculating the output of the system described by the formula for a given sample n. The key property of the difference equation is its ability to help easily find the transform, H(*z*), of a system. In the following two subsections, we will look at the general form of the difference equation and the general conversion to a *z*-transform directly from the difference equation.

Difference Equation

The general form of a linear, constant-coefficient difference equation (LCCDE), is shown below :

$$\sum_{k=0}^{N}(a_k y[n-k]) = \sum_{k=0}^{M}(b_k x[n-k]) \qquad ...(8.45)$$

We can also write the general form to easily express a recursive output, which looks like this :

$$y[n] = -\left(\sum_{k=1}^{N}(a_k y[n-k])\right) + \sum_{k=0}^{M}(b_k x[n-k]) \qquad ...(8.46)$$

Note the lower limit is K = 1 for the *y*-term.

From this equation, note that $y[n-k]$ represents the outputs and $x[n-k]$ represents the inputs. The value of N represents the **order** of the difference equation and corresponds to the memory of the system being represented. Because this equation relies on past values of the output, in order to compute a numerical solution, certain past values are to be kept stored in memory.

Conversion to Z-Transform

Using the above Eq. (8.45) we can easily generalize the **transfer function**, H(*z*), for any difference equation. Below are the steps taken to convert any difference equation into its transfer function, i.e., z-transform. The first step involves taking the Fourier Transform of all the terms in (8.45). Then we use the linearity property to pull the transform inside the summation and the time-shifting property of the *z*-transform to change the time-shifting terms to exponentials. Once this is done, we arrive at the following equation : $a_0 = 1$.

$$Y(z) = -\left|\sum_{k}^{N}(a_k Y(z)z^{-k})\right| + \sum^{M}(b_k X(z)z^{-k}) \qquad ...(8.47)$$

$$H(z) = \frac{Y(z)}{X(z)} = \frac{\sum_{k=0}^{M}(b_k z^{-k})}{1+\sum_{k=1}^{N}(a_k z^{-k})} \qquad ...(8.48)$$

Conversion to Frequency Response

Once the z-transform has been calculated from the difference equation, we can go one step further to define the frequency response of the system, or filter, that is being represented by the difference equation.

Note : Remember that the reason we are dealing with these formulas is to be able to aid us in filter design. A LCCDE is one of the easiest ways to represent FIR (finite impulse response) filters. By being able to find the frequency response, we will be able to look at the basic properties of any filter represented by a simple LCCDE.

Below is the general formula for the frequency response of a z-transform. The conversion is simple a matter of taking the z-transform formula, H(z), and replacing every instance of z with $e^{j\omega}$.

$$H(w) = H(z)\big|_{z,\, z = e^{j\omega}}$$

$$= \frac{\sum_{k=0}^{M}(b_k e^{-(j\omega k)})}{\sum_{k=0}^{N}(a_k e^{-(j\omega k)})} \qquad ...(8.49)$$

Example 1 : Finding Difference Equation

Below is a basic example showing the opposite of the steps above : given a transfer function one can easily calculate the system's difference equation.

$$H(z) = \frac{(z+1)^2}{\left(z-\frac{1}{2}\right)\left(z+\frac{3}{4}\right)} \qquad ...(8.50)$$

Given this transfer function of a time-domain filter, we want to find the difference equation. To begin with, expand both polynomials and divide them by the highest order z.

$$H(z) = \frac{(z+1)(z+1)}{\left(z-\frac{1}{2}\right)\left(z+\frac{3}{4}\right)} = \frac{z^2+2z+1}{z^2+2z+1-\frac{3}{8}}$$

$$= \frac{1+2z^{-1}+z^{-2}}{1+\frac{1}{4}z^{-1}-\frac{3}{8}z^{-2}} \qquad ...(8.51)$$

From this transfer function, the coefficients of the two polynomials will be our a_k and b_k values found in the general difference equation formula, (8.46). Using these coefficients and the above form of the transfer function, we can easily write the difference equation :

$$x[n]+2x[n-1]+x[n-2] = y[n]+\frac{1}{4}y[n-1]-\frac{3}{8}y[n-2] \qquad ...(8.52)$$

In our final step, we can rewrite the difference equation in its more common form showing the recursive nature of the system.

$$y[n] = x[n]+2x[n-1]+x[n-2]+\frac{-1}{4}y[n-1]+\frac{3}{8}y[n-2] \qquad ...(8.53)$$

Solving a LCCDE

In order for a linear constant-coefficient difference equation to be useful in analyzing a LTI system, we must be able to find the systems output based upon a known input, *x(n)*, and a set of initial conditions. Two common methods exist for solving a LCCDE : the **direct method** and the **indirect method**, the later being based on the *z*-transform. Below we will briefly discuss the formulas for solving a LCCDE using each of these methods.

Direct Method

The final solution to the output based on the direct method is the sum of two parts, expressed in the following equation :

$$y(n) = y_h(n) + y_p(n) \qquad ...(8.54)$$

The first part, $y_h(n)$, is referred to as the **homogeneous solution** and the second part, $y_h(n)$, is referred to as **particular solution**. The following method is very similar to that used to solve many differential equations, for those who took a differential calculus course or used differential equations before then this should seem very familiar.

Homogeneous Solution

We begin by assuming that the input is zero, $x(n) = 0$. Now we simply need to solve the homogeneous difference equation :

$$\sum_{k=0}^{N}(a_k y[n-k]) = 0 \qquad ...(8.55)$$

In order to solve this, we will make the assumption that the solution is in the form of an exponential. We will use lambda, λ, to represent our exponential terms. We now have to solve the following equation :

$$\sum_{k=0}^{N}(a_k \lambda^{n-k}) = 0 \qquad ...(8.56)$$

We can expand this equation out and factor out all of the lambda terms. This will give us a large polynomial in parenthesis, which is referred to as the **characteristic polynomial**. The roots of this polynomial will be the key to solving the homogeneous equation. If there are all distinct roots, then the general solution to the equation will be as follows :

$$y_h(n) = C_1(\lambda_1)^n + C_2(\lambda_2)^n + \ldots + C_N(\lambda_N)^n \qquad ...(8.57)$$

However, if the characteristic equation contains multiple roots then the above general solution will be slightly different. Below we have the modified version for an equation where λ_1 has K multiple roots :

$$y_h(n) = C_1(\lambda_1)^n + C_1 n(\lambda_1)^n + C_1 n^2(\lambda_1)^n + \ldots + C_1 n^{K-1}(\lambda_1)^n + C_2(\lambda_2)^n + \ldots + C_N(\lambda_N)^n \quad \ldots(8.58)$$

Particular Solution

The particular solution, $y_p(n)$, will be any solution that will solve the general difference equation :

$$\sum_{k=0}^{N}(a_k y_p(n-k)) = \sum_{k=0}^{M}(b_k x(n-k)) \quad \ldots(8.59)$$

In order to solve, our guess for the solution to $y_p(n)$ will take on the form of the input, $x(n)$. After guessing at a solution to the above equation involving the particular solution, one only needs to plug the solution into the difference equation and solve it out.

Indirect Method

1. Convert the difference equation into z-transform polynomial equation. See Eqns. (8.46) to (8.47).
2. Find the inverse transform as in (8.42) (a) and (b). The solution has been got.

FILTER DESIGN USING THE POLE/ZERO PLOT OF A Z-TRANSFORM

Estimating Frequency Response from Z-Plane

One of the motivating factors for analyzing the pole/zero plots is due to their relationship to the frequency response of the system. Based on the position of the poles and zeros, one can quickly determine the frequency response. This is a result of the correspondence between the frequency response and the transfer function evaluated on the unit circle in the pole/zero plots. The frequency response, or DTFT (DT = Discrete Time) of the system is defined as :

$$H(\omega) = H(z)\Big|_{z,\, z = e^{jw}}$$

$$= \frac{\sum_{k=0}^{M}(b_k e^{-(jwk)})}{\sum_{k=0}^{N}(a_k e^{-(jwk)})} \quad \ldots(8.60)$$

Next, by factoring the transfer function into poles and zeros and multiplying the numerator and denominator by e^{jw} we arrive at the following equations :

$$H(\omega) = \left|\frac{b_0}{a_0}\right| \frac{\prod_{k=1}^{M}\left(\left|e^{jw} - c_k\right|\right)}{\prod_{k=1}^{N}\left(\left|e^{jw} - d_k\right|\right)} \quad \ldots(8.61)$$

From (8.61) we have the frequency response in a form that can be used to interpret physical characteristics about the filter's frequency response. The numerator and denominator contain a product of terms of the form $\left|e^{jw} - h\right|$, where h is either a zero, denoted by c_k or a pole, denoted by d_k. Vectors are commonly used to represent the term and its parts on the complex plane. The pole or zero, h, is a vector from the origin to its location anywhere on the complex plane and e^{jw} is a vector from the origin to its location on the unit circle. The vector connecting these two points, $\left|e^{jw} - h\right|$, connects the pole or zero location to a place on the unit circle dependent on the value of w. From this, we can begin to understand how the magnitude of the frequency response is a ratio of the distances to the poles and zeros present in the z-plane as ω goes from zero to pi. These characteristics allow us to interpret $|H(\omega)|$ as follows :

$$|H(\omega)| = \left|\frac{b_0}{a_0}\right| \frac{\prod \text{“distances from zeros”}}{\prod \text{“distances from poles”}} \qquad ...(8.62)$$

In conclusion, using the distances from the unit circle to the poles and zeros, we can plot the frequency response of the system. As ω goes from 0 to 2π, the following two properties, taken from the above equations, specify how one should draw $|H(\omega)|$.

While moving around the unit circle...

1. if close to a zero, then the magnitude is small. If a zero is on the unit circle, then the frequency response is zero at that point.
2. if close to a pole, then the magnitude is large. If a pole is on the unit circle, then the frequency response goes to infinity at that point.

Drawing Frequency Response from Pole/Zero Plot

Let us now look at several examples of determining the magnitude of the frequency response from the pole/zero plot of a z-transform.

EXAMPLE 1

In this first example we will take a look at the very simple z-transform shown below:

$$H(z) = z + 1 = 1 + z^{-1}$$

$$H(w) = 1 + e^{-(jw)}$$

For this example, some of the vectors represented by $\left|e^{jw} - h\right|$, for random values of ω, are explicitly drawn onto the complex plane shown in the Fig. 22 below. These vectors show how the amplitude of the frequency response changes as ω goes from 0 to 2π, and also show the physical meaning of the terms in (8.62) above. One can see that when ω = 0, the vector is the longest and thus the frequency response will have its largest amplitude here. As ω approaches π, the length of the vectors decrease as does the amplitude of $|H(\omega)|$. Since there are no poles in the transform, there is only this one vector term rather than a ratio as seen in (8.62).

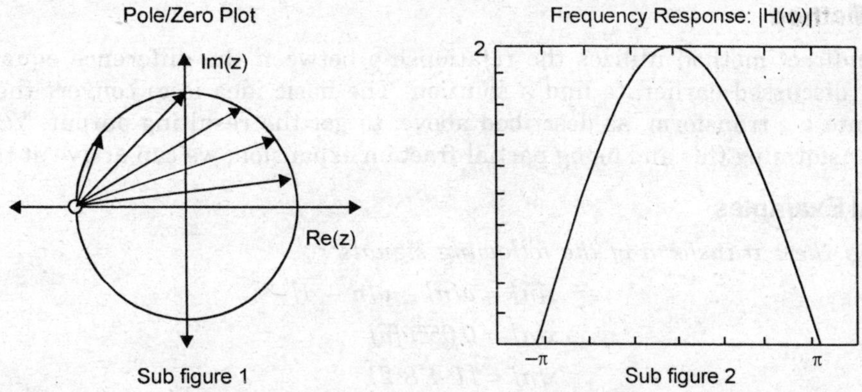

Fig. 22. The first figure represents the pole/zero plot with a few representative vectors graphed while the second shows the frequency response with a peak at +2 and graphed between plus and minus π.

Example 2

For this example, a more complex transfer function is analyzed in order to represent the system's frequency response.

$$H(z) = \frac{z}{z - \frac{1}{2}} = \frac{1}{1 - \frac{1}{2}z^{-1}}$$

$$H(\omega) = \frac{1}{1 - \frac{1}{2}e^{-(jw)}} \qquad \text{...(8.63)}$$

Below we can see the two figures described by the above equations. The Fig. 23(a) represents the basic pole/zero plot of the z-transform, H(ω). Fig. 23(b) shows the magnitude of the frequency response. From the formulas and satements in the previous section, we can see that when ω = 0 the frequency will peak since it is at this value of ω that the pole is closest to the unit circle. The ratio from (8.62) helps us see the mathematics behind this conclusion and the relationship between the disances from the unit circle and the poles and zeros. As ω moves from 0 to π, we see how the zero begins to mask the effects of the pole and thus force the frequency response closer to 0.

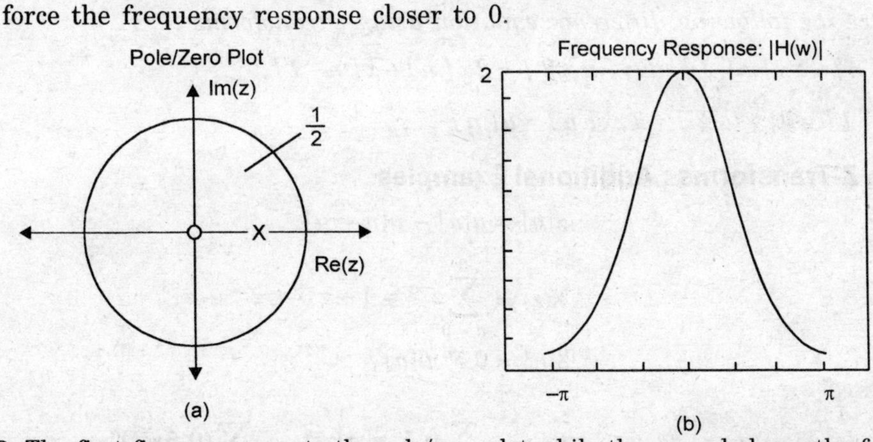

Fig. 23. The first figure represents the pole/zero plot while the second shows the frequency response with a peak at +2 and graphed between plus and minus π.

We can also calculate H(ω) from (8.63), which was shown earlier in the example 1, page 241.

Indirect Method

The indirect method utilizes the relationship between the difference equation and z-transform, discussed earlier, to find a solution. The basic idea is to convert the difference equation into a z-transform, as described above, to get the resulting output, $Y(z)$. Then by inverse transforming this and using partial-fraction expansion, we can arrive at the solution.

Additional Examples

1. *Find the z transform of the following signals :*

(a) $x[n] = u[n] - u[n-4]$

(b) $x[n] = 0.05^n u[n]$

(c) $x[n] = [1\ 4\ 8\ 2]$

(d) $x[n] = [0\ 1\ 2\ 3\ 4]$

(e) $x[n] = 2(0.8)^n u[n]$

2. *Find the inverse Z-transforms of the following signals :*

(a) $$X(z) = \frac{(z-1)(z+0.8)}{(z-0.5)(z+0.2)}$$

(b) $$X(z) = \frac{(z+0.8)}{(z-0.5)(z+0.2)}$$

(c) $$X(z) = \frac{z^3+z+1}{(z^2-0.5z+0.25)(z-1)}$$

(d) $$X(z) = \frac{(z^2-1)(z+0.8)}{(z-0.5)^2(z+0.2)}$$

3. *Use the Final Value Theorem to determine the final value of x[n] for each of the signals defined in Problem 2. Compare your answer obtained from the Final Value Theorem to the answer found by taking* $\lim_{n\to\infty} x[n]$.

4. *Solve the following difference equation using z-transforms :*

(a) $y[n]+3y[n-1]+2y[n-2] = 2x[n]-x[n-1];$

$y[-1]=0;\ y[-2]=1,\ x[n]=u[n]$

Answer to Z-Transforms : Additional Examples

1. (a) $x[n] = u[n] - u[n-u]$

$$X(z) = \sum_{n=0}^{3} z^{-n} = 1 + z^{-1} + z^{-2} + z^{-3}$$

(b) $X[n] = 0.5^n\, u[n]$

$$X(z) = \sum_{n=-\infty}^{\infty} 0.5^n\, u[n] z^{-n} = \sum_{n=0}^{\infty} (0.5z^{-1})^n$$

Use geometric series convergence :

$$\sum_{n=0}^{\infty} a^n = \frac{1}{1-a} \text{ of } |a|<1$$

$$X(z) = \frac{1}{1-0.5z^{-1}}$$

(*c*) $$X[n] = [1\ 4\ 8\ 2]$$

$$X(z) = 1 + 4z^{-1} + 8z^{-2} + 2z^{-3}$$

(*d*) $$X[n] = [0\ 1\ 2\ 3\ 4]$$

$$X(z) = z^{-1} + 2z^{-2} + 3z^{-3} + 4z^{-4}$$

(*e*) $$X[n] = 2(.8)^n u[n]$$

$$X(z) = \frac{2}{1-0.8z^{-1}}$$

2. (*a*) $$X(z) = \frac{(z-1)(z+0.8)}{(z-0.5)(z+0.2)}$$

$$\frac{X(z)}{z} = \frac{C_1}{z} + \frac{C_2}{z-.5} + \frac{C_3}{z+0.2}$$

$$C_1 = \frac{X(z)}{z} z\Big|_{z=0} = 8$$

$$C_2 = \frac{X(z)}{z}(z-.5)\Big|_{z=.5} = \frac{(z-1)(z+.8)}{z(z+.2)}\Big|_{z=.5} = -1.857$$

$$C_3 = \frac{X(z)}{z}(z+.2|_{z=-.2} = 5.143$$

$$X(z) = 8 - \frac{-1.857z}{z-.5} - \frac{5.143z}{z+0.2}$$

$$X[n] = 8[n] - 1.857\,(0.5)^n u[n] - 5.143\,(-0.2)^n u[n]$$

(*b*) $$X(z) = \frac{z+0.8}{(z-0.5)(z+.2)}$$

$$X(z) = \frac{1.857}{z-.5} - \frac{0.8571}{z+.2},$$

let $$X_1(z) = \frac{1.857z}{z-.5} - \frac{0.8571z}{z+.2}$$

then $$X_1[n] = \left[1.857(.5)^n - .8571\,(-0.2)^n\right] u[n]$$

$$X(z) = z^{-1}$$

$X(z) \rightleftharpoons$ $$x[n] = \left[1.857(0.5)^{n-1} - .8571\,(-0.2)^{n-1}\right]u[n-1]$$

(c)
$$X(z) = \frac{z^3+z+1}{(z^2-.5z+.25)(z-1)}$$

$$\frac{X(z)}{z} = \frac{-4}{z} + \frac{4}{(z-1)} + \frac{C_1 z + C_2}{z^2-.5z+.25}$$

$$z^3+z+1 = (-4z+4)(z^2-.5z+.25) + 4z(z^2-.5z+.25)$$
$$+(C_1 z + C_2)(z^2 - z)$$

$$z^3 : 1 = -4 + 4 + C_1 \Rightarrow C_1 = 1$$
$$z^1 : 1 = -1 - 2 + 1 - C_2 \Rightarrow C_2 = -3$$

$$X(z) = -4 + \frac{4z}{z-1} + \frac{z^2-3z}{z^2-0.5z+.25}$$

let $a = 0.5,\ -2a \cos(\Omega) = -0.5,\ 2a \sin \Omega = 2.75/3$

$\Rightarrow$ $\Omega = 1.047$ rad.

$$\frac{z^2-3z}{z^2-0.5z+0.25} = \frac{z^2 - a\cos(\Omega)z + 0.25 + 2.75}{z^2 - 2a\cos(\Omega)z + a^2}$$

$$x[n] = -4\delta[n] + 4u[n] + (0.5)^n \cos(1.047n)u[n]$$
$$+(0.5)^n 6.35 \sin(1.047n)u[n],\ (2.75 = 6.35a \sin \Omega)$$

(d)
$$X(z) = \frac{(z^2-1)(z+0.8)}{(z-0.5)^2(z+0.2)}$$

$$\frac{X(z)}{z} = \frac{C_1}{z} + \frac{C_2}{z+.2} + \frac{C_3}{z-.5} + \frac{C_4}{(z-.5)^2}$$

$$C_1 = -16,\ C_2 = 5.88,\ C_4 = -2.79$$

$$C_3 = \frac{d}{dz}\left(\frac{(z^2-1)(z+.8)}{z(z+.2)}\right)\Bigg|_{z=.5} = 11.12$$

$$x[n] = -16\delta[n] + 5.88(-0.2)^n u[n] + 11.12(0.5)^n u[n]$$
$$-2.79n(0.5)^n u[n]$$

3. (a)
$$X(z) = \frac{(z-1)(z+0.8)}{(z-0.5)(z+0.2)}$$

$$\lim_{n\to\infty} X[n] = [(z-1)X(z)]_{z=1}$$
$$= \frac{(z-1)^2(z+.8)}{(z-0.5)(z+0.2)}\Bigg|_{z=1} = 0$$

Matches 2(a)

(*b*) $$X(z) = \frac{z+.8}{(z-.5)(z+.2)}$$

$$\lim_{n\to\infty} X[n] = \left.\frac{(z-1)(z+.8)}{(z-.5)(z+.2)}\right|_{z=1} = 0$$

Matches 2(*b*)

(*c*) $$X(z) = \frac{z^3+z+1}{(z^2-.5z+.25)(z-1)}$$

$$\lim_{n\to\infty} X[n] = \left.\frac{(z^3+z+1)(z-1)}{(z^2-.5z+.25)(z-1)}\right|_{z=1} = 4$$

matches 2(*c*)

(*d*) $$X(z) = \frac{(z^2-1)(z+.8)}{(z-.5)^2(z+.2)},\quad \lim_{n\to\infty} X[n] = 0$$

matches 2(*d*)

4. (*a*) $y[n] + 3y[n-1] + 2y[n-2] = 2x[n] - x[n-1]$;

$y[-1] = 0$; $y[-2] = 1$, $x[n] = u[n]$

$$y(z) + (3z^{-1}Y(z) + y[-1]) + \partial(z^{-2}Y(z) + y[-2] + z^{-1}y[-1]) = 2X(z) - z^{-1}X(z) - x[-1]$$

$$Y(z)\,(1 + 3z^{-1} + 2z^{-2}) + 2 = \frac{2z}{z-1} - \frac{1}{z-1}$$

$$Y(z) = \frac{z^2}{z^2+3z+2}\ \frac{2z-1-2z+2}{z-1}$$

$$= \frac{z^2}{(z+1)(z+2)(z-1)}$$

$$= \frac{C_1}{z+1} + \frac{C_2}{z+2} + \frac{C_3}{z-1}$$

$$= \frac{-1/2}{z+1} + \frac{4/3}{z+2} + \frac{1/6}{z-1}$$

$$zY(z) = \frac{-1/2z}{z+1} + \frac{4/3z}{z+2} + \frac{116z}{z-1}$$

$$\left(-\frac{1}{2}(-1)^n + \frac{4}{3}(-2)^n + \frac{1}{6}\right)u[n]$$

$$y[n] = \left(-\frac{1}{2}(-1)^{n-1} + \frac{4}{3}(-2)^{n-1} + \frac{1}{6}\right)u[n-1]$$

Discrete Transfer Functions and System Response : Addl. Examples.

1. *Find the transfer functions of the following discrete-time systems :*

(a) $y[n] + 0.5y[n-1] = 2x[n]$

(b) $y[n] + 2y[n-1] - y[n-2] = 2x[n] - x[n-1] + 2x[n-2]$

(c) $y[n] + y[n-2] = 2x[n] - x[n-1]$

(d) $y[n] = x[n] - 2x[n-1] + x[n-2]$

(e) $y[n] + y[n-1] + 0.25y[n-2] = x[n] - x[n-1]$

2. *Plot the poles and zeros of the following transfer functions. Determine the stability.*

(a) $$H(z) = \frac{(z-0.5)}{z+0.75}$$

(b) $$H(z) = \frac{(z-0.5)}{z-0.75}$$

(c) $$H(z) = \frac{(z-0.5)}{z-2}$$

(d) $$H(z) = \frac{z^2+1}{z^2-0.25}$$

(e) $$H(z) = \frac{z(z-1)}{z^2-0.5z-0.5}$$

(f) $$H(z) = \frac{z^2+1}{z^2-1.5z-1}$$

(g) $$H(z) = \frac{(z-0.5)(z+0.5)}{z^2+z+0.74}$$

(h) $$H(z) = \frac{(z-0.5)(z+0.5)}{z^2+z+4.25}$$

3. *Give the general form of the transient response for each of the transfer functions given in Problem 2.*

4. *Give the general form of the transient response for each of the transfer functions given below.*

(a) $$H(z) = \frac{(z-0.5)(z-0.1)}{(z+0.75)^2}$$

(b) $$H(z) = \frac{(z-0.5)(z+0.5)(z+1)}{(z^2+z+0.74)(z-0.75)}$$

(c) $$H(z) = \frac{(z-0.5)^2(z+0.5)(z+1)^2}{(z^2+z+0.74)^2(z-0.75)}$$

5. *Determine the unit step response for each of the transfer functions given in Problem 2.*

6. *Given the following sysem :*

$$y[n] = x[n] - x[n-1] + x[n-2]$$

(a) Find the transfer function.

(b) Give the impulse response.

(c) Determine the stability.

(d) Sketch the frequency response and determine the type of filter.

Answers to the Discrete Transfer Functions : Addl. Examples

1. (*a*)

$$y[n] + 0.5\,y[n-1] = 2x[n]$$

$$Y(z)(1+.5z^{-1}) = 2X(z)$$

$$H(z) = \frac{2}{1+.5z^{-1}} = \frac{2z}{z+.5}$$

(*b*)

$$y[n]+2y[n-1]-y[n-2] = 2x[n]-x[n-1]+2x[n-2]$$

$$y(z)(1+2z^{-1}-z^{-2}) = X(z)(2-z^{-1}+2z^{-2})$$

$$H(z) = \frac{2-z^{-1}+2z^{-2}}{1+2z^{-1}-z^{-2}} = \frac{2z^2-z+2}{z^2+2z-1}$$

(*c*)

$$y[n]+y[n-2] = 2x[n]-x[n-1]$$

$$Y(z)(1+z^{-2}) = X(z)(2-z^{-1})$$

$$H(z) = \frac{2-z^{-1}}{1+z^{-2}} = \frac{2z^2-z}{z^2-1}$$

(*d*)

$$y[n] = x[n]-2x[n-1]+x[n-2]$$

$$Y(z) = X(z)(1-2z^{-1}+z^{-2})$$

$$H(z) = 1-2z^{-1}+z^{-2} = \frac{z^2-2z+1}{z^2}$$

2. (*a*)

$$H(z) = \frac{z-0.5}{z+0.75}$$

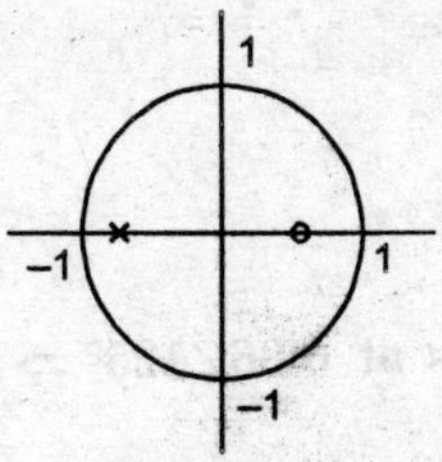

Stable

(*b*)

$$H(z) = \frac{z-0.5}{z-0.75}$$

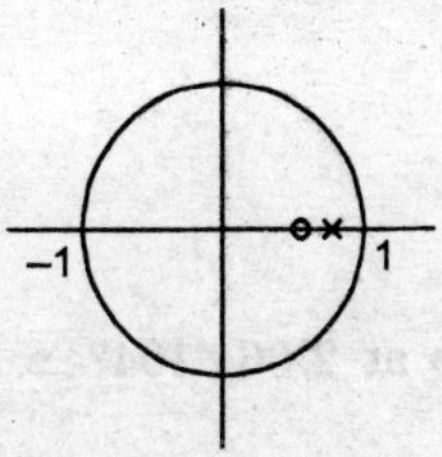

Stable

(*c*)

$$H(z) = \frac{z-0.5}{z+2}$$

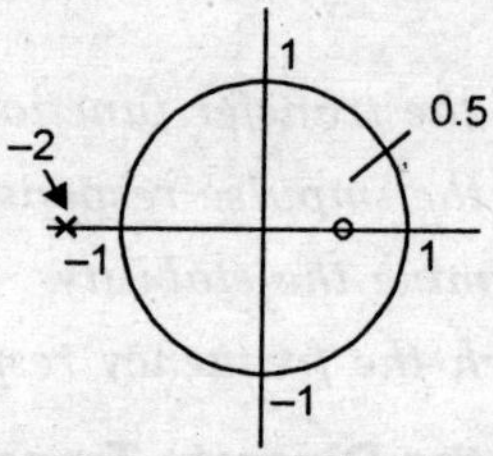

(*d*)

$$H(z) = \frac{z^2+1}{z^2-0.25}$$

$$= \frac{(z-j)(z+j)}{(z-\tfrac{1}{2})(z+\tfrac{1}{2})}$$

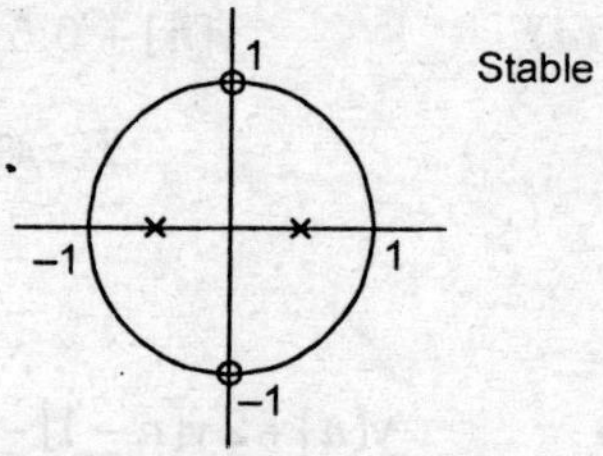

(*e*)

$$H(z) = \frac{z(z-1)}{z^2-0.5z-0.5}$$

$$= \frac{z(z-1)}{(z-1)(z+0.5)}$$

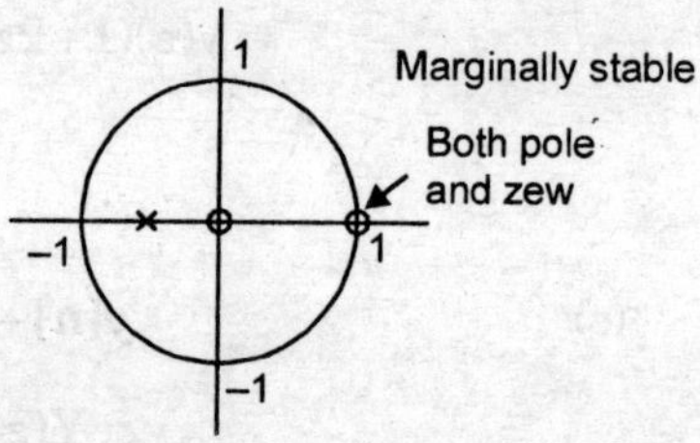

(*f*)

$$H(z) = \frac{z^2+1}{z^2-1.5z-1}$$

$$= \frac{(z+j)(z-j)}{(z-2)(z+0.5)}$$

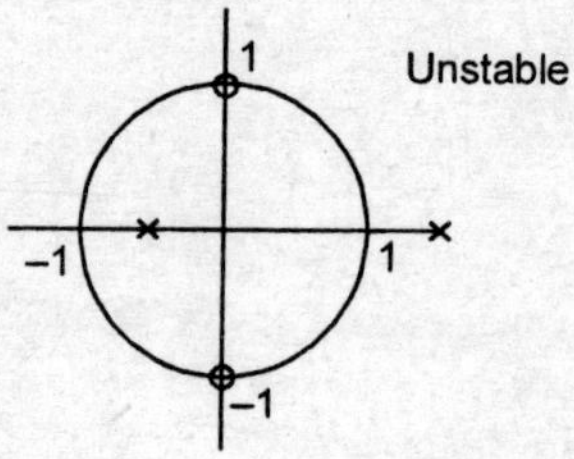

(*g*)

$$H(z) = \frac{(z-0.5)(z+0.5)}{z^2+z+.74}$$

$$= \frac{(z-0.5)(z+0.5)}{(z+0.5+0.7j)(z+0.5-0.7j)}$$

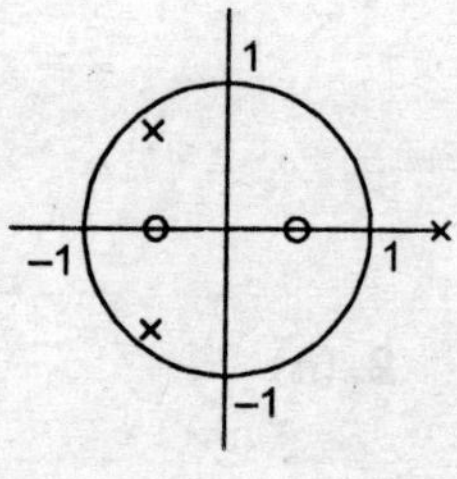

poles at $0.86\angle 125° \Rightarrow$ stable

(*h*)

$$H(z) = \frac{(z-0.5)(z+0.5)}{z^2+z+4.25}$$

$$= \frac{(z-0.5)(z+0.5)}{(z+0.5+2j)(z+0.5-2j)}$$

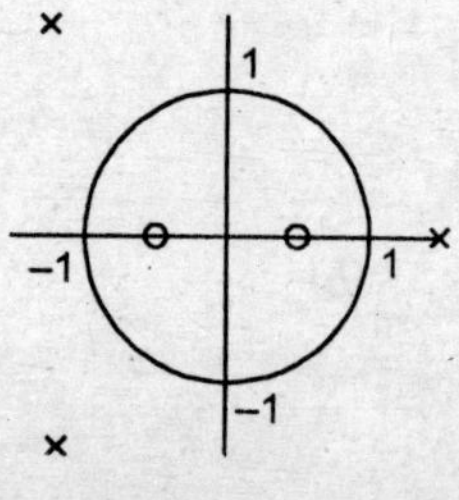

poles at $2.06\angle 104° \Rightarrow$ unstable

3. (*a*)
$$H(z) = \frac{z-0.5}{z+0.75}, \text{ pole at } -0.75$$
$$y[n] = C_1(-0.75)^n\, u[n] + \delta[n]C_2$$

(*b*)
$$H(z) = \frac{z-0.5}{z-0.75}, \text{ pole at } 0.75$$
$$y[n] = C_1(0.75)^n\, u[n] + C_2\, \delta[n]$$

(*c*)
$$H(z) = \frac{z-0.5}{z+2}, \text{ pole at } -2$$
$$y[n] = C_1(-2)^n\, u[n] + C_2\, \delta[n]$$

(*d*)
$$H(z) = \frac{z^2+1}{z^2-0.25}, \text{ poles at } 0.5, -0.5$$
$$y[n] = C_1(0.5)^n\, u[n] + C_2(-0.5)^n\, u[n] + C_3\, \delta[n]$$

(*e*)
$$H(z) = \frac{z(z-1)}{z^2-0.5z-0.5}, \text{ poles at } 1, -0.5$$
$$y[n] = C_1 u[n] + C_2\,(-0.5)^n\, u[n] + C_3\delta[n]$$

(*f*)
$$H(z) = \frac{z^2+1}{z^2-1.5z-1}, \text{ poles at } 2, -0.5$$
$$y[n] = C_1\, 2^n\, u[n] + C_2(-0.5)^n\, u[n] + C_3\, \delta[n]$$

(*g*)
$$H(z) = \frac{(z-0.5)(z+0.5)}{z^2+z+0.74}, \text{ poles at } 0.86\angle 125^\circ \;\; 2.19 \text{ rad.}$$
$$y[n] = C(0.86)^n \cos\,(2.19\, n + \theta)\, u[n] + C_2\delta[n]$$

(*h*)
$$H(z) = \frac{(z-0.5)(z+0.5)}{z^2+z+4.25}, \text{ poles at } 2.06\angle 104^\circ \;\; 1.81 \text{ rad}$$
$$y[n] = C(2.06)^n \cos\,(1.81n + \theta)\, u[n] + C_2\, \delta[n]$$

4. (*a*)
$$H(z) = \frac{(z-0.5)(z-0.1)}{(z+0.75)^2}$$
$$y[n] = C_1\,(-0.75)^n\, u[n] + C_2 n\,(-0.75)^n\, u[n] + C_3\delta[n]$$

(*b*)
$$H(z) = \frac{(z-0.5)(z+0.5)(z+1)}{(z^2+z+0.74)(z-0.75)}$$

poles at 0.75, $0.86\angle 2.19$ rad

$$y[n] = C_1(0.75)^n\, u[n] + C_2\,(0.86)^n \cos\,(2.19n + \theta)\, u[n] + C_3\, \delta[n]$$

(*c*)
$$H(z) = \frac{(z-0.5)^2(z+0.5)(z+1)^2}{(z^2+z+0.74)^2(z-0.75)}$$

poles at 0.75, $0.86\angle 2.19$, $0.86\angle 2.19$

$$y[n] = C_1 (0.75)^n u[n] + C_2 (0.86)^n \cos(2.19 n + \theta_1) u[n] + C_3 (0.86)^n n \cos(2.19n + \theta_2) u[n] + C_4 \delta[n]$$

5. (*a*)

$$H(z) = \frac{z-0.5}{z+0.75}, \; X(z) = \frac{z}{z-1}$$

$$y(z) = H(z)X(z) = \frac{(z-0.5)z}{(z+0.75)(z-1)}$$

$$\frac{y(z)}{z} = \frac{0.71}{z+0.75} + \frac{0.285}{z-1}$$

$$y(z) = \frac{0.71z}{z+0.75} + \frac{0.285z}{z-1}$$

$$y[n] = \left(0.71(-0.75)^n + 0.285\right) u[n]$$

(*b*)

$$H(z) = \frac{z-0.5}{z-0.75}$$

$$\frac{Y(z)}{z} = \frac{z-0.5}{(z-0.75)(z-1)}$$

$$Y(z) = \frac{-z}{(z-0.75)} + \frac{2z}{z-1}$$

$$y[n] = \left(-(0.75)^n + 2\right) u[n]$$

(*c*)

$$H(z) = \frac{z-0.5}{z+2}$$

$$\frac{Y(z)}{z} = \frac{z-0.5}{(z+2)(z-1)}$$

$$Y(z) = \frac{0.8337}{z+2} + \frac{0.167z}{z-1}$$

$$y[n] = \left(0.833(-2)^n + 0.167\right) u[n]$$

(*d*)

$$H(z) = \frac{z^2+1}{z^2-0.25}$$

$$\frac{Y(z)}{z} = \frac{z^2+1}{(z-0.5)(z+0.5)(z-1)}$$

$$Y(z) = \frac{-2.5z}{z-0.5} + \frac{0.833}{z+0.5} + \frac{2.67}{z-1}$$

$$y[n] = \left(-2.5(0.5)^n + 0.833(-0.5)^n + 2.67\right) u[n]$$

(e)

$$H(z) = \frac{z(z-1)}{z^2 - 0.5z - 0.5}$$

$$\frac{Y(z)}{z} = \frac{z(z-1)}{(z-1)(z+0.5)(z-1)} = \frac{z}{(z+0.5)(z-1)}$$

$$Y(z) = \frac{0.313z}{z+0.5} + \frac{0.672}{z-1}$$

$$y[n] = \left(0.33(-0.5)^n + 0.67\right) u[n]$$

(f)

$$H(z) = \frac{z^2+1}{z^2 - 1.5z - 1} = \frac{z^2+1}{(z-2)(z+0.5)}$$

$$\frac{Y(z)}{z} = \frac{z^2+1}{(z-2)(z-1)(z+0.5)}$$

$$Y(z) = \frac{2z}{z-2} + \frac{-1337}{z-1} + \frac{0.33z}{z+0.5}$$

$$y[n] = \left((2)^{n+1} - 1.33 + 0.33(-0.5)^n\right) u[n]$$

(g)

$$H(z) = \frac{(z-0.5)(z+0.5)}{z^2 + z + 0.74}$$

$$\frac{Y(z)}{z} = \frac{(z-0.5)(z+0.5)}{(z^2+z+0.74)(z-1)}$$

$$= \frac{C_1}{z-1} + \frac{C_2 z + C_3}{z^2 + z + 0.74}$$

$$C_1 = 0.273$$

$$z_2 - 0.25 = 0.273\ (z_2 + z + 0.74) + (z - 1)\ (C_2 z + C_3)$$

$$C_2 = 0.727, C_3 = +0.452$$

$$y(z) = \frac{0.273z}{z-1} + \frac{0.727(z^2 + 0.62z)}{z^2 - 2a\cos(\Omega)z + a^2}$$

$$a = 0.86,\ \Omega = 2.19 \text{ rad.}$$

$$= \frac{0.2737}{z-1} + \frac{0.727(z^2 - a\cos(\Omega)z)}{z^2 - 2a\cos(\Omega) + a^2} + \frac{0.087z}{z^2 - 2a\cos(\Omega) + a^2}$$

where $0.087 = 0.125\ a \sin \Omega$

$$y[n] = 0.273 + 0.727\ (0.86)^n \cos(2.19n) + 0.125\ (0.86)^n \sin(2.19n)$$

$$n \geq 0$$

(*h*)
$$H(z) = \frac{(z-0.5)(z+0.5)}{z^2+z+4.25}$$

$$\frac{Y(z)}{z} = \frac{z^2-0.25}{(z^2-2a\cos\Omega\, z+a^2)(z-1)},$$

$$a = 2.04$$

$$\Omega = 1.816$$

$$\frac{y(z)}{z} = \frac{0.12}{z-1} + \frac{C_1 z + C_2}{z^2-2a\cos\Omega\, z+a^2}$$

$$z^2 - 0.25 = 0.12\ (z_2 + z + 4.25) + (z-1)\ (C_1 z + C_2)$$

$$C_1 = 0.88,\ C_2 = 0.76$$

$$y(z) = \frac{0.12z}{z-1} + \frac{0.88z^2+0.76z}{z^2-2a\cos(\Omega)\,z+a^2}$$

$$\frac{0.88(z^2-a\cos(\Omega)z)}{z^2-2a\cos(\Omega)\,z+a^2} + 0.16a\sin(\Omega).z$$

since ($0.16a \sin\Omega = 0.32$)

$$y[n] = \{(0.12 + 0.88(2.06)^n\cos(1.816n)$$

$$+0.16\sin(1.816n)(2.06)^n\}\ u[n]$$

(*a*)
$$Y(z) = (1 - z^{-1} + z^{-2})\ X(z)$$

$$H(z) = 1 - z^{-1} + z^{-2} = \frac{z^2-z^{-1}+1}{z^2}$$

(*b*)
$$h[n] = \delta[n] - \delta[n-1] + \delta[n-2]$$

(*c*) This is an FIR filter $\Rightarrow h[n] \to 0$ *in* finite number of steps $\Rightarrow$ it is stable (also note that the poles are inside the unit circle).

d)
$$H(\Omega) = H(z)\big|_{z=e^{j\Omega}} = 1 - e^{-j\Omega} + e^{-2j\Omega}$$

$$= e^{-j\Omega}(e^{j\Omega} - 1 + e^{-j\Omega})$$

$$= e^{-je}(2\cos(\Omega) - 1)$$

$$[H(\Omega)] = |2\cos(\Omega) - 1|$$

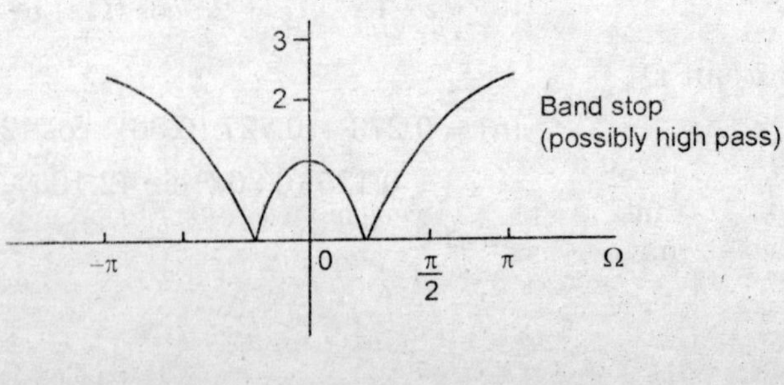

EXERCISES

1. (*a*) Determine the *z*-transform of the following samples of a time signal.

$$x(n) = \{3, 0, 0, 7, 6, 5, 4, -9\}$$

[**Ans.** $3 + 7z^{-3} + 6z^{-4} + 5z^{-5} + 4z^{-6} - 9z^{-7}$]

(*b*) If $x(n) = (0.25)^n$ for $n \geq 4$, and zero for $n < 4$, find $x(z)$.

[**Ans.** $0.253z^{-3}/(1 - 0.25z^{-1})$]

2. If $x(n)$ can be written by the formula $x(n) = (0.5)n$, express X(z).

[$1 + 0.5z^{-1} + 0.25z^{-2} + 0.125z^{-3}$...]

3. If $x(n) = 1$ for n lying between 0 to N–1 and is zero elsewhere, find X(z). Also mark the region of convergence of X(z), the z-transform of $x(n)$. Also mark the pole and zeros.

[$X(z) = x_0 + x_1z^{-1} + x_2z^{-2} + ...x_{N-1}\, z^{-(N-1)}$ and since all x values $x_0 ...x_{N-1}$ are 1.

Only if the series is infinite, this summation can be summed up only if $z^{-1} < 1$ or $z > 1$, which means the region of convergence is outside the unit circle. But here the series is finite! Hence it can be summed up for all values of z, except $z = 0$! The region of convergence is the entire z plane (except z = 0).

The pole is a $z = 1$ and zeros are at $z = 0$.]

4. Determine the inverse z transform of the following X(z).

$$X(z) = \frac{1}{\{1 + \tfrac{1}{2}z^{-1}\}} \text{ with region of convergence } |z| > 1/2$$

[$(-1)^n\, (\tfrac{1}{2})^n\, u(n)$]

5. Find the inverse z transform of

$$X(z) = \{3 - 4z^{-1}\}/\ [1 - 3.5z^{-1} + 1.5z^{-2}]$$

specifying the region of convergence and find x(n) for the following condition :

System is causal.

[The partial fraction expansion is found and hence the poles as z = ½and z = 3. The system is causal means that values of $|z| > 3$. $x(n) = (1/2)^n u(n) - 2\ (3)^n u(n)$.

6. Given the impulse response of a system as

$h(n) = 1$ for $n = -1$,

2 for $n = 0, 1$

for $n = 1$, –1 1 for $n = 2$, 0 for all other n.

For a given input signal $x(n) = \{1, 2, 3, 1, 0...\}$, find the response.

[$y(n) = \{..., 0, 0, 1, 4, 8, 8, 3, -2, -1, 0, 0, ...\}$.

7. Find the impulse response for the discrete time systems

(*a*) $y[n] + 1.2\, y[n - 1] = 2x[n - 1]$

(*b*) $y[n] = x[n] + 0.5x[n - 1] + x[n - 2]$

[**Ans.** (*i*) $-1.2^{n-1}\ .2$ for $n \geq 1$

(*ii*) [1 .5 1], $h(n) = 0$ for all other n]

8. Given two 8-point sequences as :

{1, 3, 5, 7, 2, 4, 6, 3} and {0.5, 1.5, 3.5, 5.5, 1, 2, 4, 6}

Find the Fourier transform of the two sequences (using Computer program), and thereby find the convolution.

[F.T1 = 31, –4.5–1.12j, –8 + 3j, 2.5 – 3.12j, –3, 2.5 + 3.12j, –8 –3j, –4.5 + 1.12j
F.T2 = 24, –0.5 + 1.2j, –6 + 8j, –0.5 + 0.02j, –6, –0.5–0.02j, –6–8j, –0.5–1.2j.
The IDFT of the product : –46.5, –49.5, 43.5, 70.5, –31.5, –34.5, 58.5, 85.5

[**Ans.** 0.5, 3, 10.5, 27, 46.5, 62, 69, 70, 101, 109]

9. Find the inverse z-transform of

$$X(z) = \frac{(z^2 - 1)(z + 2)}{(z - 3)^2(z + 4)}$$

[**Ans.** –1/18 $\delta(n)$ – (30/49)4^n $\delta(n)$ + 0.901$(3)^n$ $\delta(n)$ + (40/7)(3^n) $n\delta(n)$]

10. Find the circular convolution of the following two sequences :

$x1(n)$ = {2, 1, 2, 1} and $x2(n)$ = {1, 2, 3, 4}

[**Ans.** 14, 16, 14, 16]

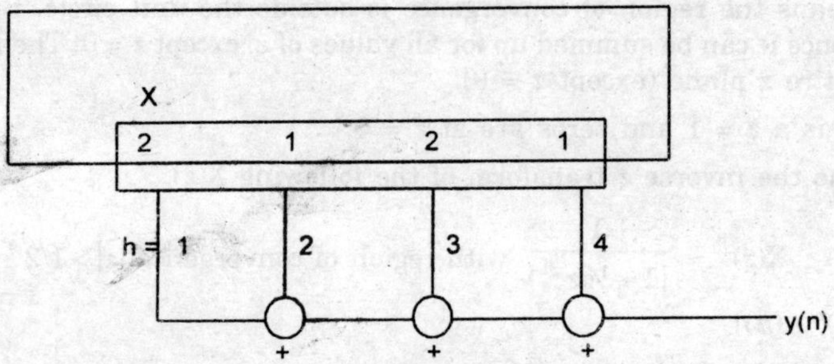

Fig. 24

11. A certain analog signal sampled at 20 kHz has the following ten samples.
500, –335, 184, 0, 666, –7.3, 0.05, 0.25, –0.04.

Using the computer program, find the spectrum of the sample over 90 equidistant frequency points on the unit semi circle of the z plane.

Then find the z transform on semi circles of radii r_0 = 0.8 and 0.6. What is the inference based on these two contours?

[A pole is lurkingly seen in the unit semi-circle spectrum near 3 kHz. In the 0.6 radius circle, the pole is exactly noted at 3300 Hz.]

INDEX

D

E

W

Z